THE LEGO® BOOST IDEA BOOK

95 Simple Robots and Hints for Making More!

YOSHIHITO ISOGAWA

The LEGO® BOOST Idea Book.

Printed in USA
First printing

22 21 20 19 18   1 2 3 4 5 6 7 8 9

ISBN-10: 1-59327-984-1
ISBN-13: 978-1-59327-984-4

Publisher: William Pollock
Production Editor: Riley Hoffman
Cover Design: Mimi Heft
Photographer: Yoshihito Isogawa
Author Photo: Sumiko Hirano
Developmental Editor: Annie Choi
Technical Reviewer: Sumiko Hirano
Proofreader: Paula L. Fleming

For information on distribution, translations, or bulk sales, please contact No Starch Press, Inc. directly:
No Starch Press, Inc.
245 8th Street, San Francisco, CA 94103
phone: 1.415.863.9900; info@nostarch.com
www.nostarch.com

Library of Congress Cataloging-in-Publication Data
Names: Isogawa, Yoshihito, 1962- author.
Title: The LEGO BOOST idea book : 95 simple robots and hints for making more! /
Yoshihito Isogawa.
Description: San Francisco : No Starch Press, Inc., [2018].
Identifiers: LCCN 2018028586 (print) | LCCN 2018033120 (ebook) | ISBN
9781593279851 (epub) | ISBN 159327985X (epub) | ISBN 9781593279844 (print)
| ISBN 1593279841 (print) | ISBN 9781593279851 (ebook) | ISBN 159327985X
(ebook)
Subjects: LCSH: Robots--Design and construction--Amateurs' manuals. |
Robotics--Popular works. | Robots--Popular works. | LEGO toys.
Classification: LCC TJ211 (ebook) | LCC TJ211 .I844 2018 (print) | DDC
629.8/92--dc23
LC record available at https://lccn.loc.gov/2018028586

Contents

PART 1 • Moving with the Move Hub

PART 2 • Using the Interactive Motor

PART 3 • More Exciting Ideas!

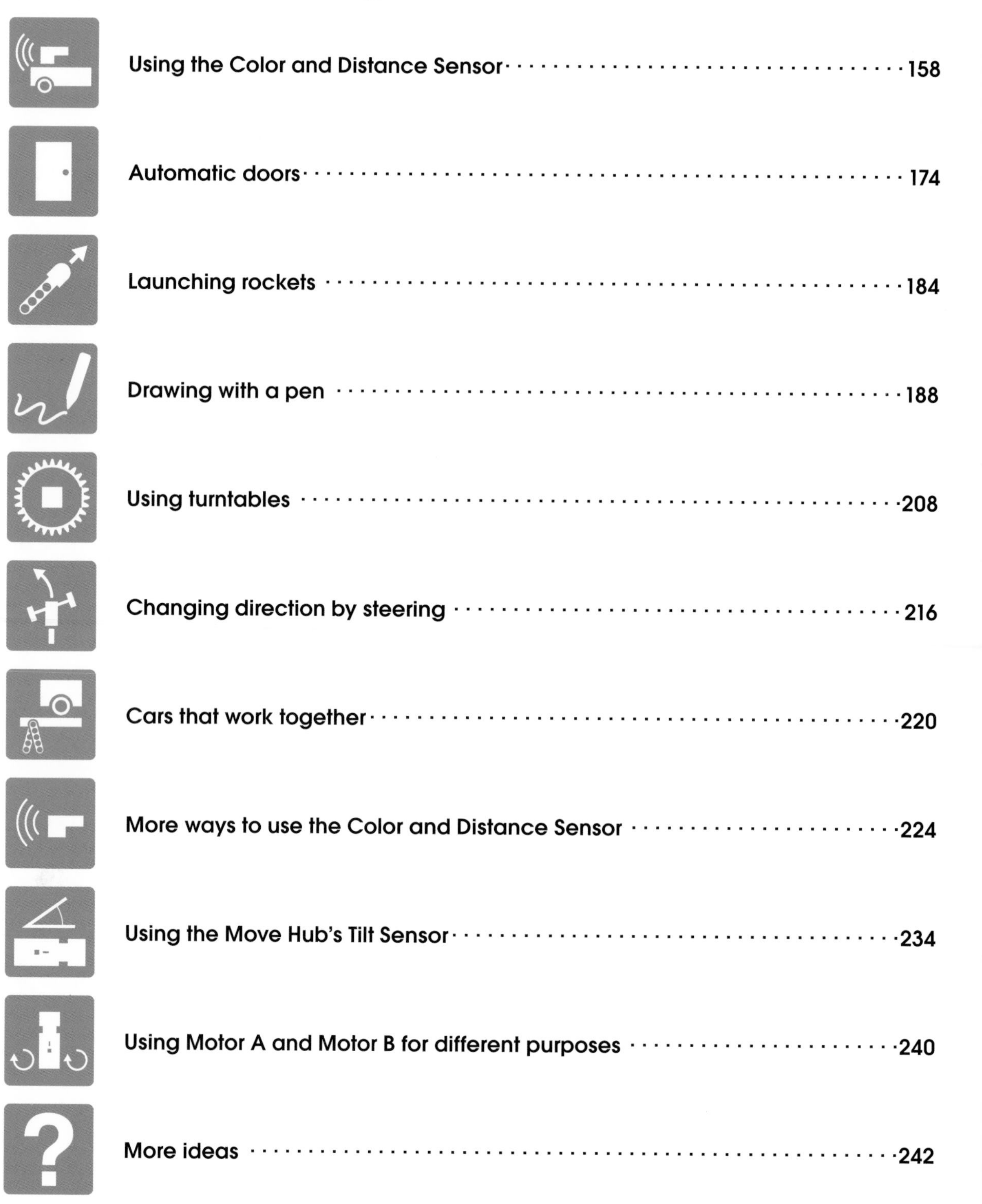

WELCOME!

Introduction

This book is not a beginner's guide to LEGO BOOST. It's also not a book for building robots like the ones included in the LEGO BOOST Creative Tool app. If you've already tried building and programming with BOOST and you're ready for more ideas to challenge yourself, this book will help you do that.

To build the models in this book, all you need is the LEGO BOOST Creative Toolbox set (#17101).

How to Use This Book

Most of the models in this book are small, simple mechanisms, and the programs you need to control them are also simple. When you build the models and get them moving, you'll understand those mechanisms and programs that much better. As you develop and evolve them more and more, you might even create your own builds. It would be a good idea to combine some mechanisms as well. Feel free to remodel, reinforce, and decorate. Your creativity has no limits.

You don't have to make these models in the order they appear. Flip through the pages and then try making models you find interesting. You might want to start with relatively simple models first.

Recommended Books

For a beginner's guide to BOOST, check out *The LEGO BOOST Activity Book* by Daniele Benedettelli.

If you would like to try even more mechanisms, check out two of my other books: *The LEGO Power Functions Idea Book, Vol. 1* and *The LEGO Power Functions Idea Book, Vol. 2*.

Acknowledgments

LDraw data and the LPub application were used to create the illustrations in this book. I would like to thank those involved in the development of those programs.

Programming BOOST

In the LEGO BOOST Creative Toolbox app, you can create the programs shown in this book by tapping the Creative Canvas icon on the right side of the menu. If the screen is rolled down, tap the screen to roll it up and display the project screen. When the project screen opens, tap the **+** icon in the top-left corner. A new screen where you can make programs will be displayed.

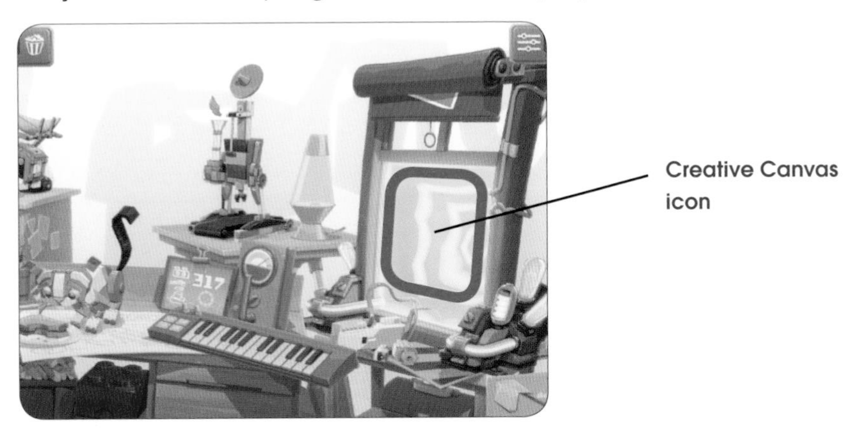

Note that this Creative Canvas icon becomes available only after you try making some programs for the robot projects in the LEGO BOOST Creative Toolbox app.

You can set the difficulty level for programming blocks to one of three levels. In this book, we'll use the standard level 2. You can select the right level like this:

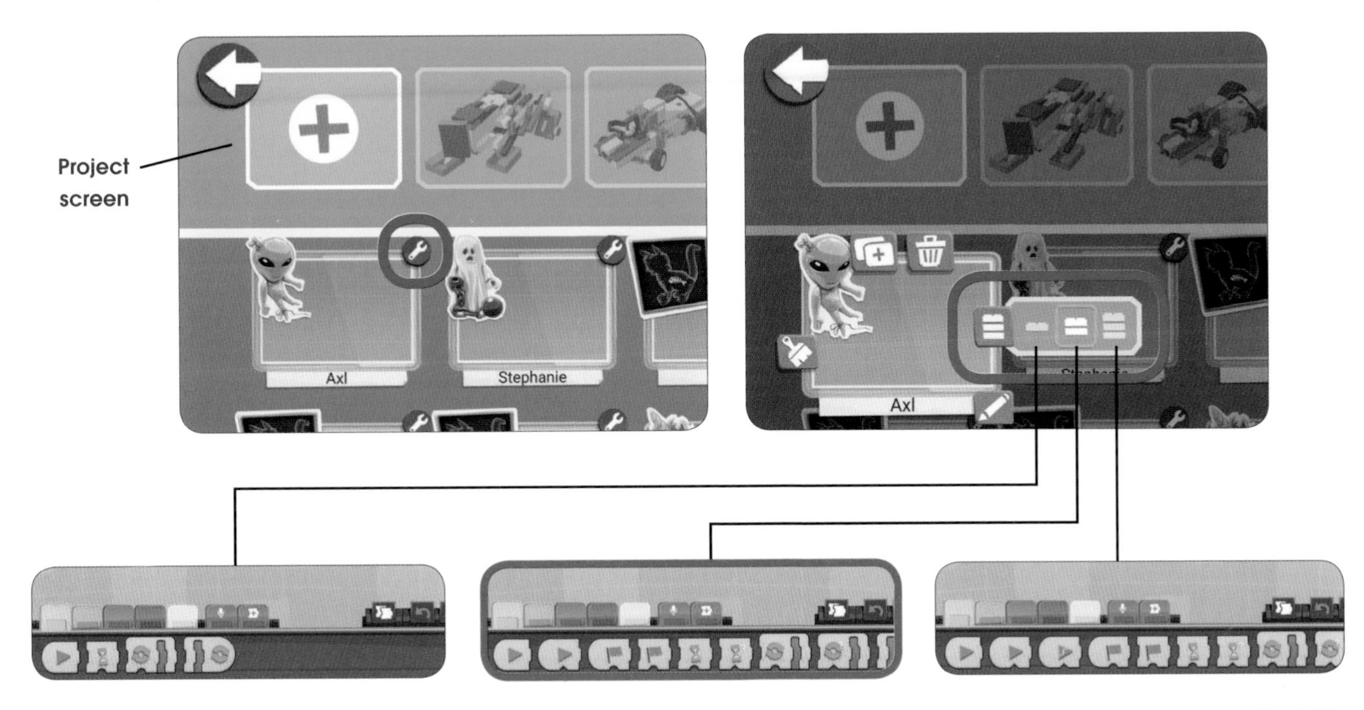

The programs in this book were created with the LEGO BOOST Creative Toolbox app version 1.5.0.

Warm-Up

You won't find step-by-step building instructions in this book. Instead, use the photographs taken from various angles to try to reproduce the model. Building in this way is like putting together a puzzle. You'll soon get the hang of this process and learn to enjoy it!

Let's practice first.

1 — This is the number of the model.

All the parts you need for this model are shown in the box below. Find them in your BOOST set and start building!

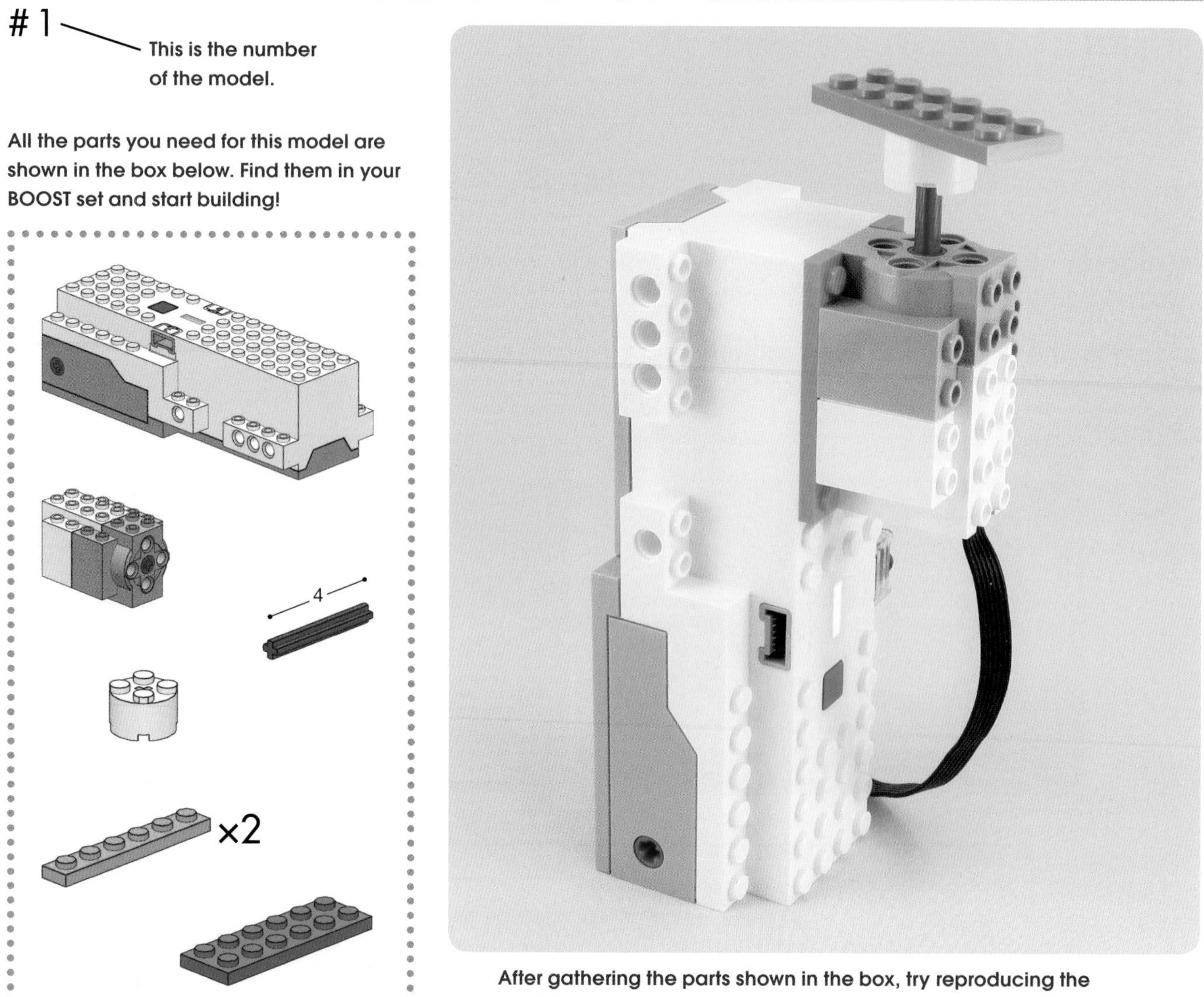

After gathering the parts shown in the box, try reproducing the model using the photos on this page and the next.

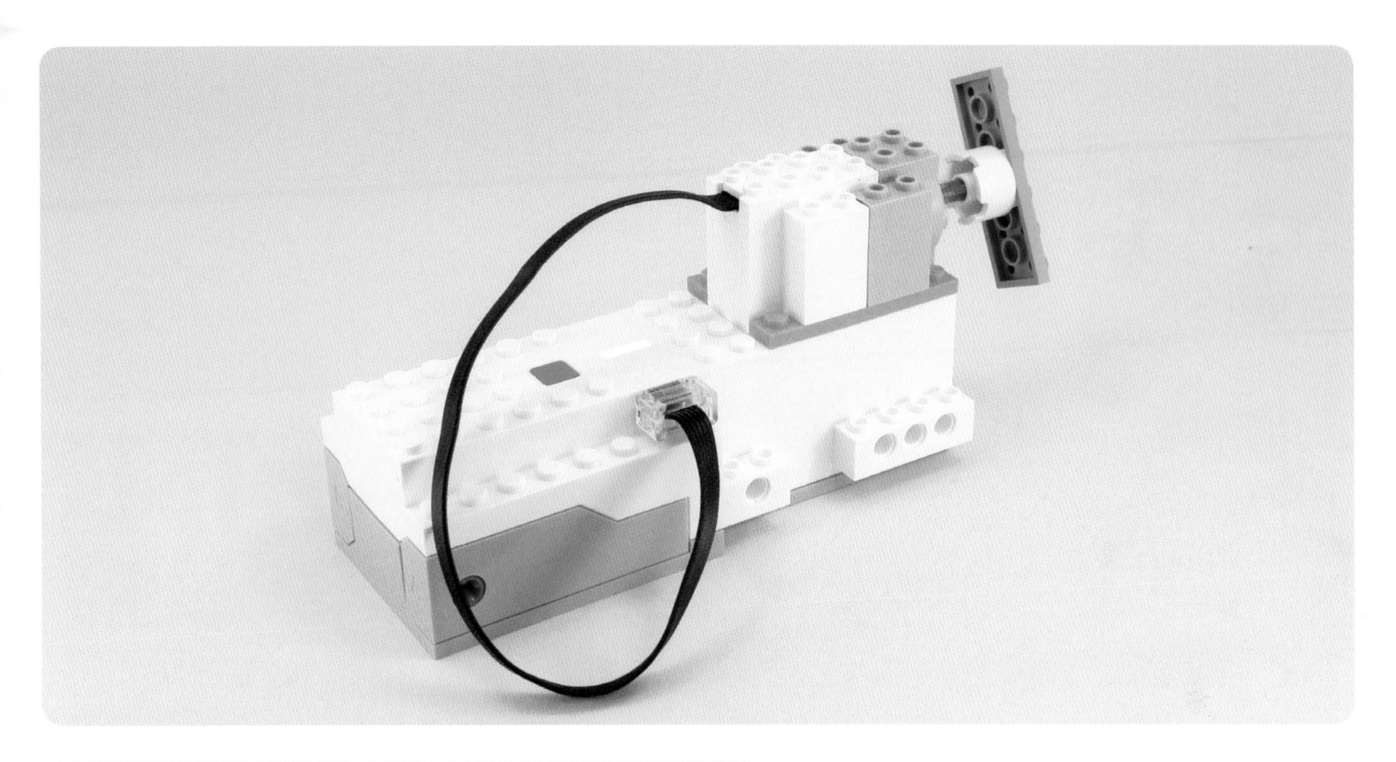

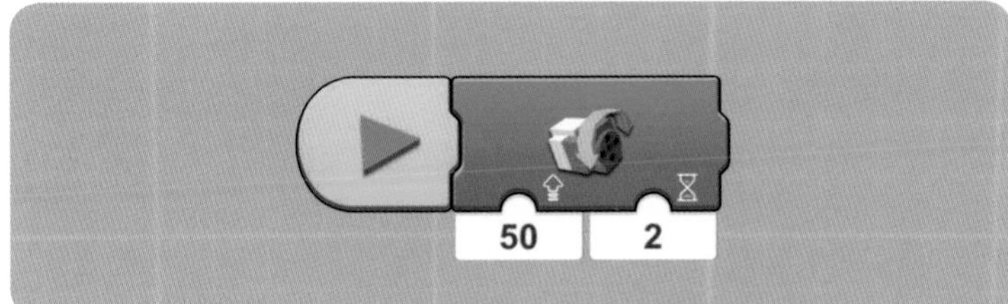

This is a sample program that you can use to set this model in motion.

This is the "hint" icon, which suggests other ways of building and programming. Try to create your own unique and fun models using these tips.

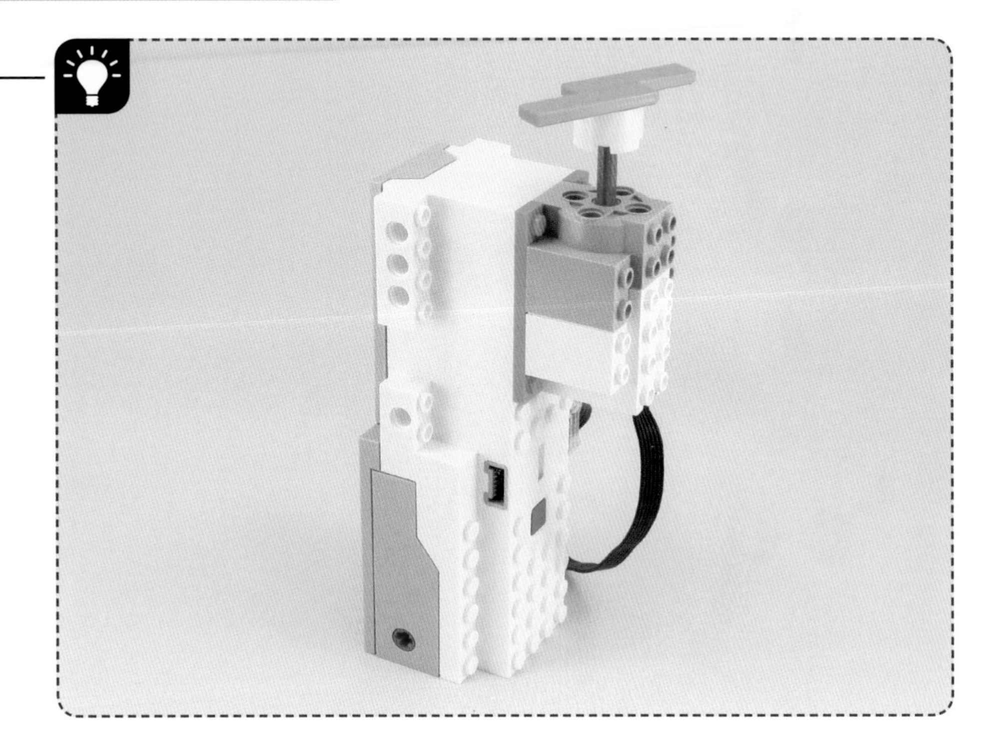

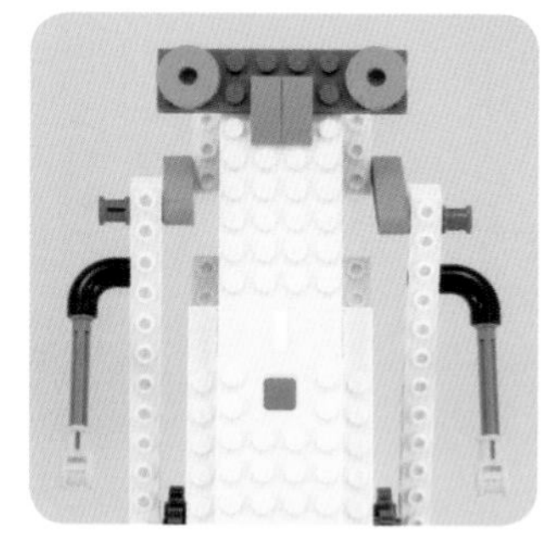

PART 1

Moving with the Move Hub

Moving on wheels

1

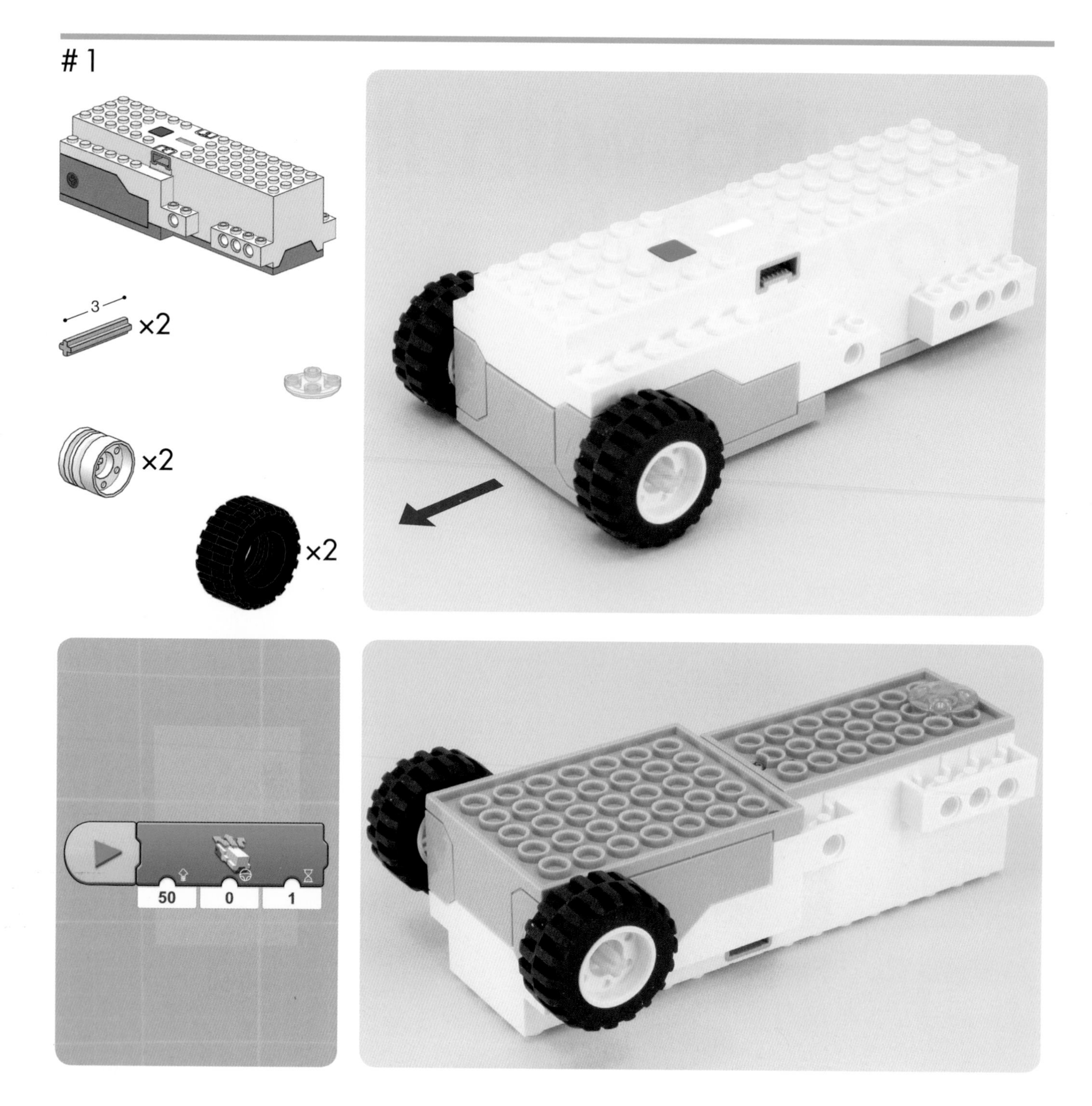

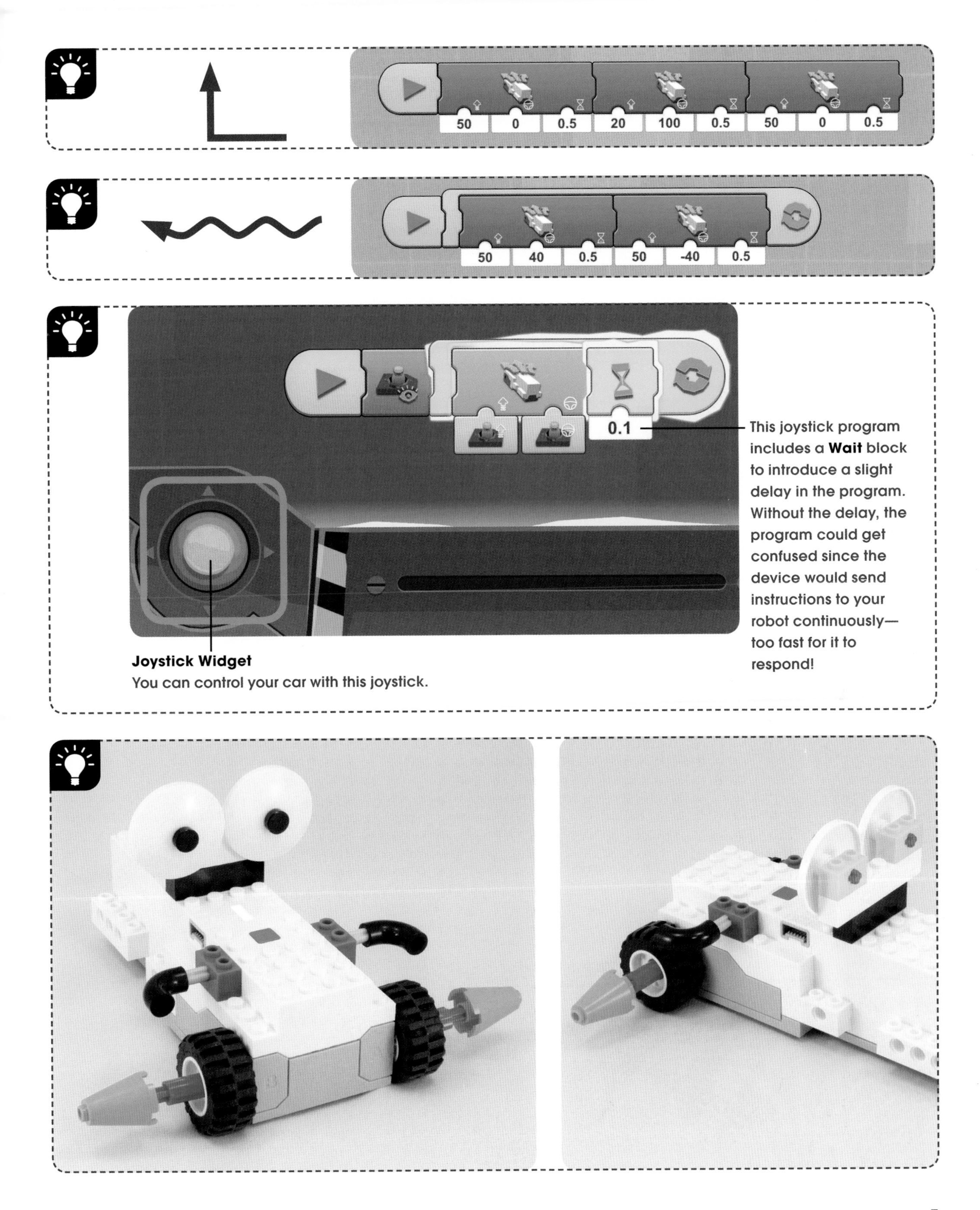

This joystick program includes a **Wait** block to introduce a slight delay in the program. Without the delay, the program could get confused since the device would send instructions to your robot continuously—too fast for it to respond!

Joystick Widget
You can control your car with this joystick.

#2

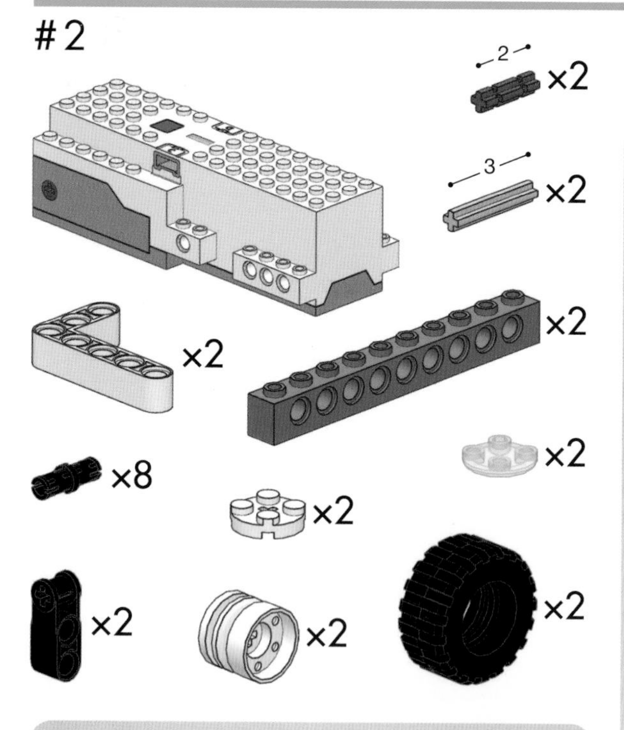

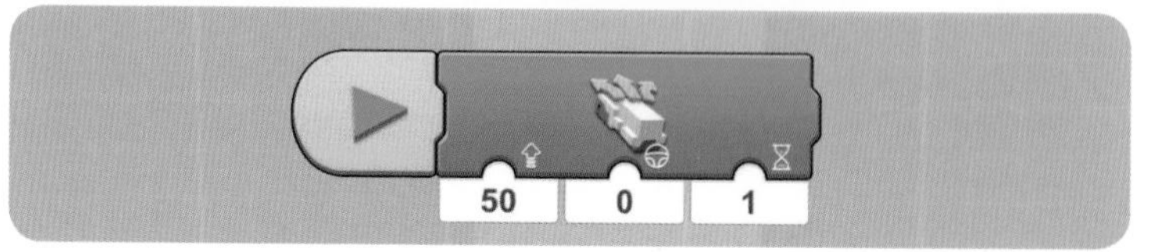

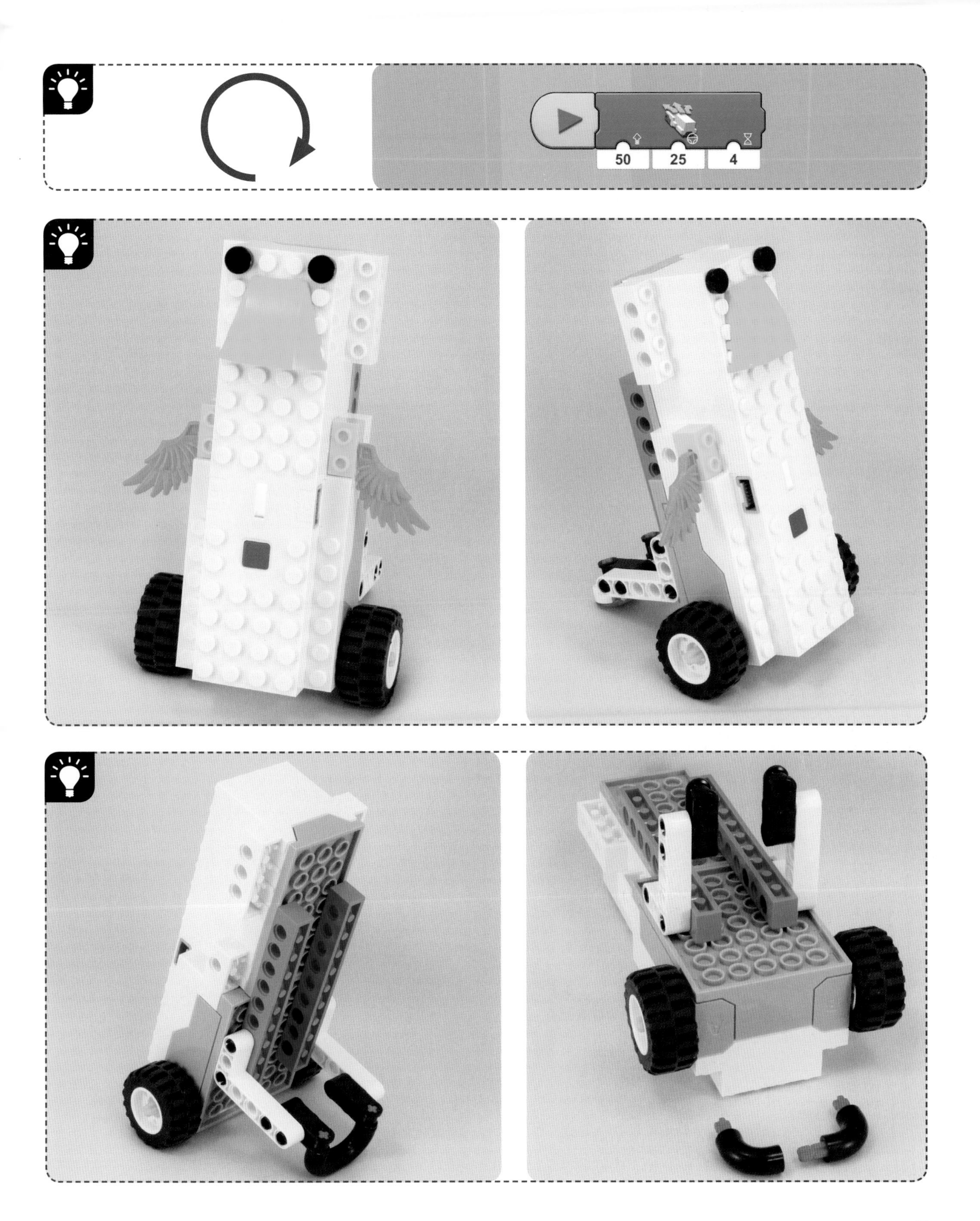
50
25
4

3

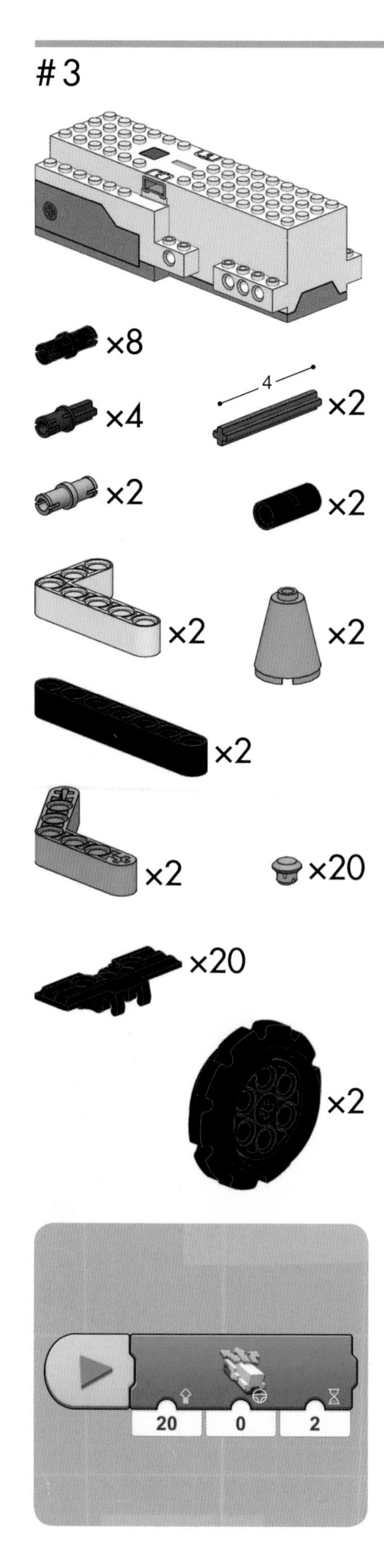

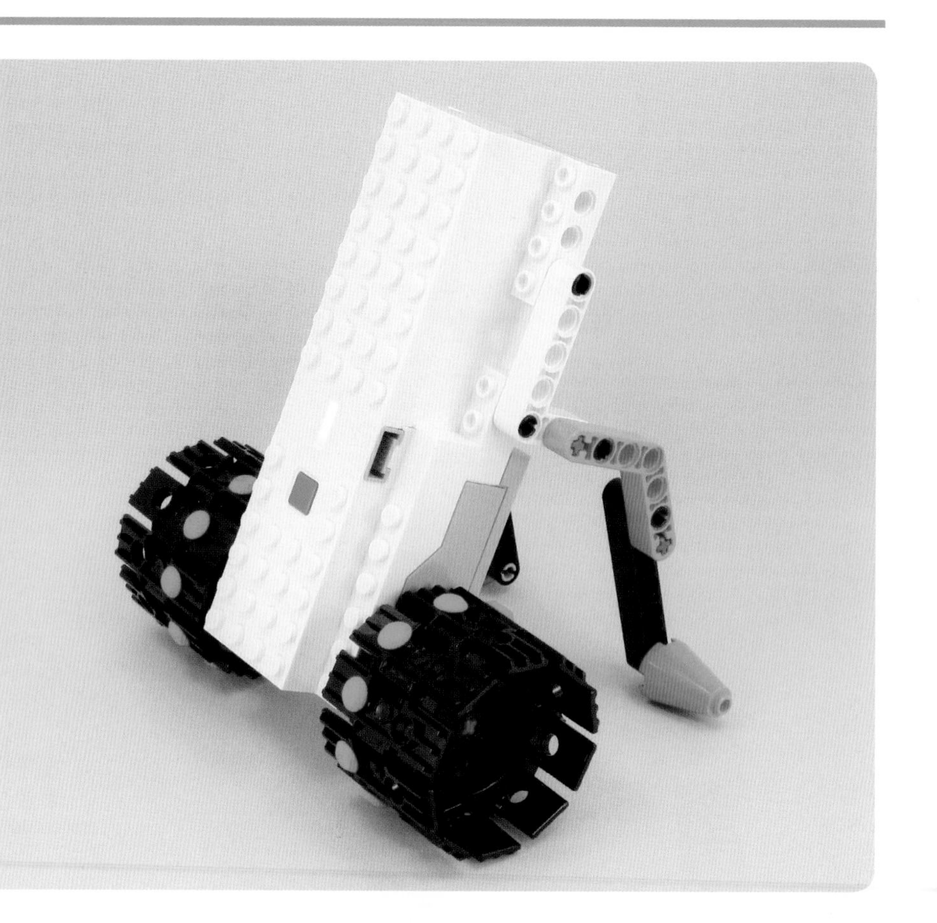

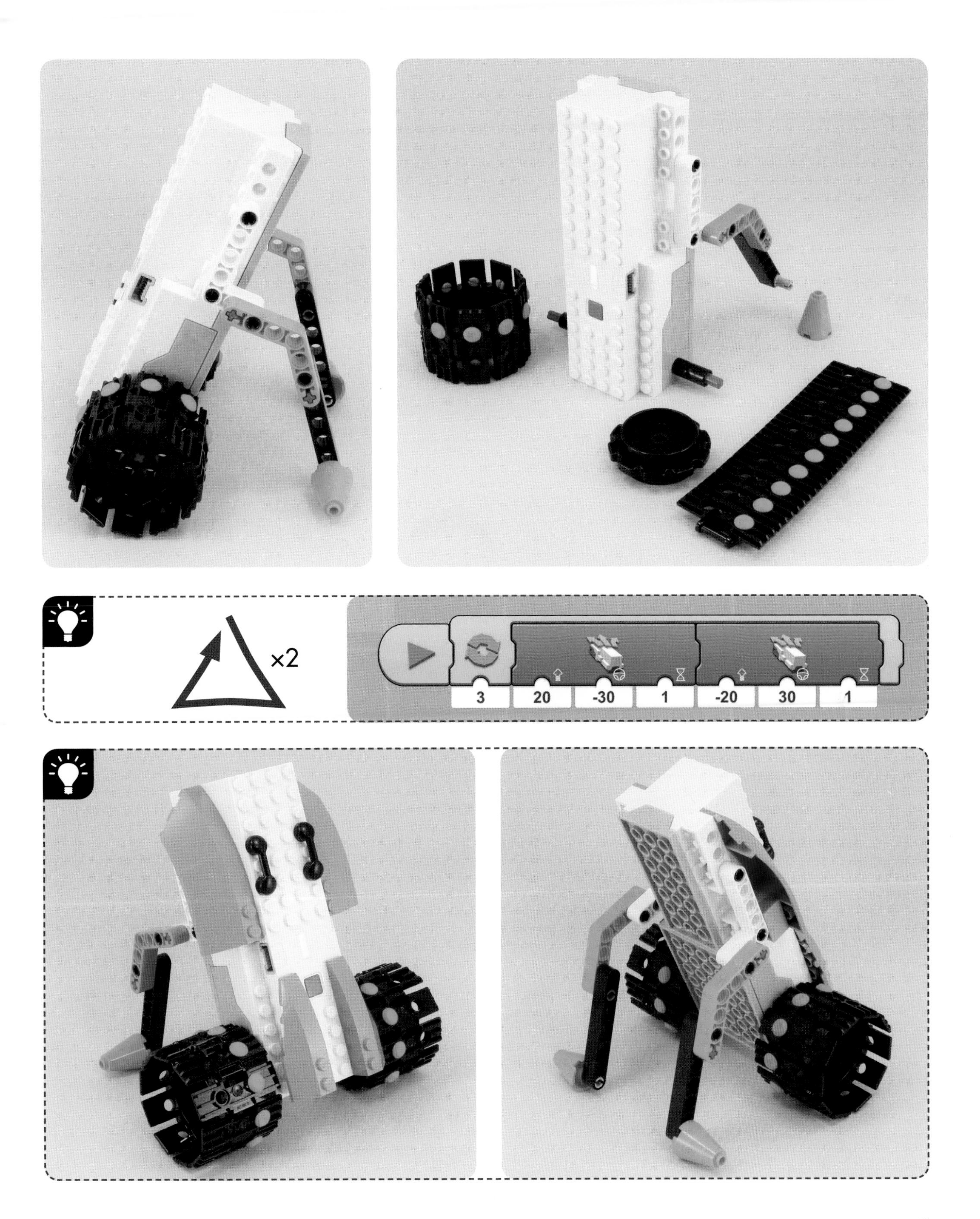
×2
3
20
-30
1
-20
30
1

#4

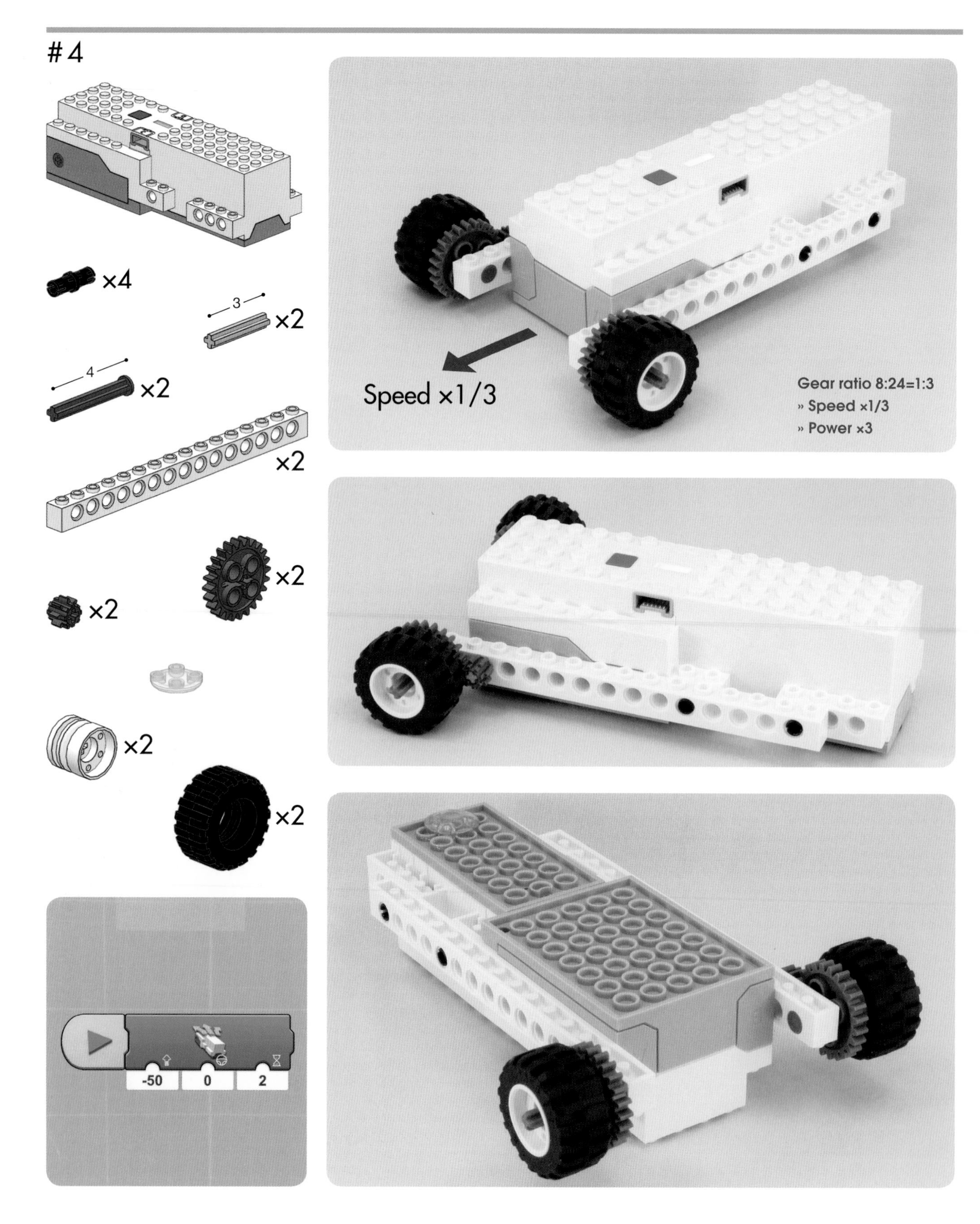

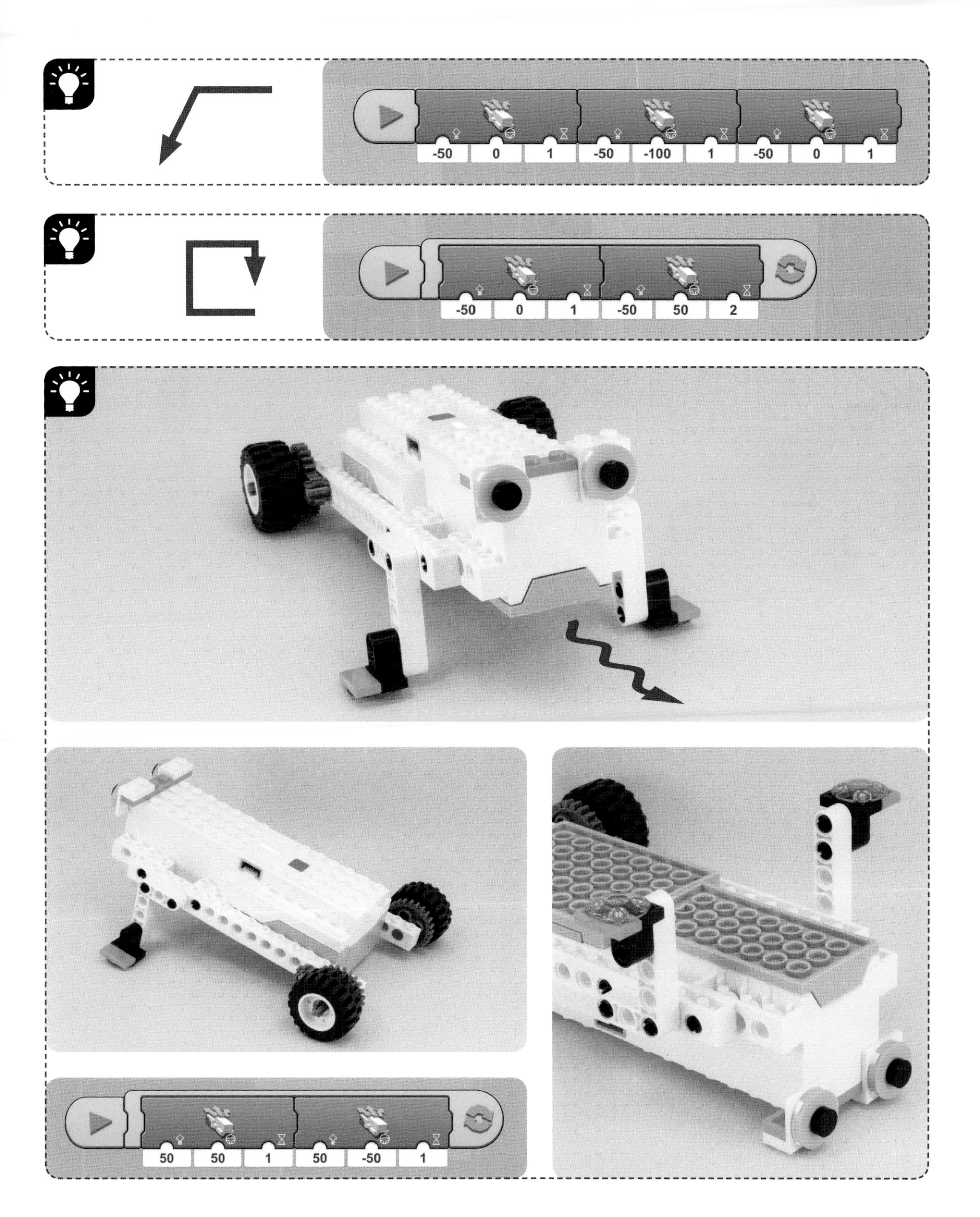
-50 0 1 -50 -100 1 -50 0 1
-50 0 1 -50 50 2
50 50 1 50 -50 1

#5

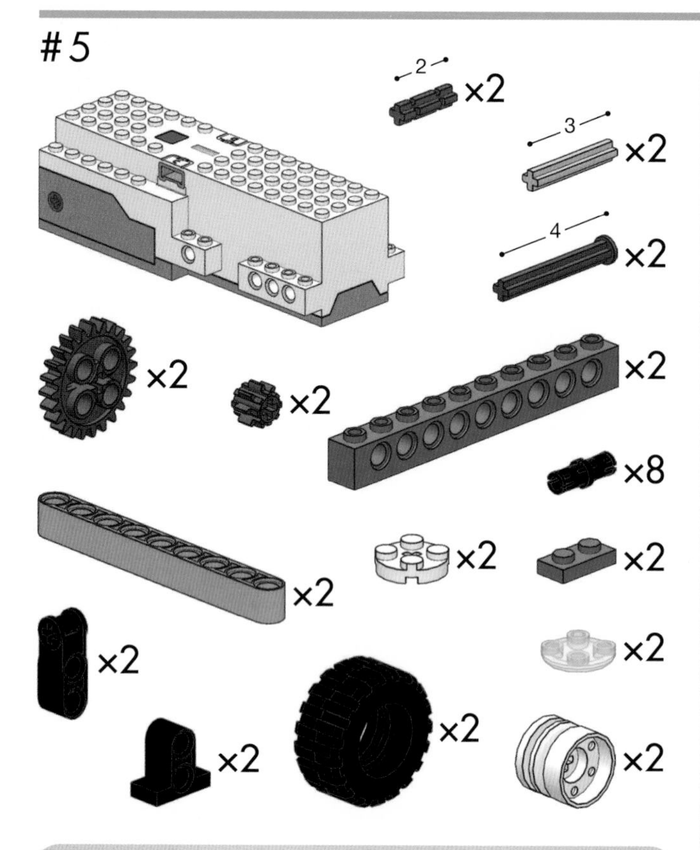

Gear ratio 8:24=1:3
» Speed ×1/3
» Power ×3

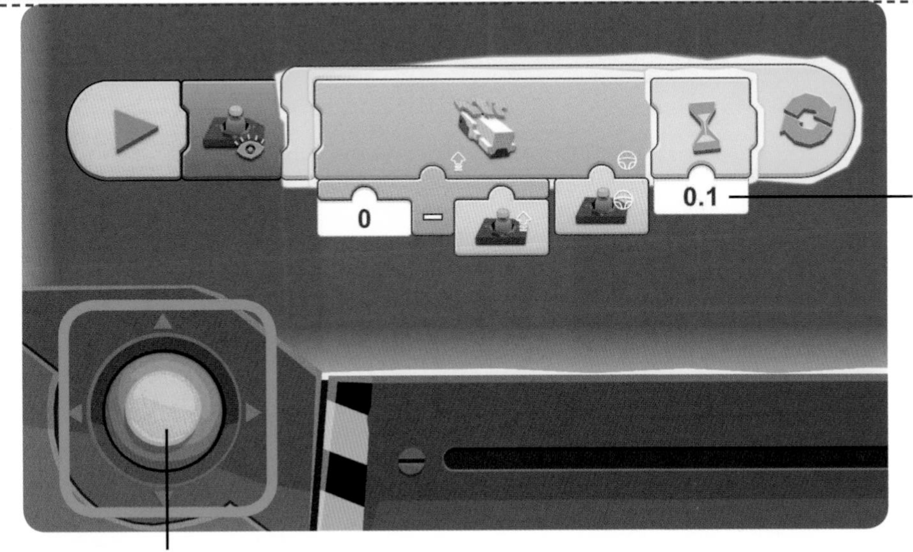

This joystick program includes a **Wait** block to introduce a slight delay in the program. Without the delay, the program could get confused since the device would send instructions to your robot continuously—too fast for it to respond!

Joystick Widget
You can control your car with this joystick.

#6

×4
3
×2
3
×2
12
×2
×2
×2
×2
×2
×2
50
2
Speed ×1/3
Gear ratio 8:24=1:3
» Speed ×1/3
» Power ×3

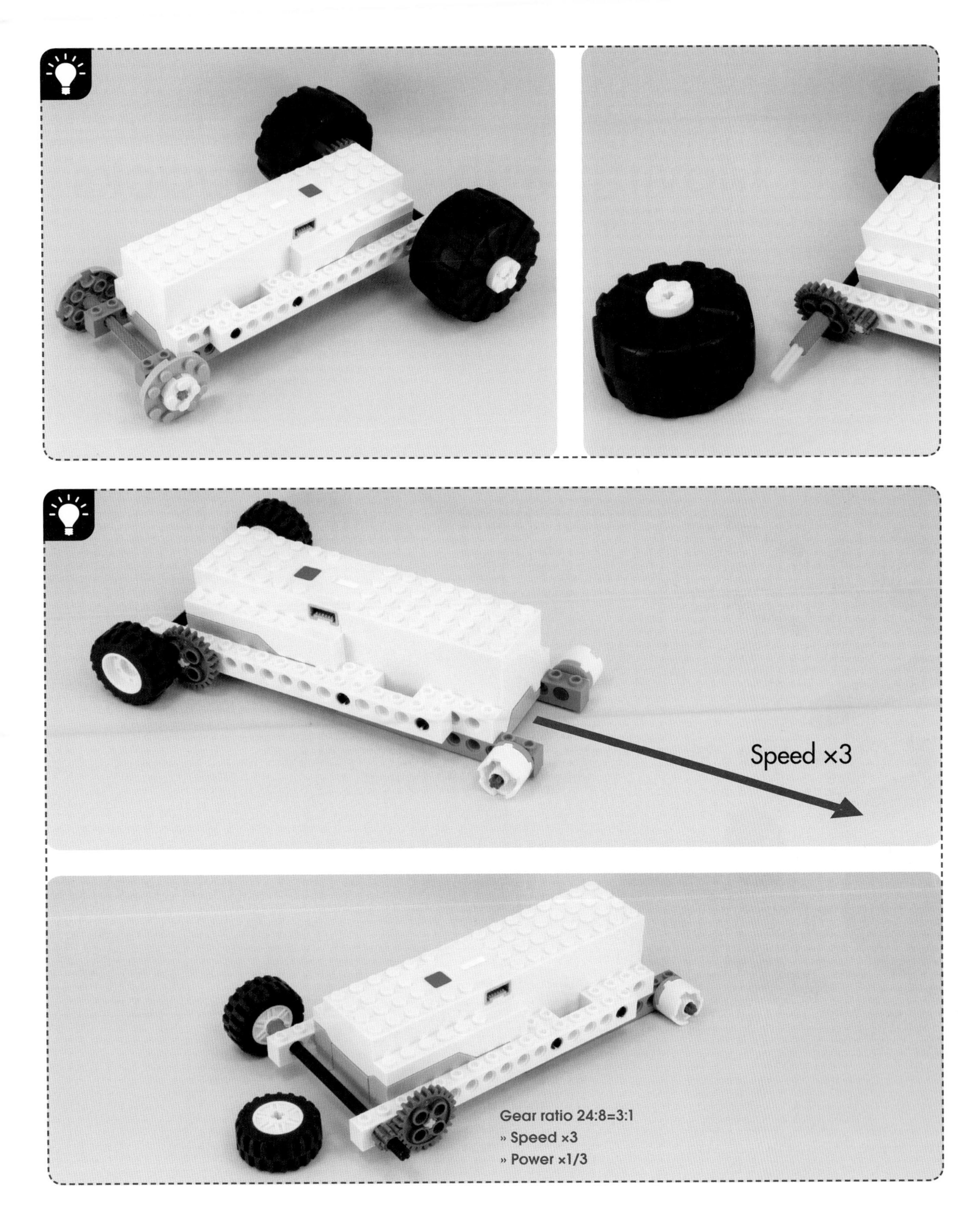
Speed ×3
Gear ratio 24:8=3:1
» Speed ×3
» Power ×1/3

Moving with crawler tracks

#7

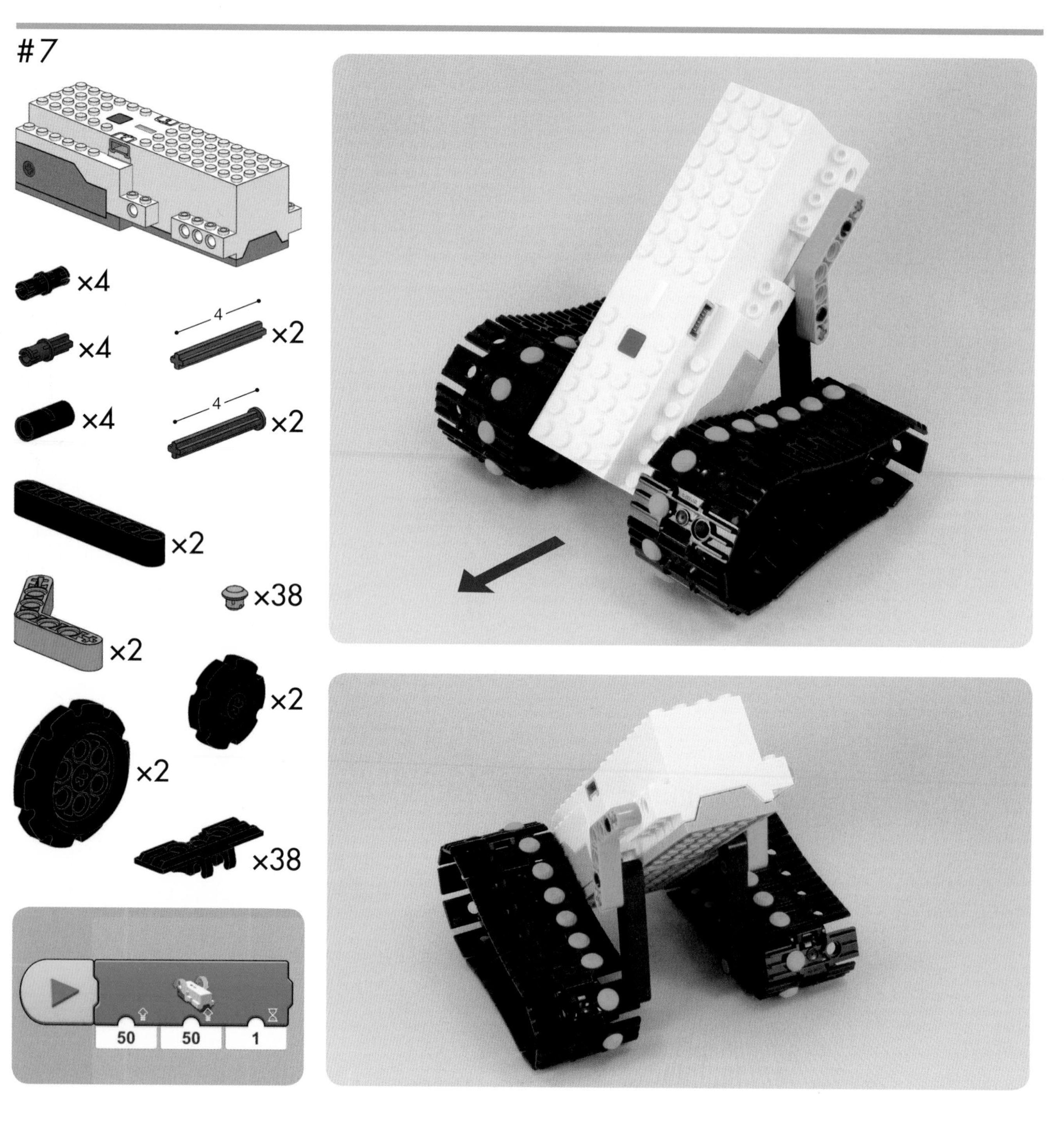

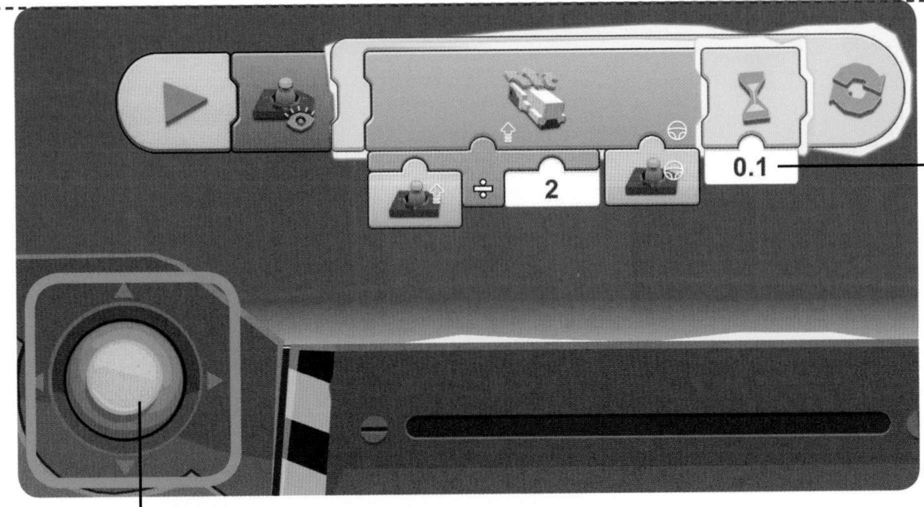

This joystick program includes a **Wait** block to introduce a slight delay in the program. Without the delay, the program could get confused since the device would send instructions to your robot continuously—too fast for it to respond!

Joystick Widget
You can control your car with this joystick.

8

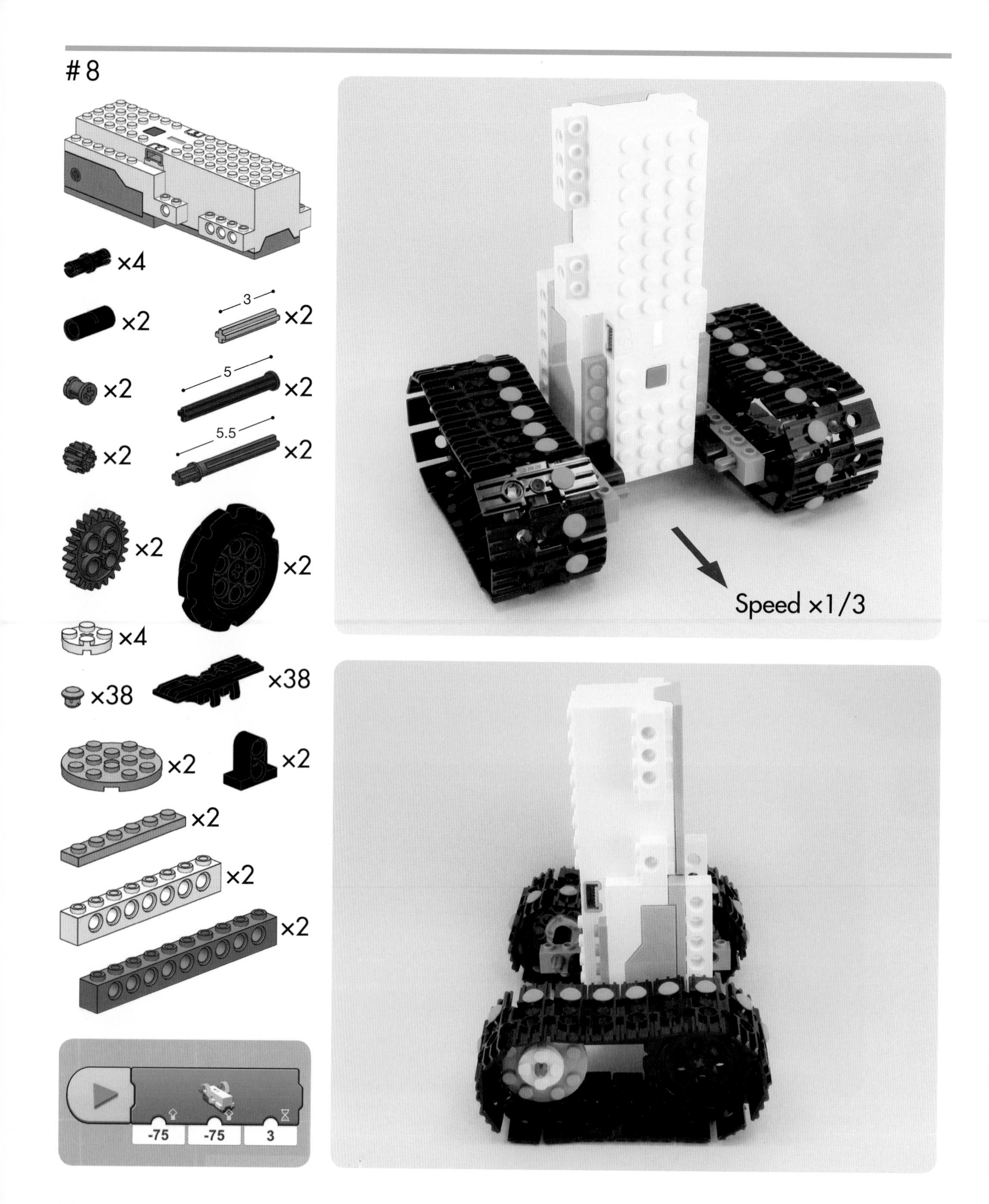

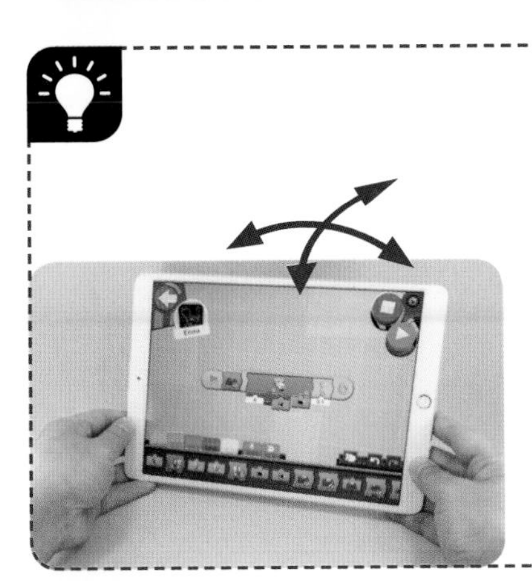

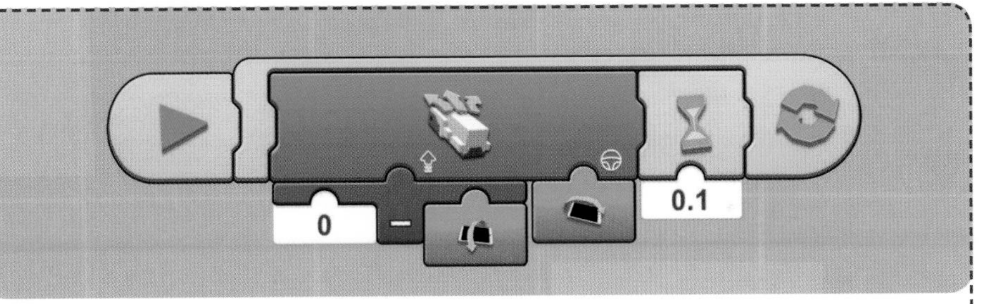

You can control this model by tilting your tablet or smartphone back and forth or left and right. (Warning: This might not work for every device!)

Suspended vehicles

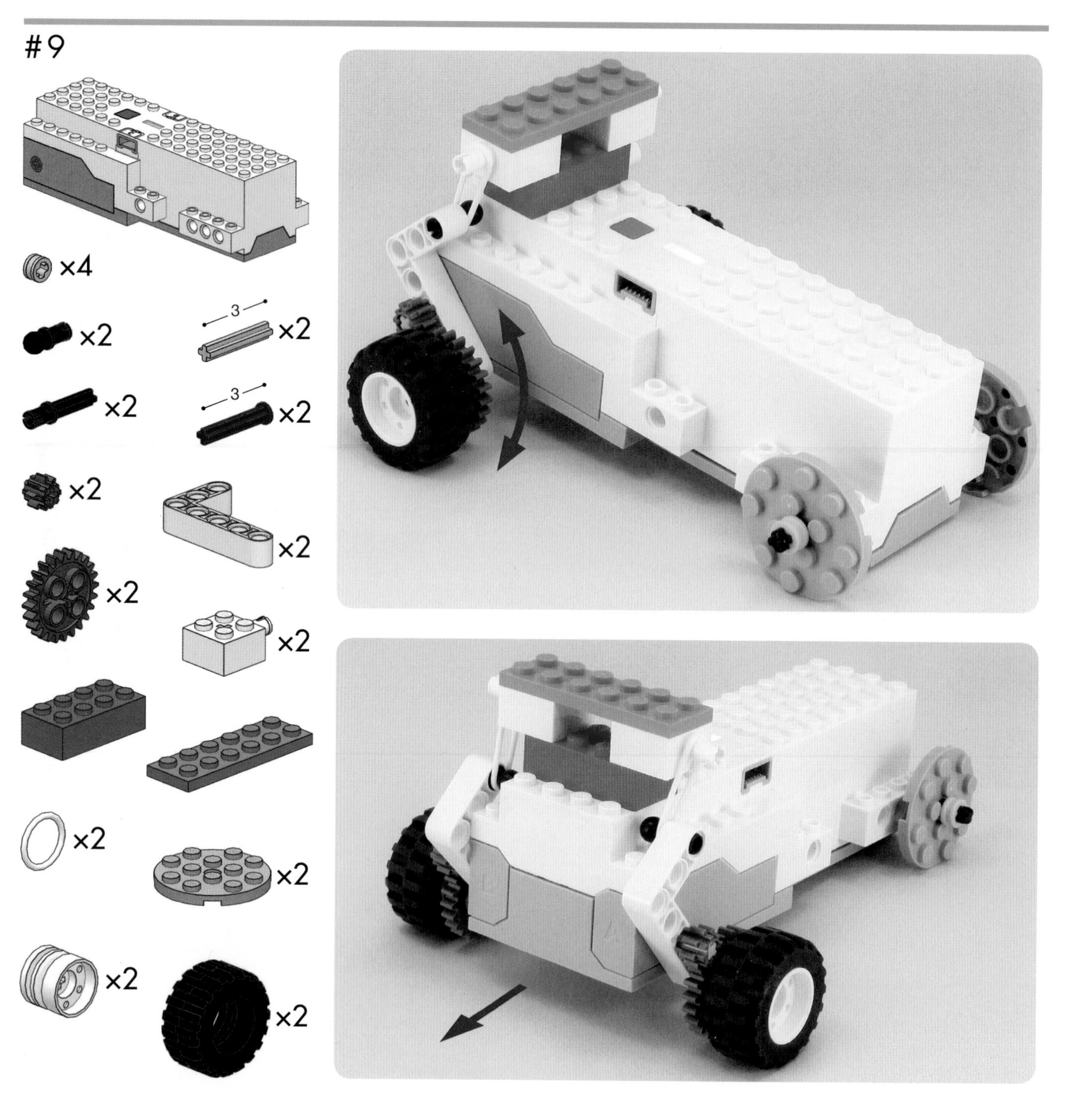

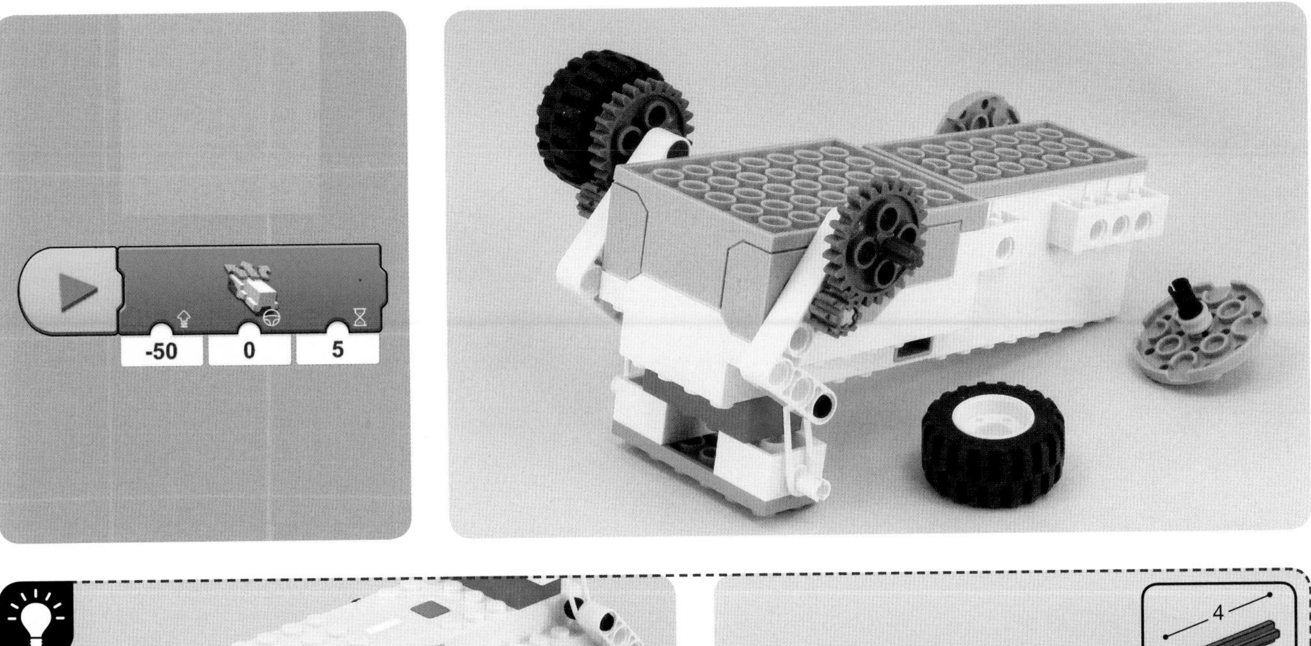
-50
0
5

7
4

10

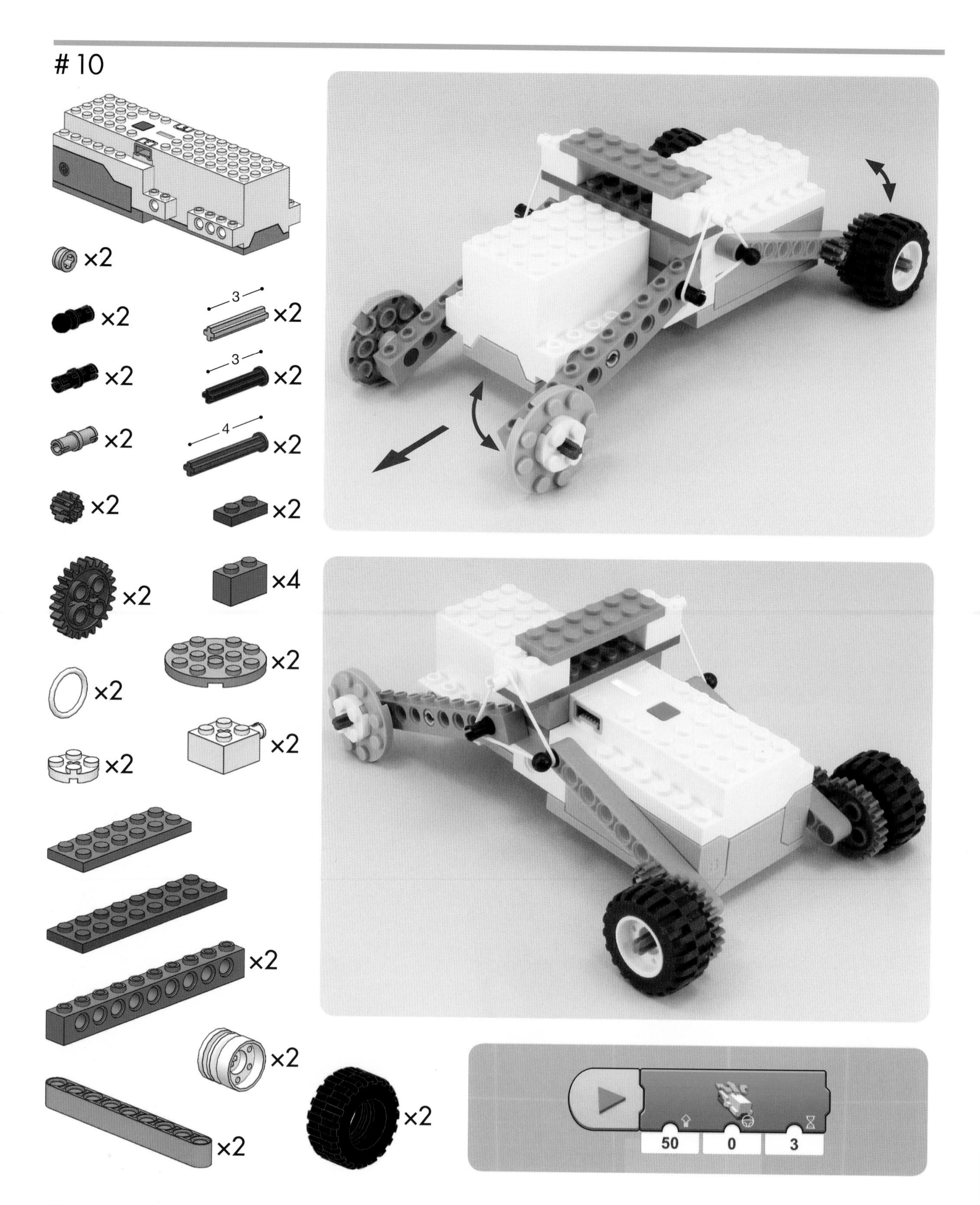

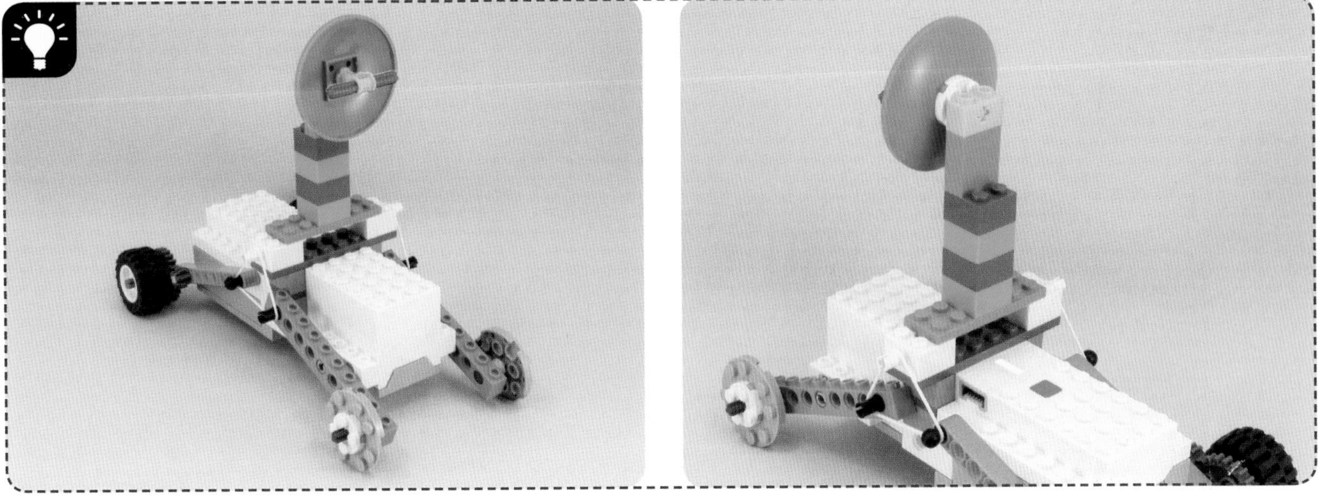

11

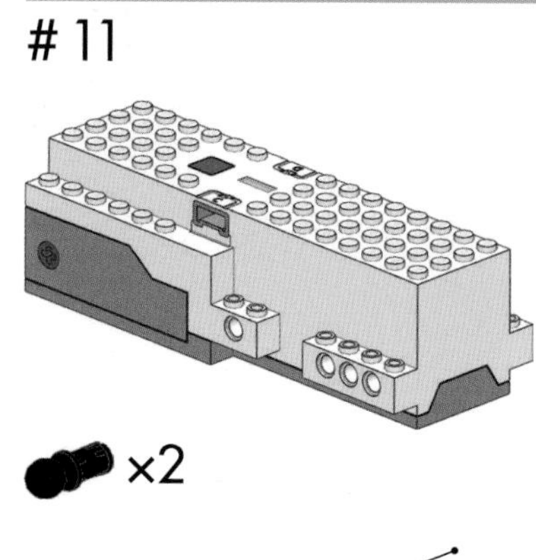

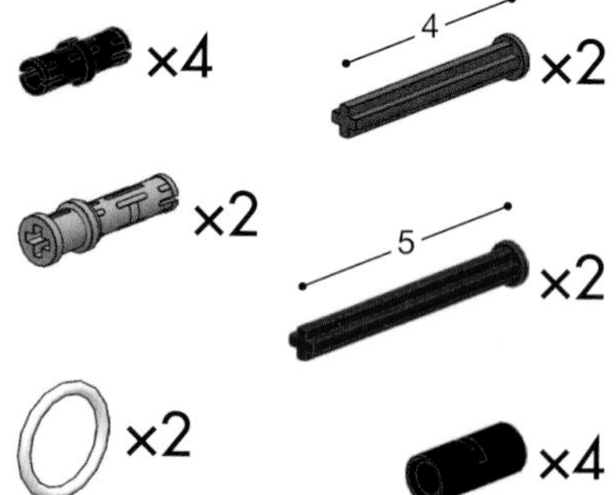

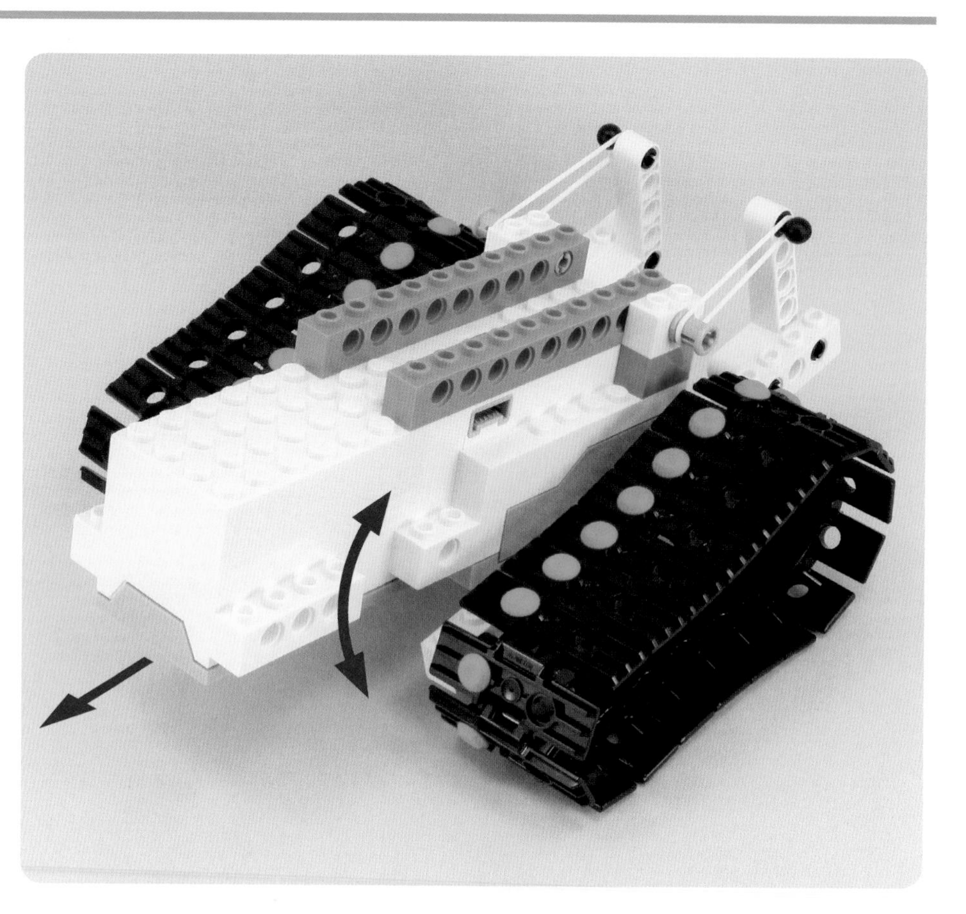

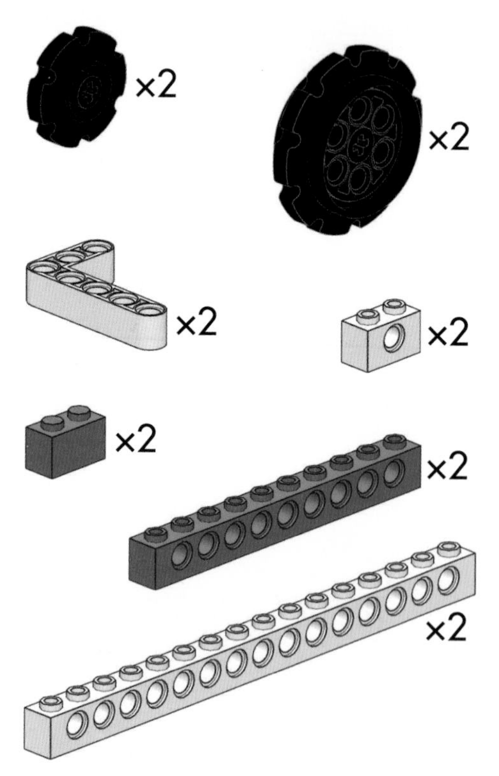

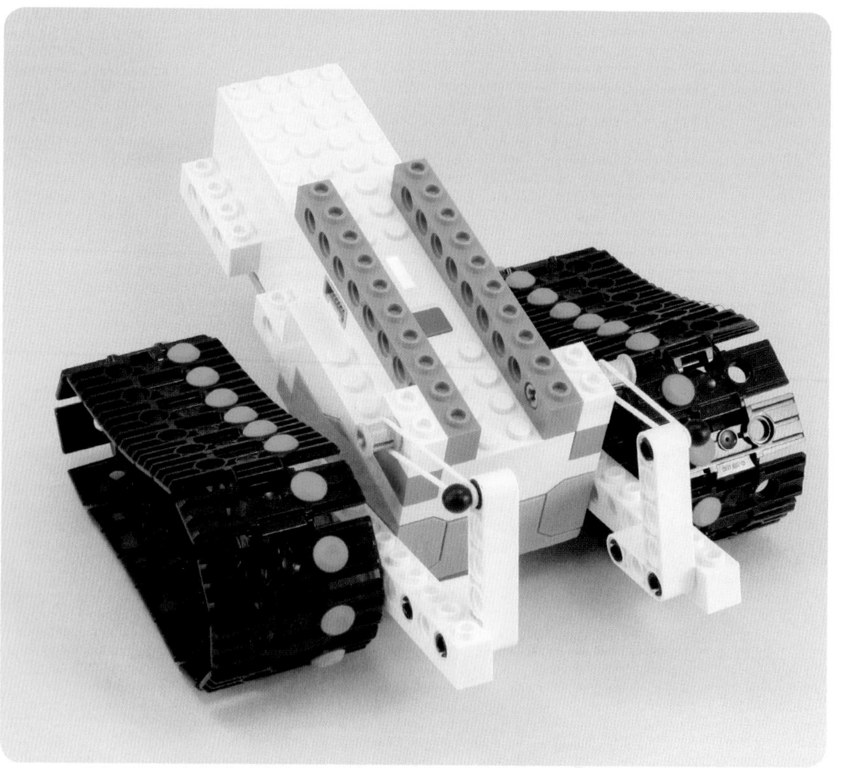

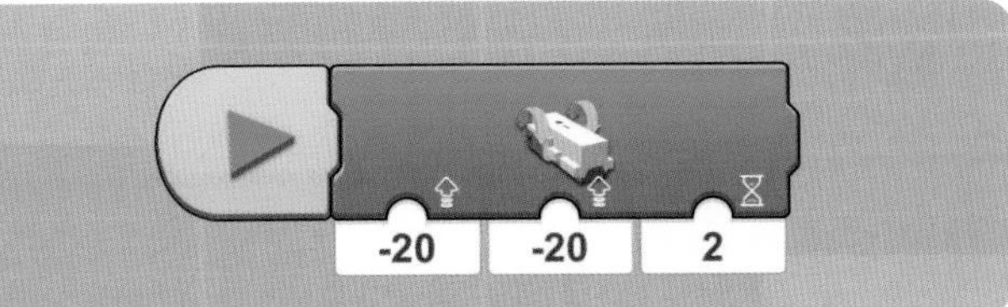
-20
-20
2

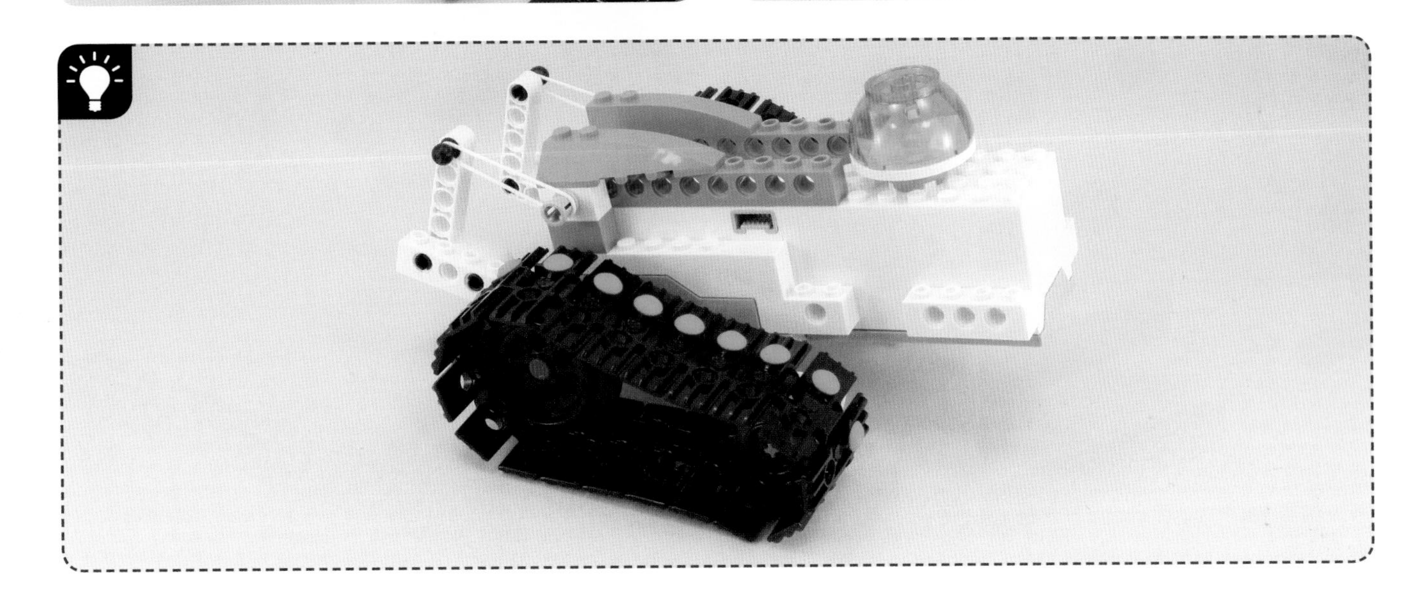

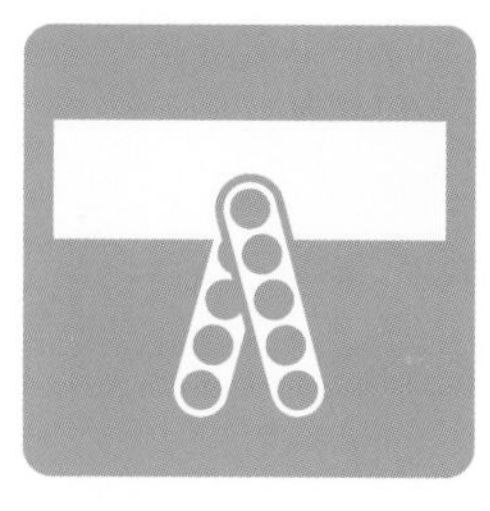

Walking machines

12

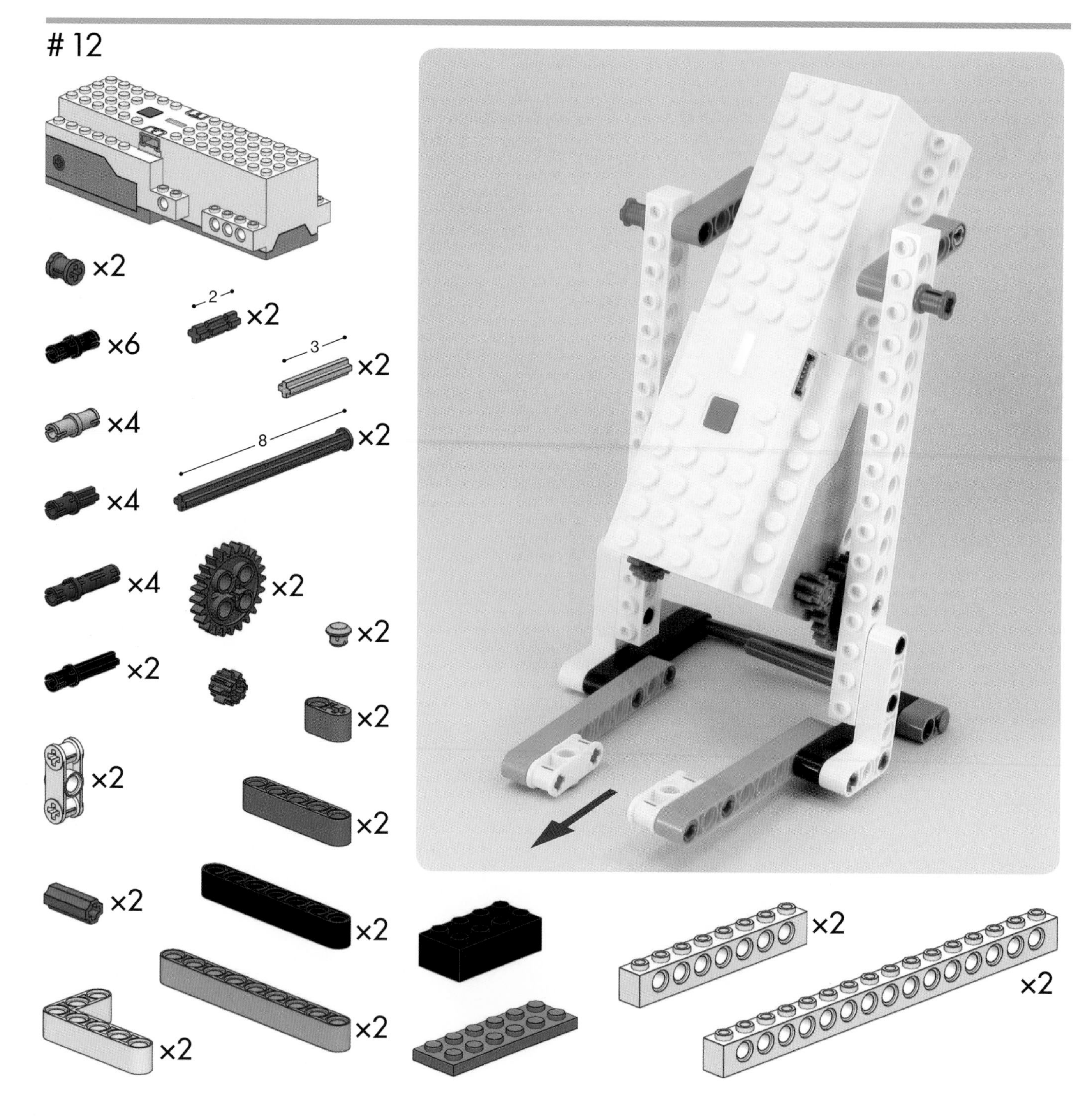

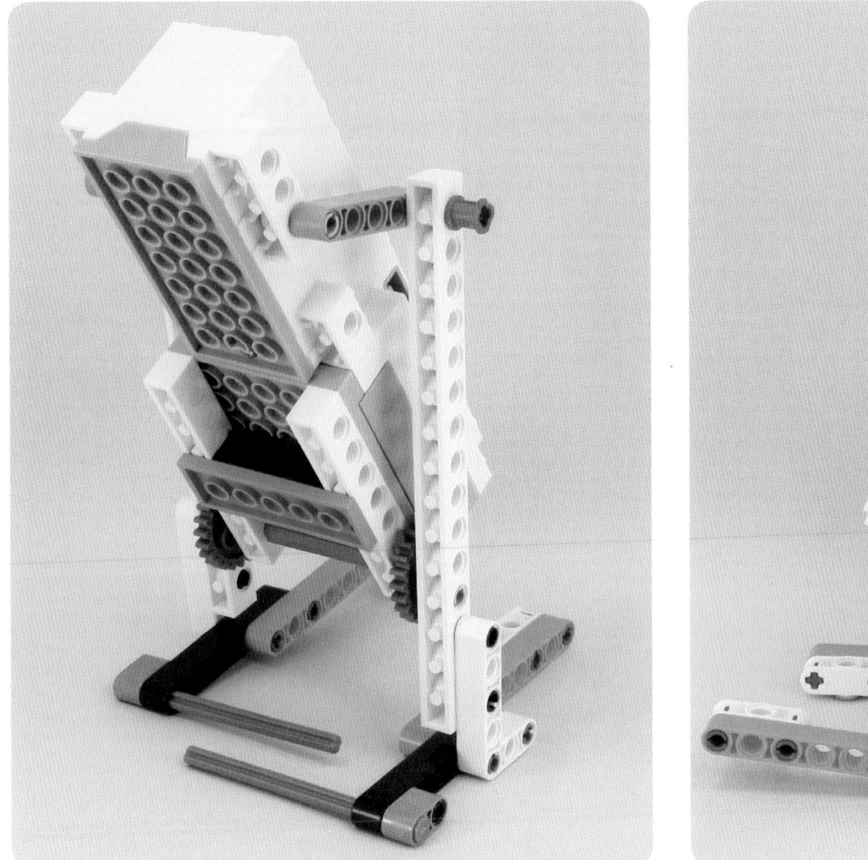

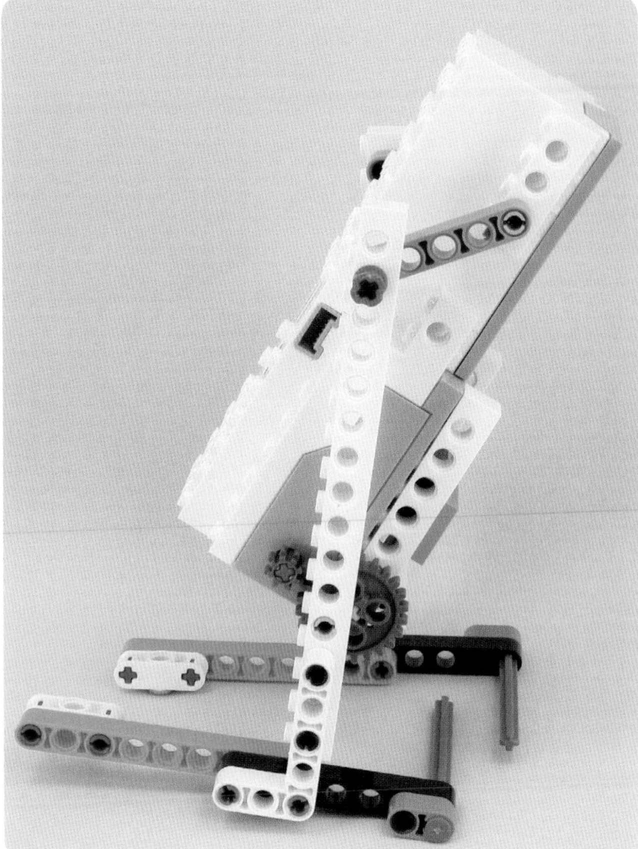

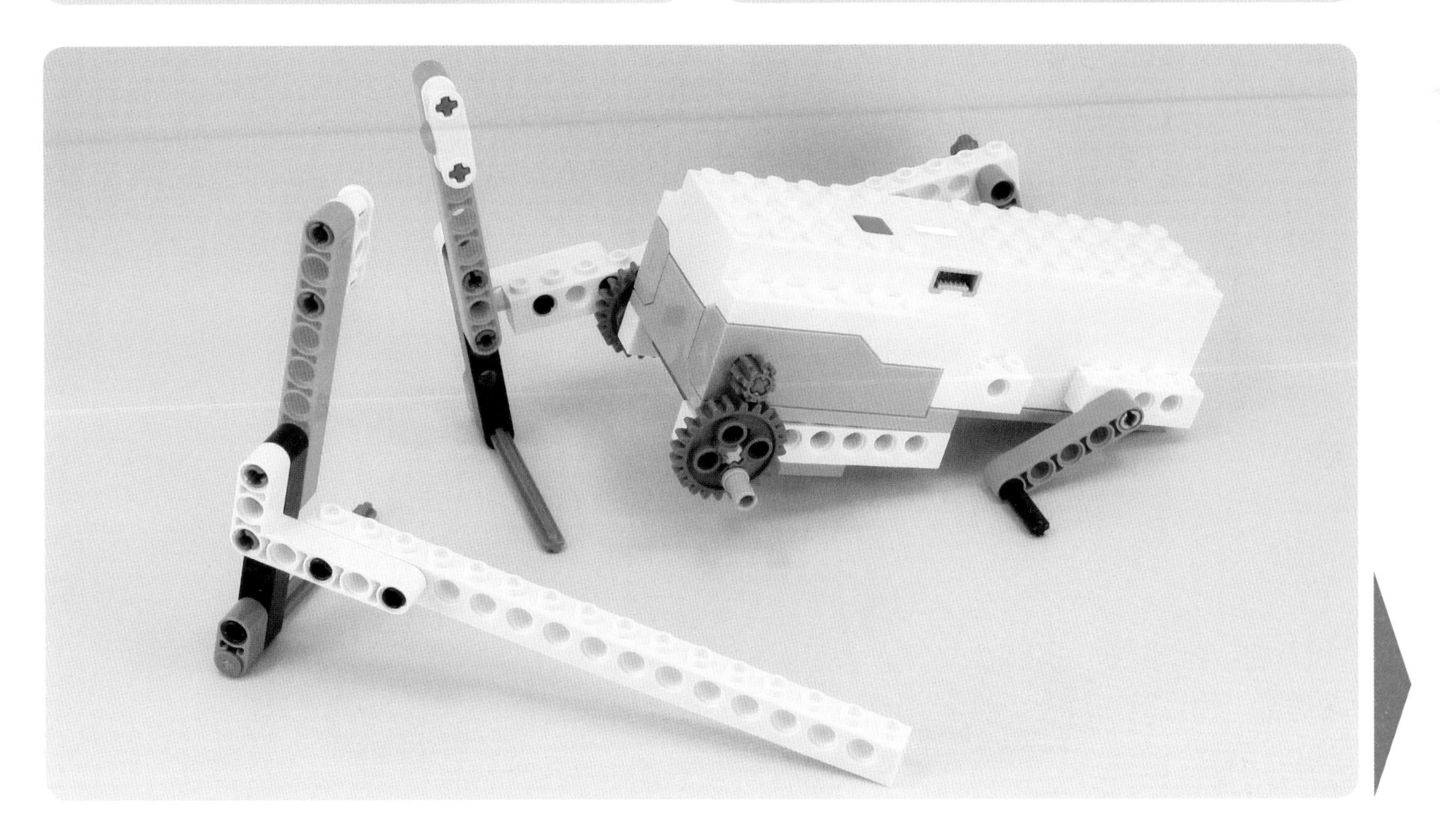

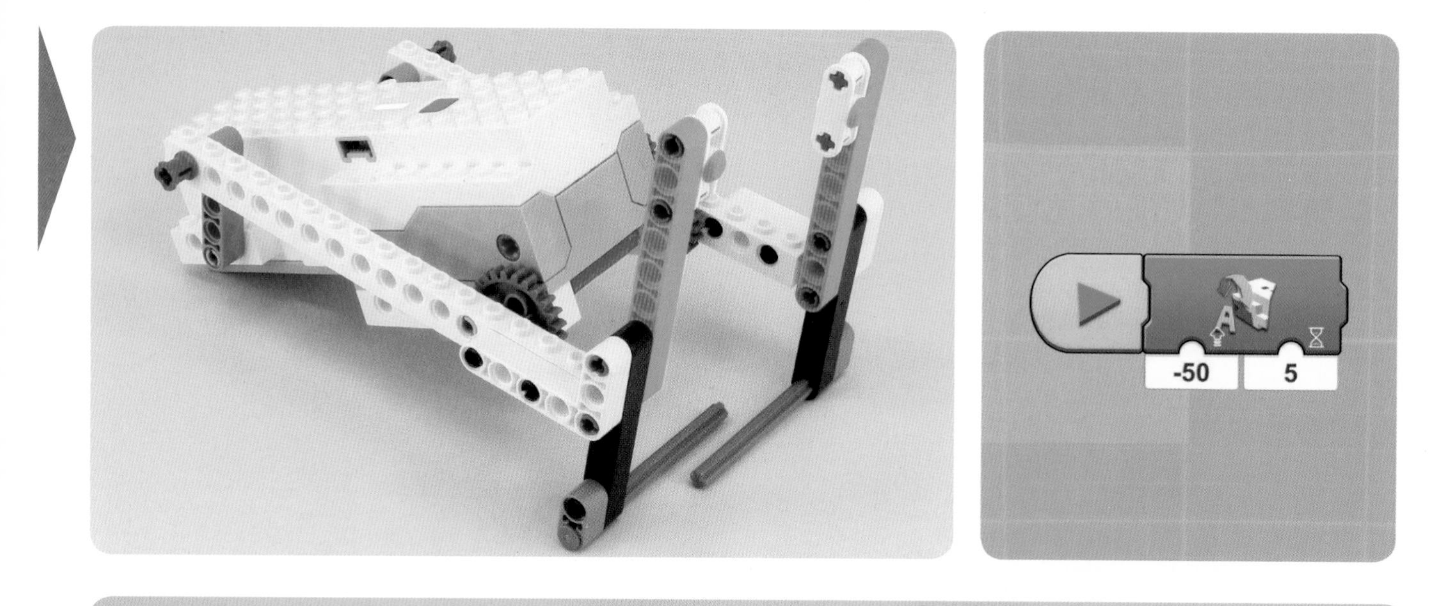
-50
5

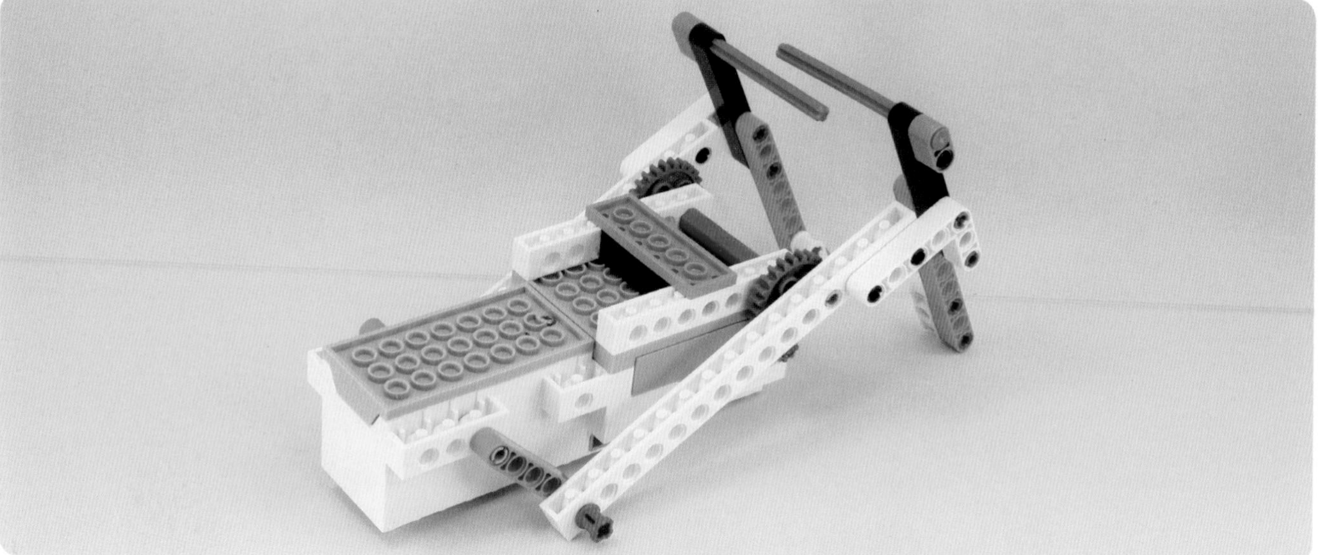

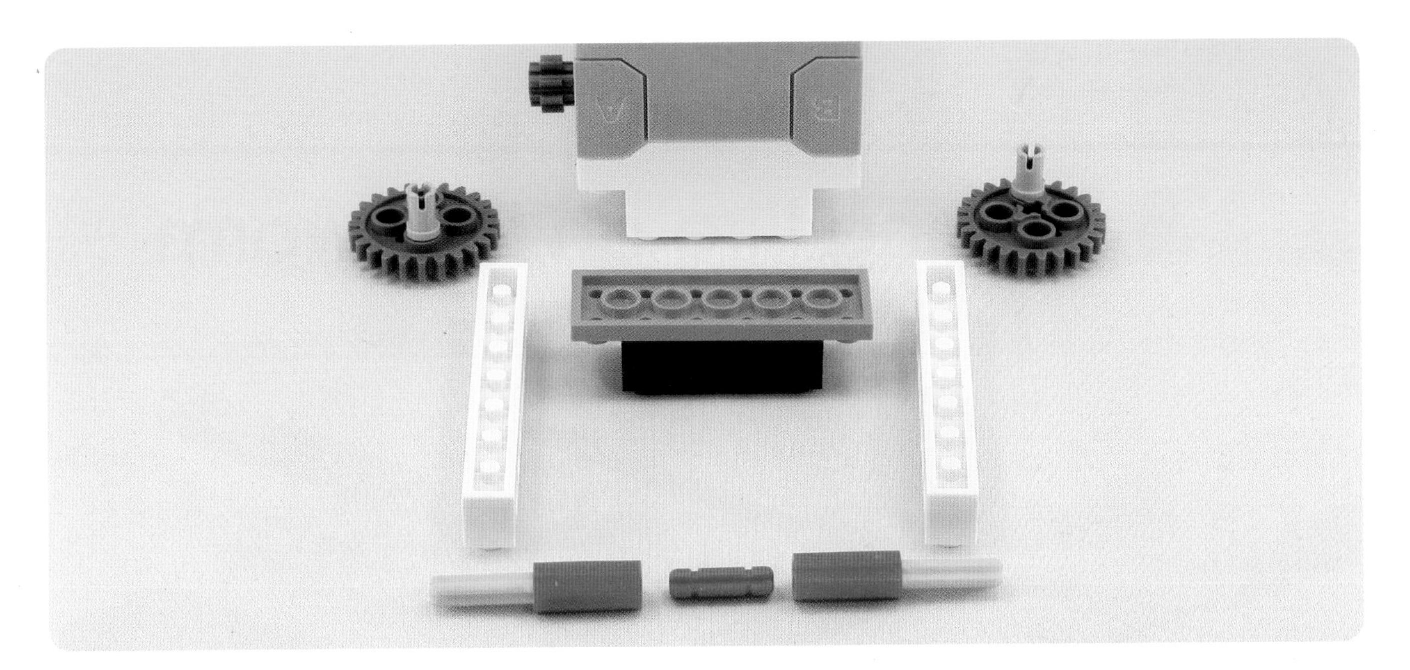

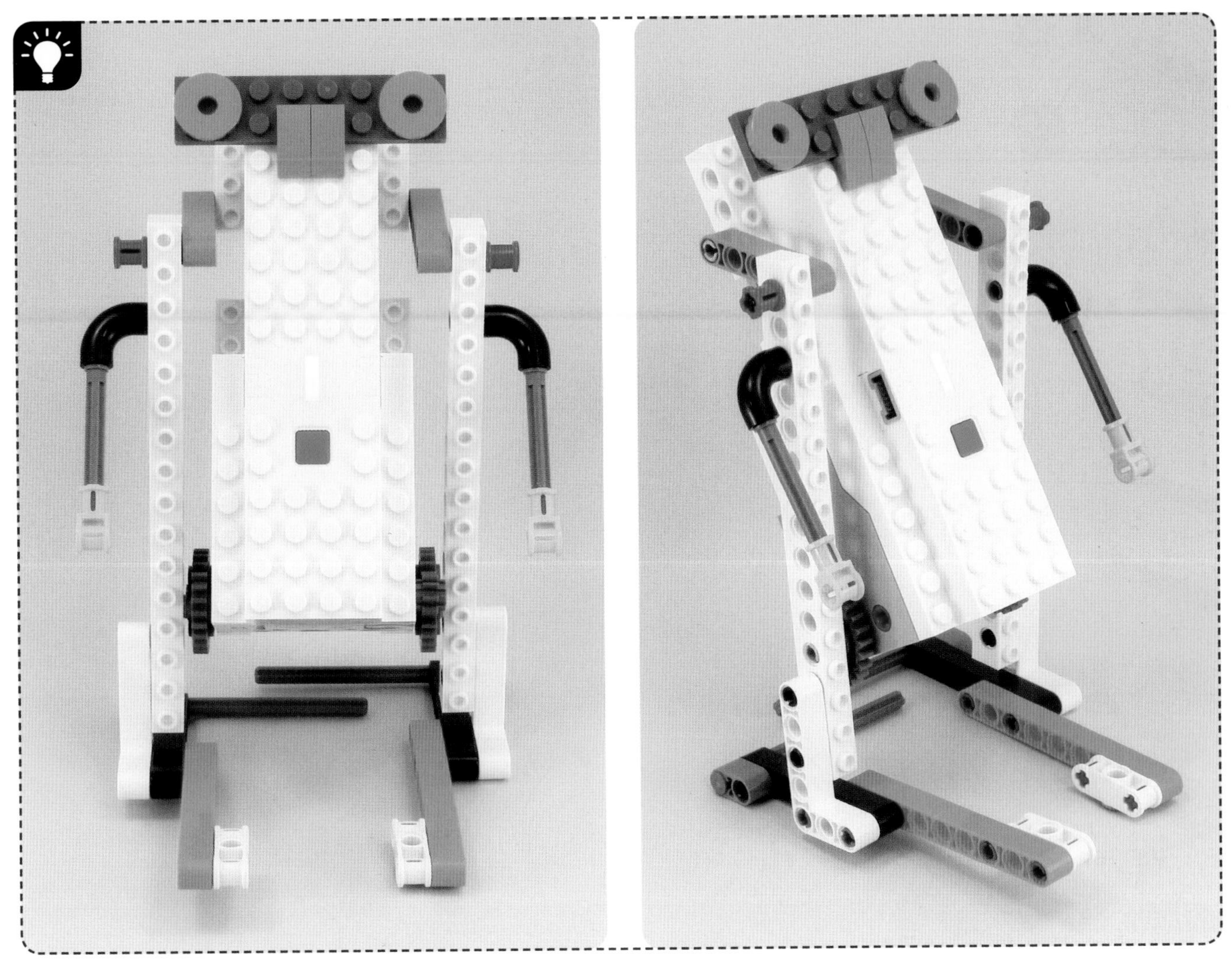

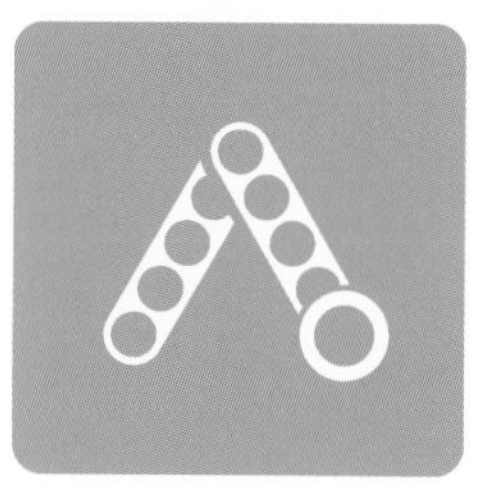

Moving like an inchworm

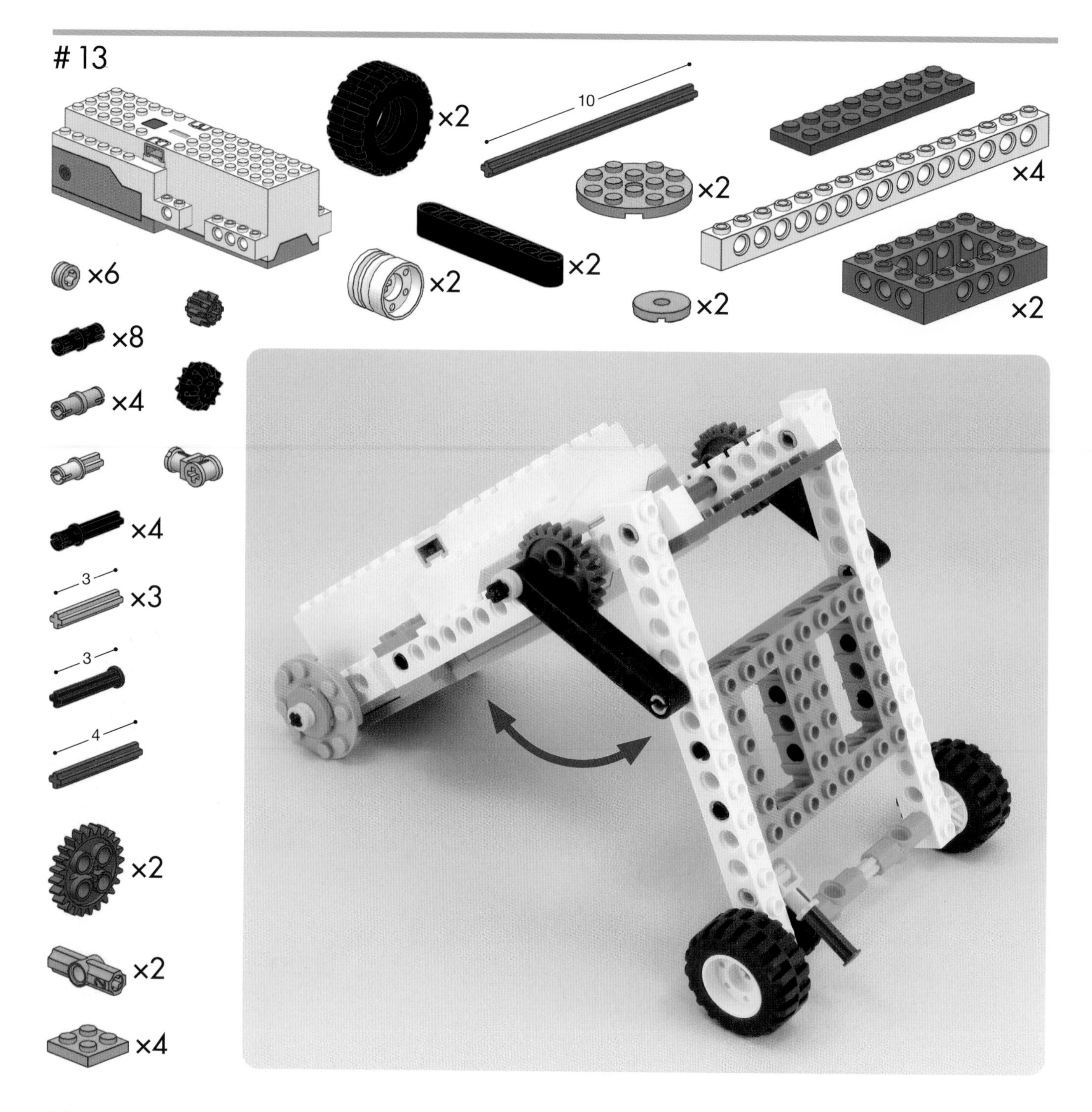

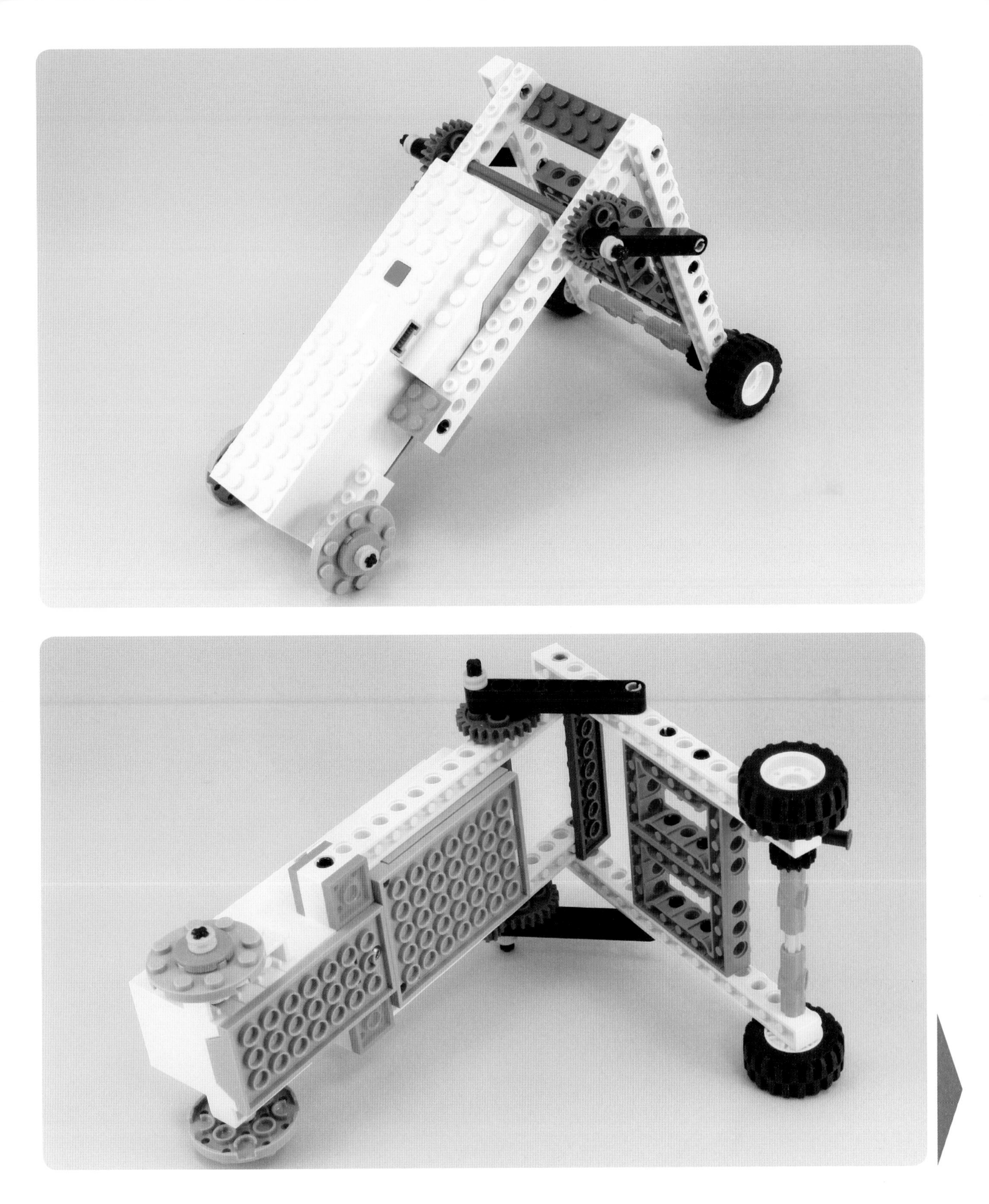

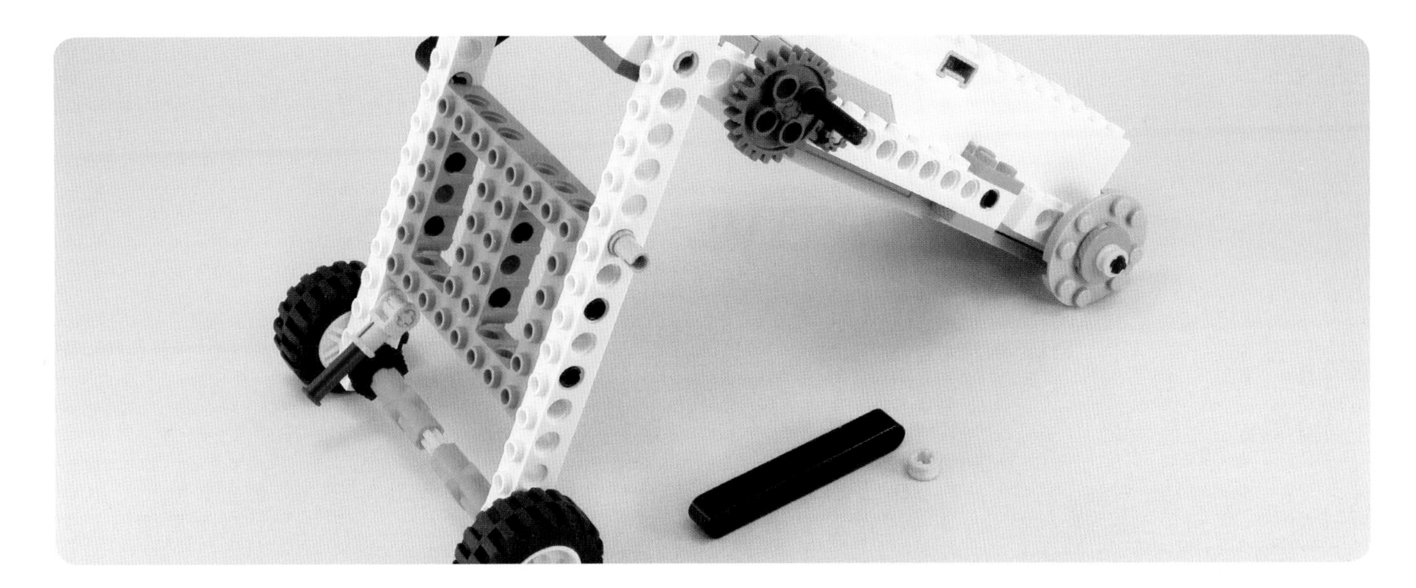

100
5

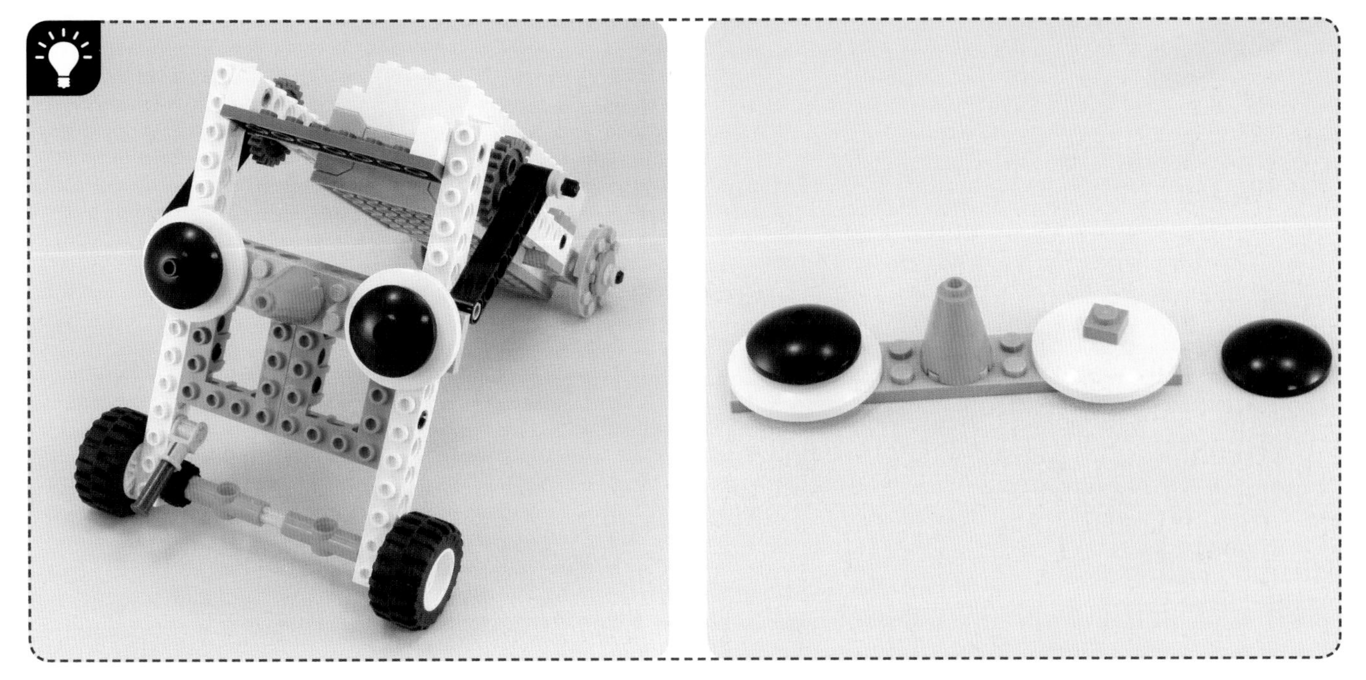

Other ways to move

14

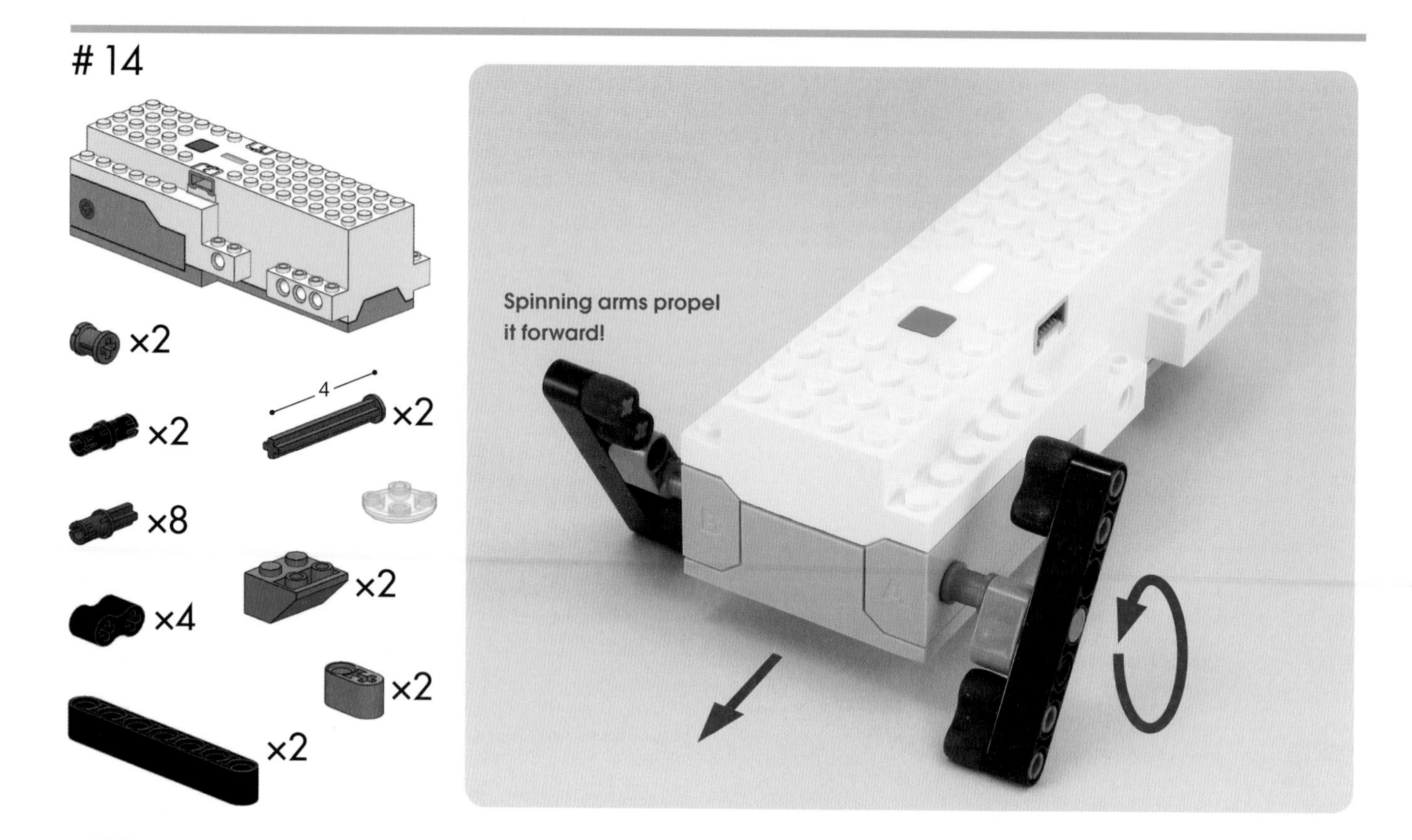

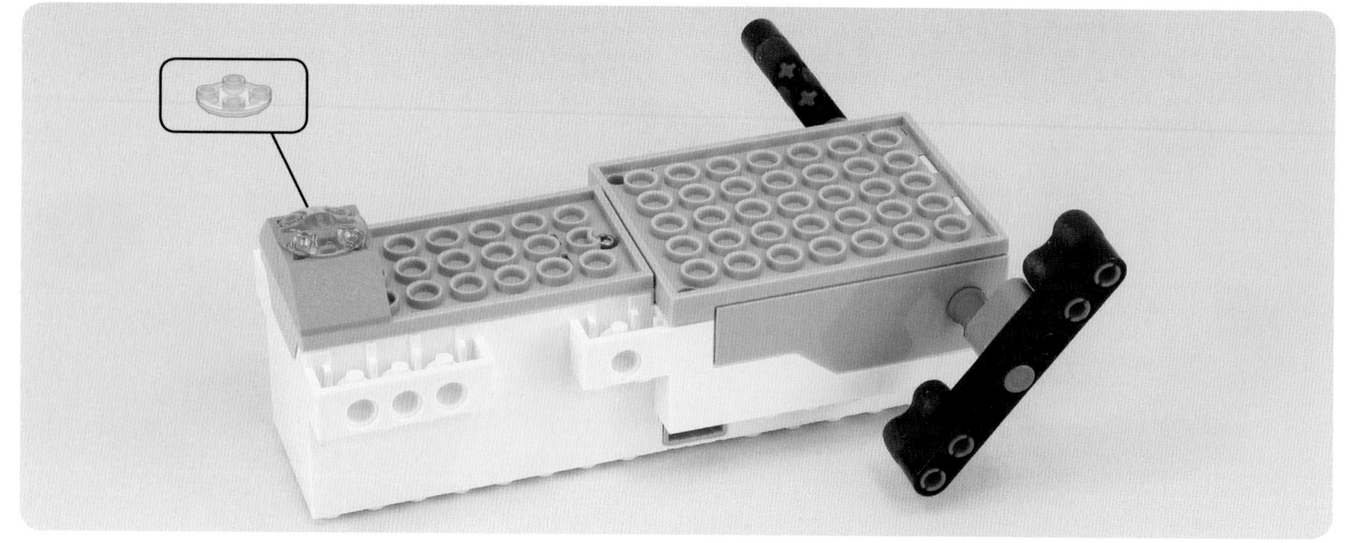

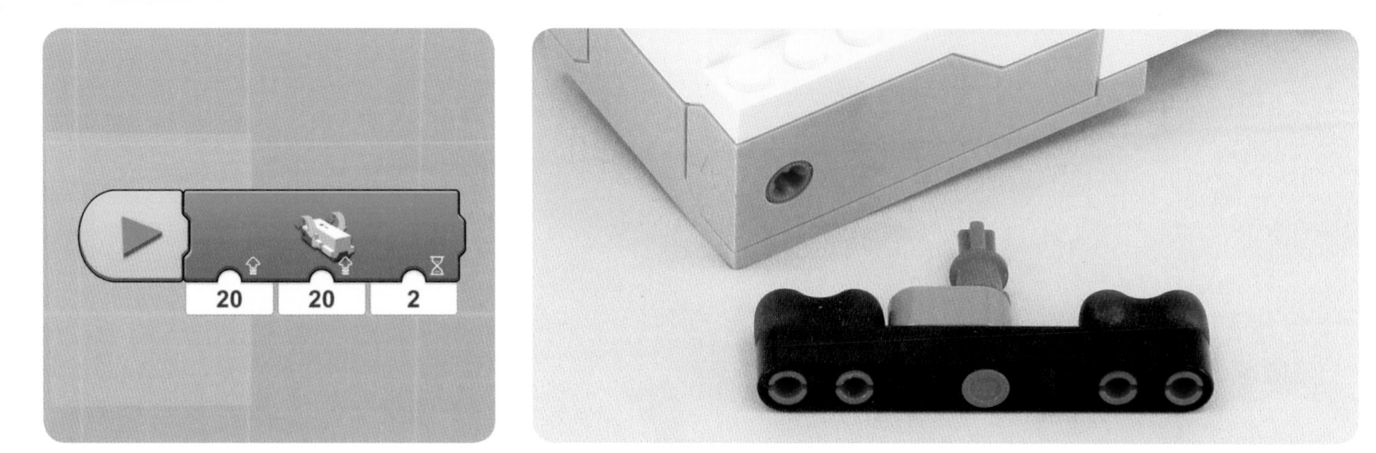
20
20
2

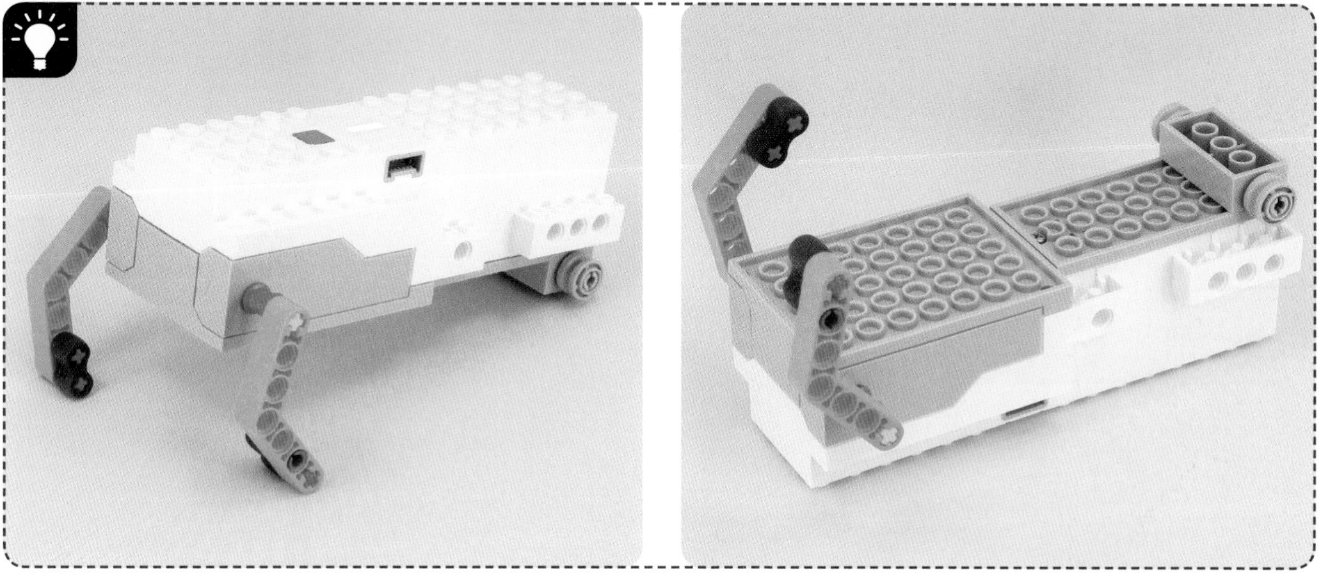

15

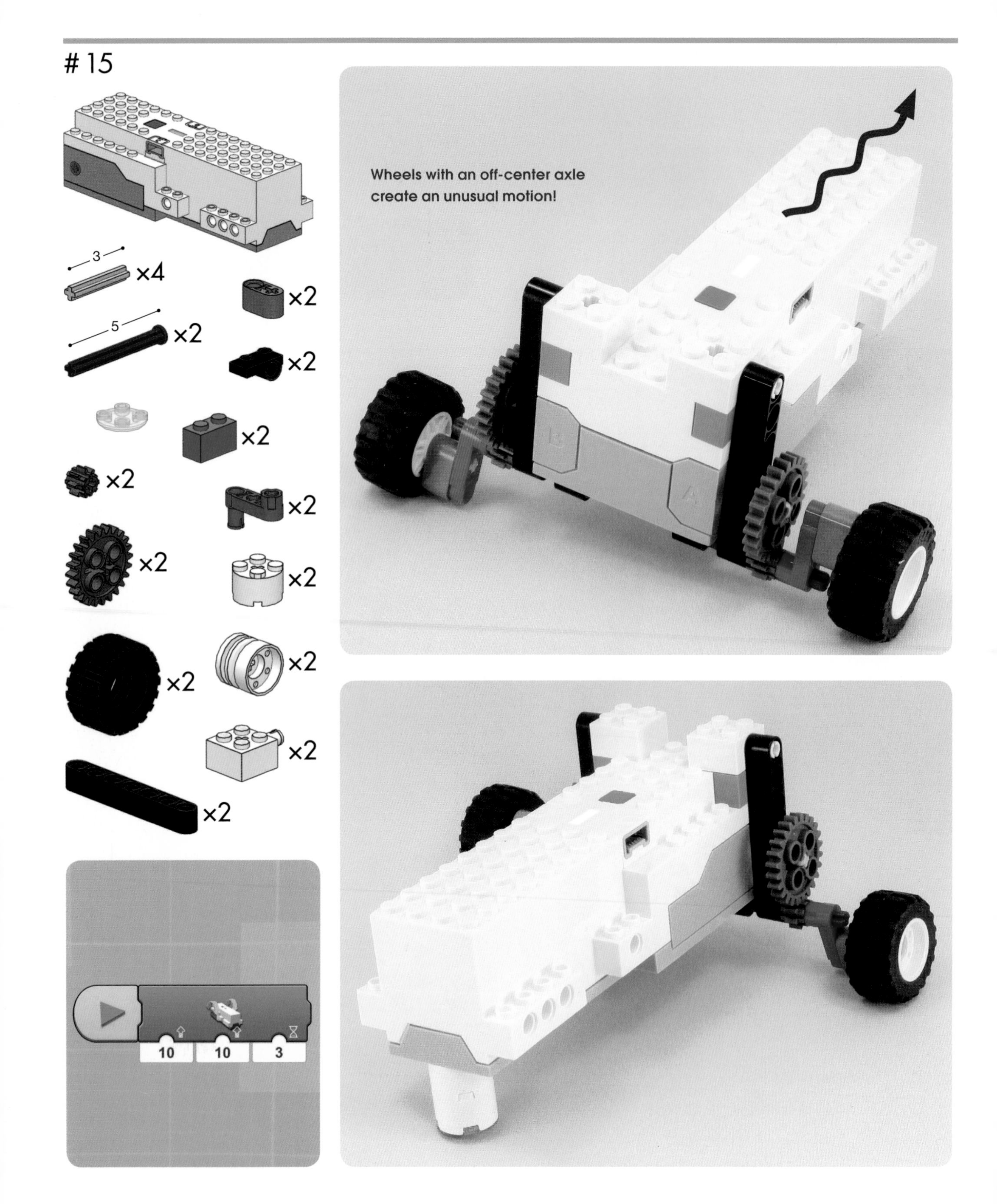

×4
×2
×2
×2
×2
×2
×2
×2
×2
×2
×2
×2
×2
×2
10
10
3
Wheels with an off-center axle create an unusual motion!

Joystick Widget You can control your car with this joystick.

This joystick program includes a **Wait** block to introduce a slight delay in the program. Without the delay, the program could get confused since the device would send instructions to your robot continuously—too fast for it to respond!

16

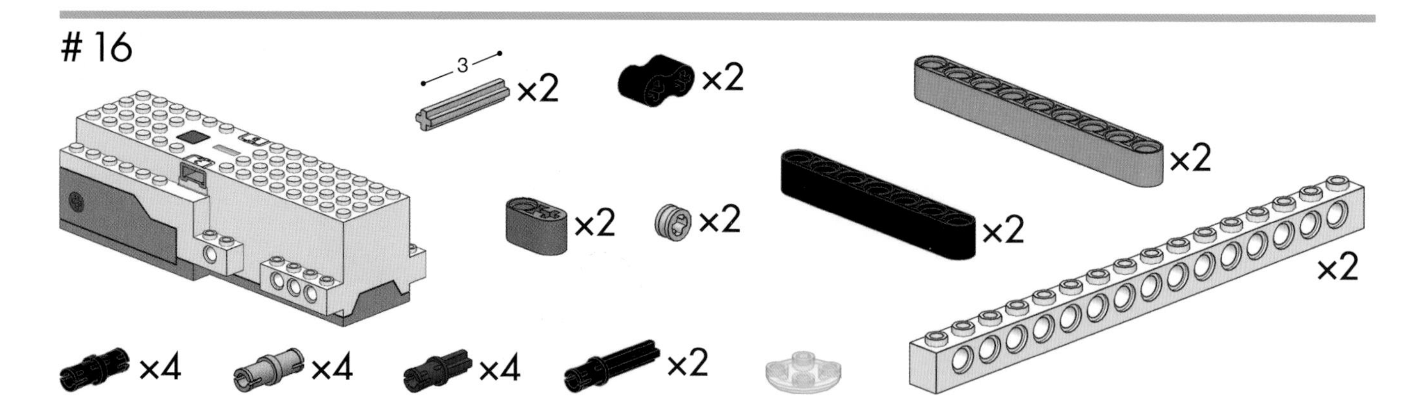
3
×2
×2
×2
×2
×2
×2
×2
×4
×4
×4
×2

Going forward by moving arms back and forth

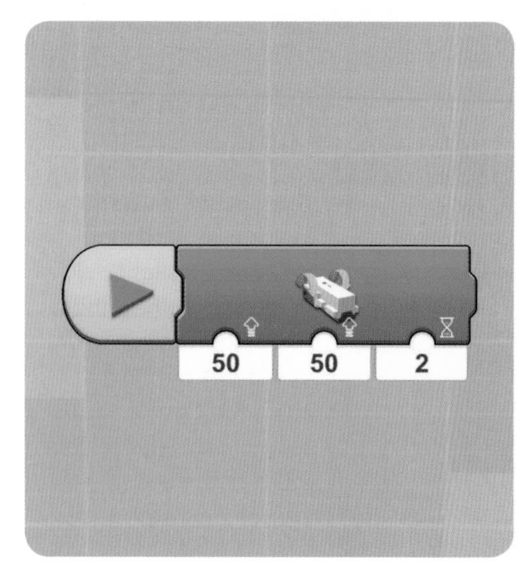
50
50
2

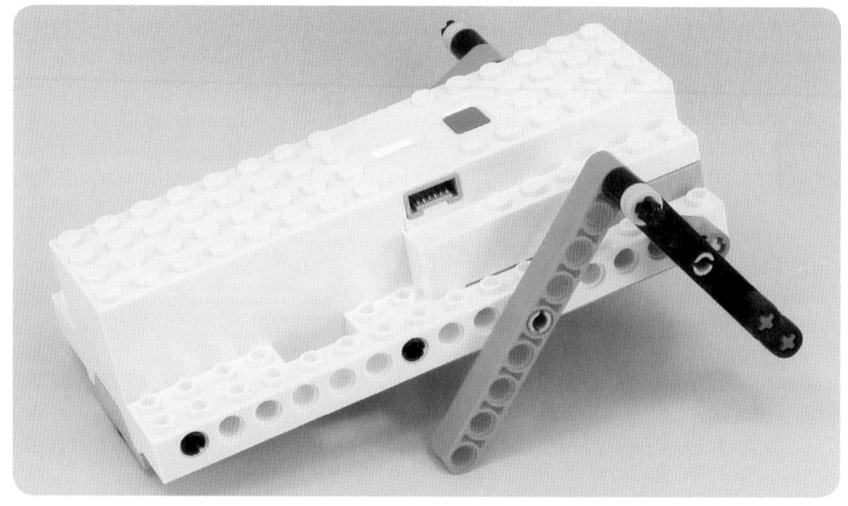

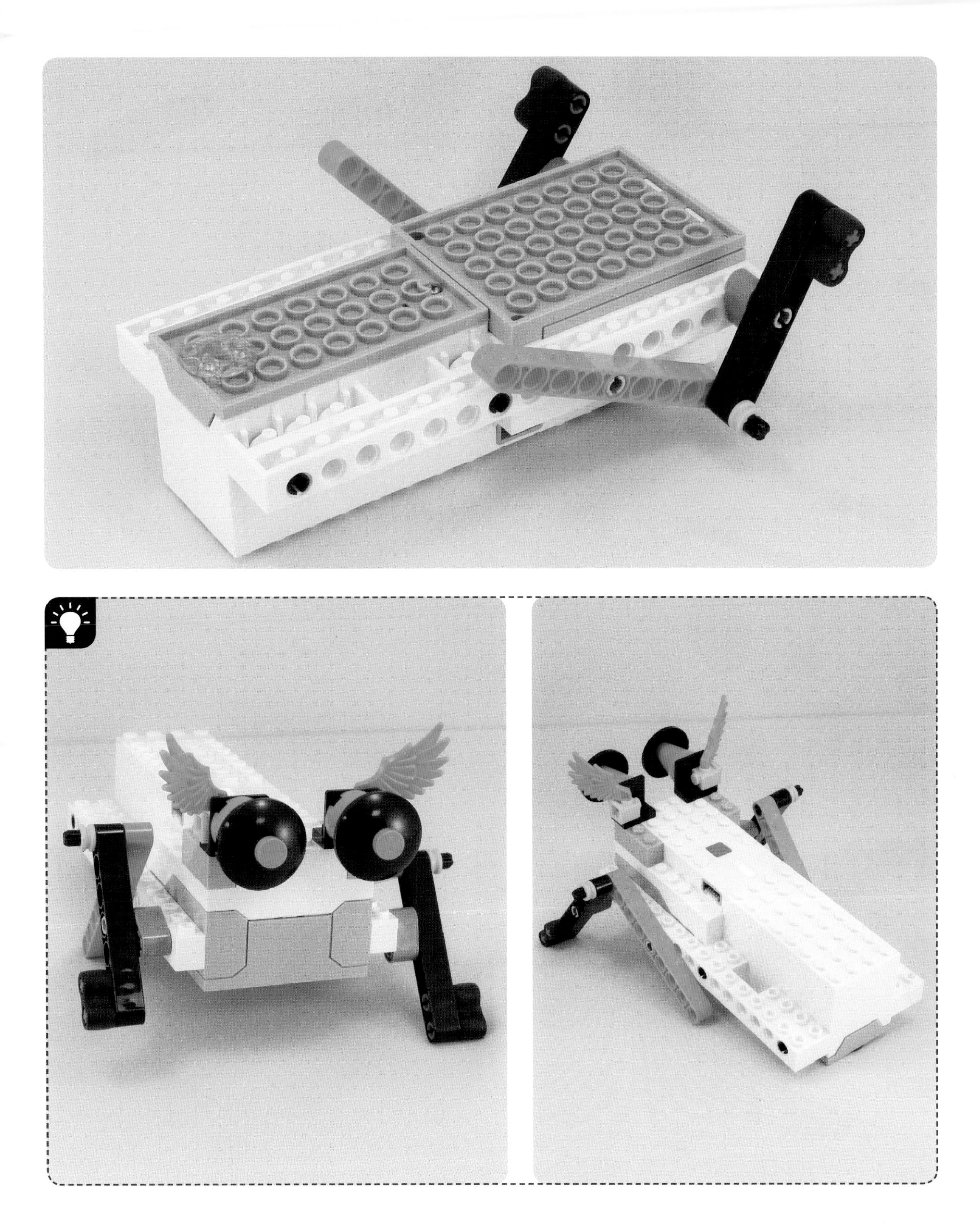

17

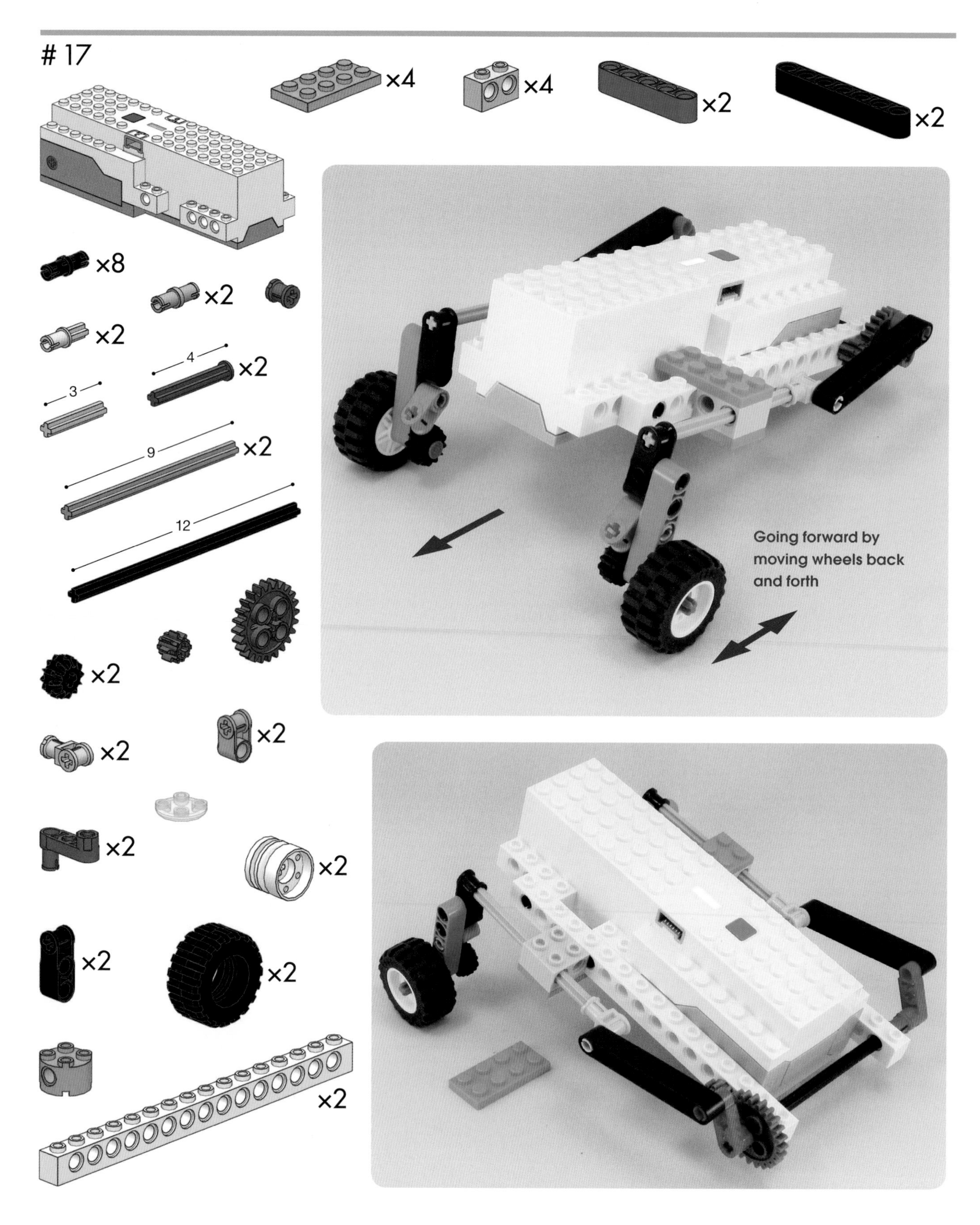

×4
×4
×2
×2
×8
×2
×2
4
×2
3
9
×2
12
×2
×2
×2
×2
×2
×2
×2
×2
Going forward by moving wheels back and forth

B
60
5

18

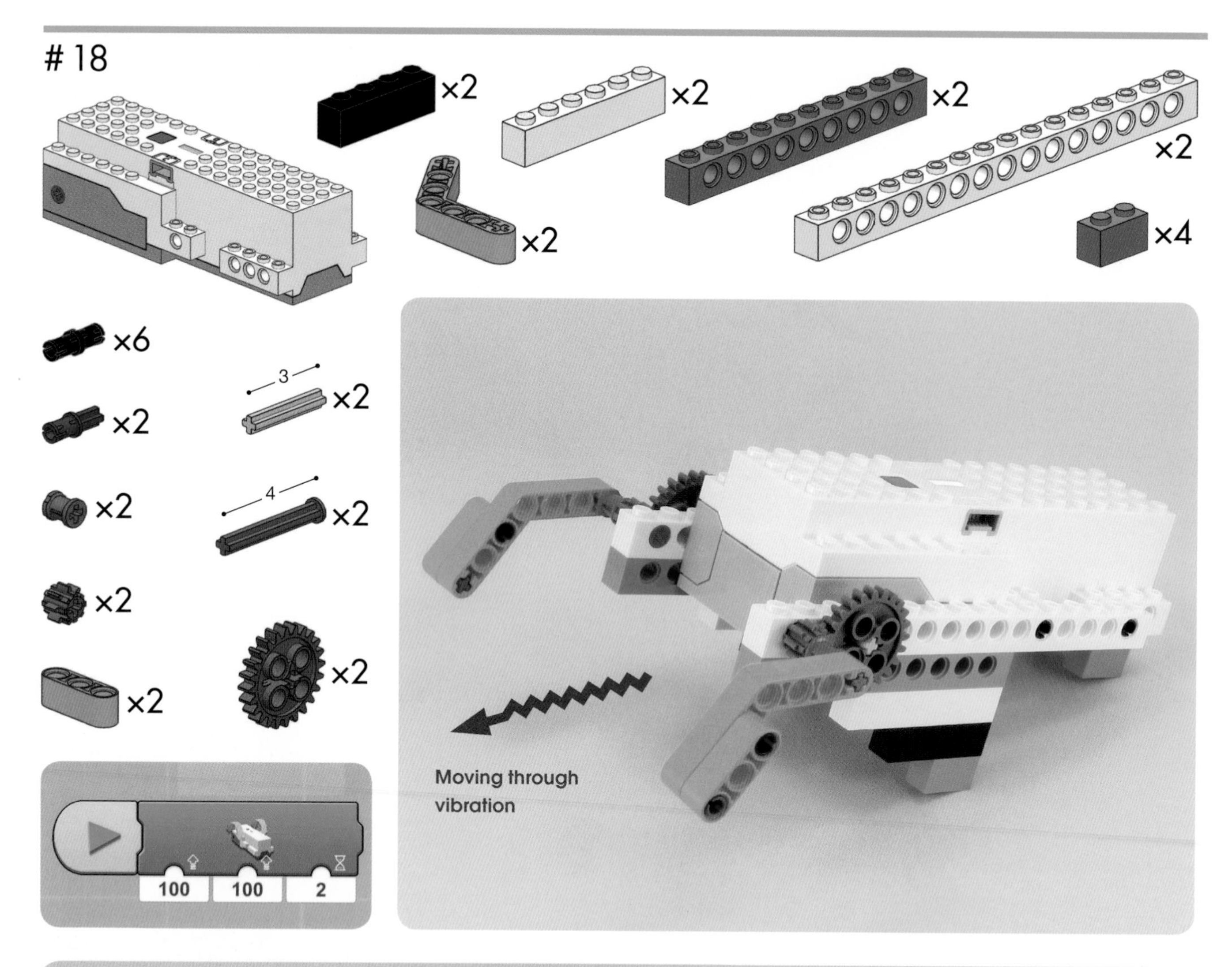

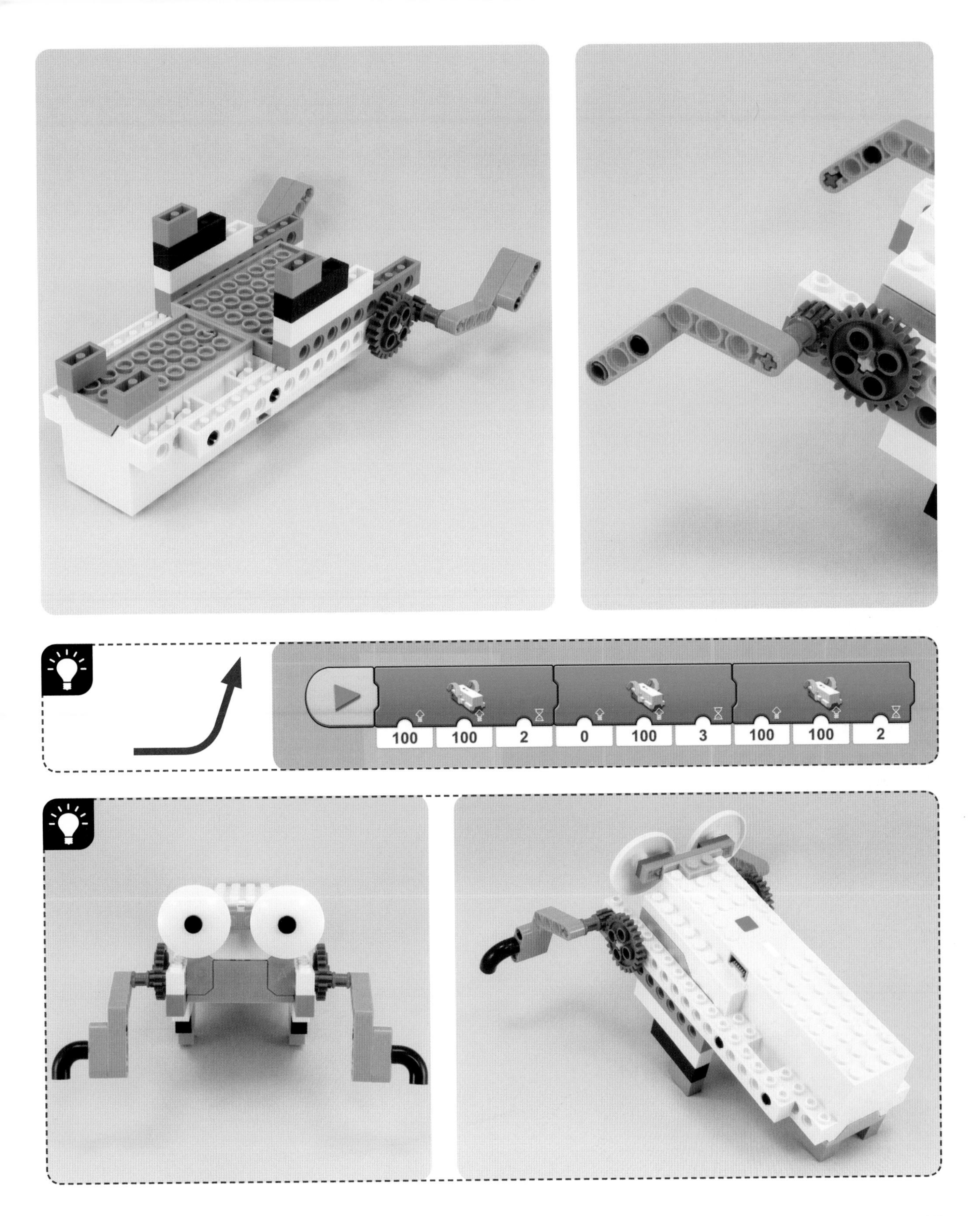
100
100
2
0
100
3
100
100
2

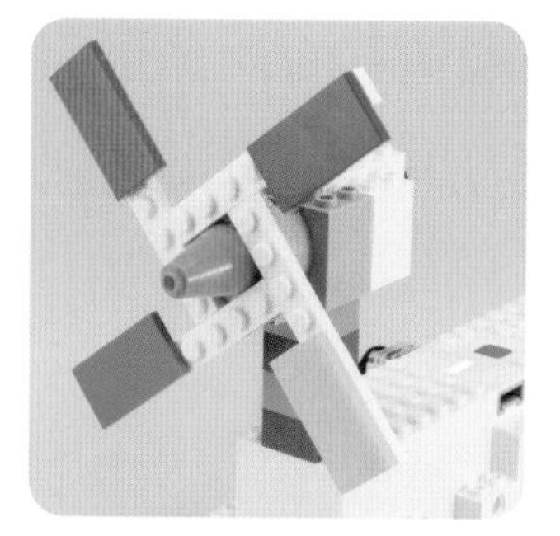

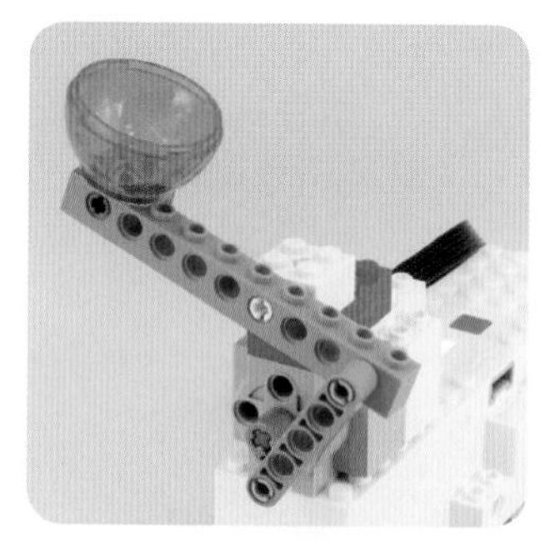

PART 2

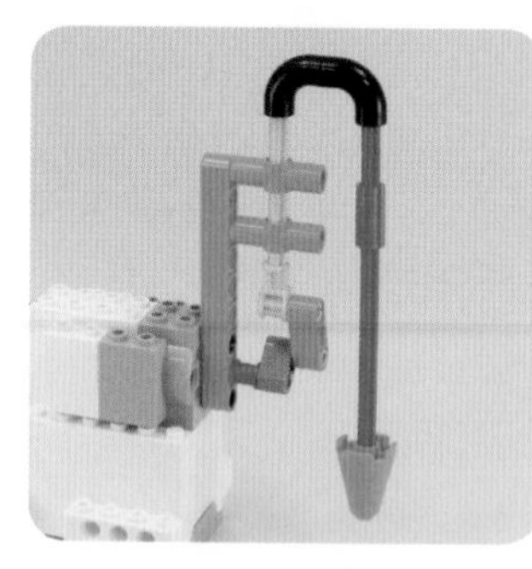

Using the Interactive Motor

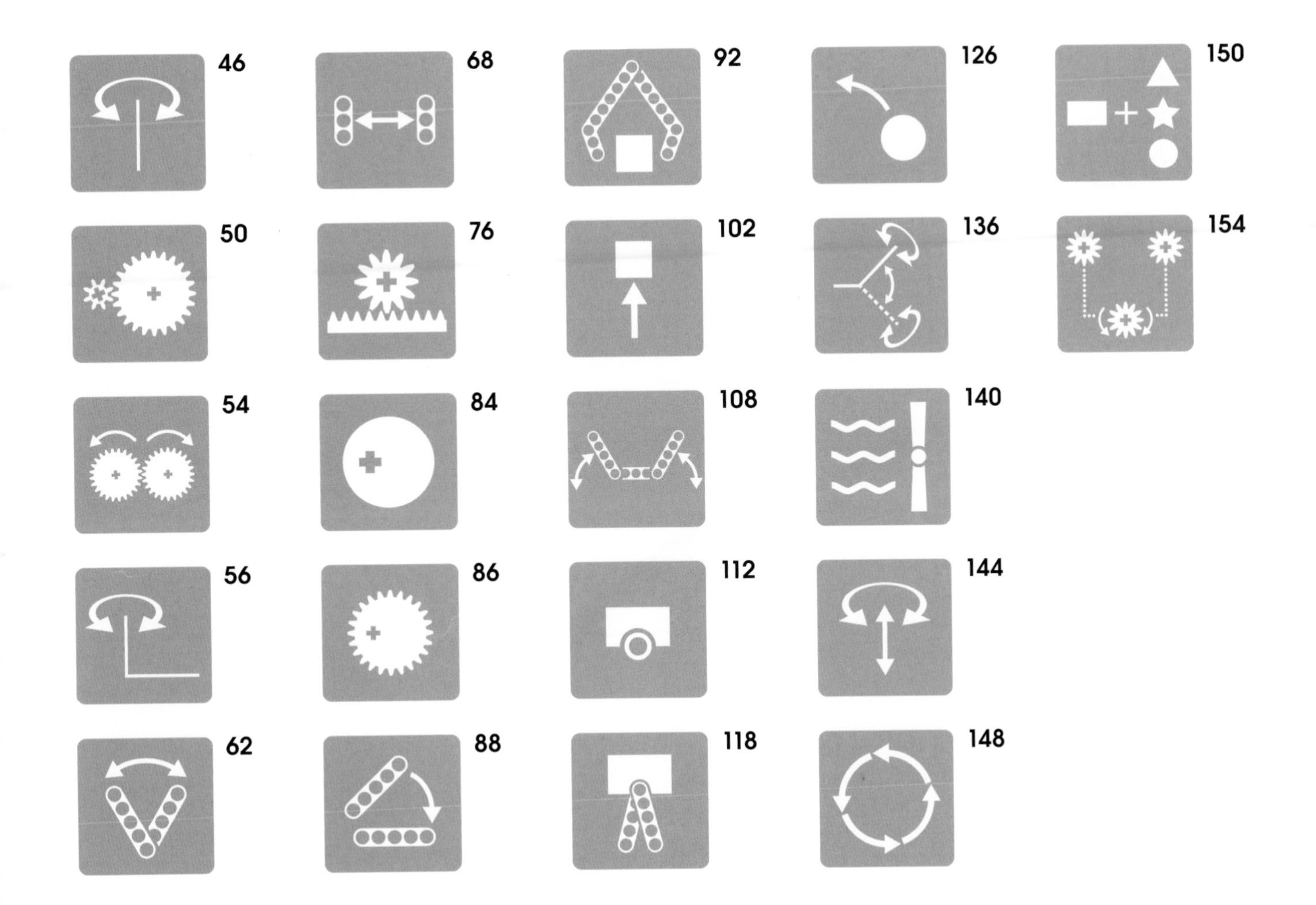

Spinning things

19

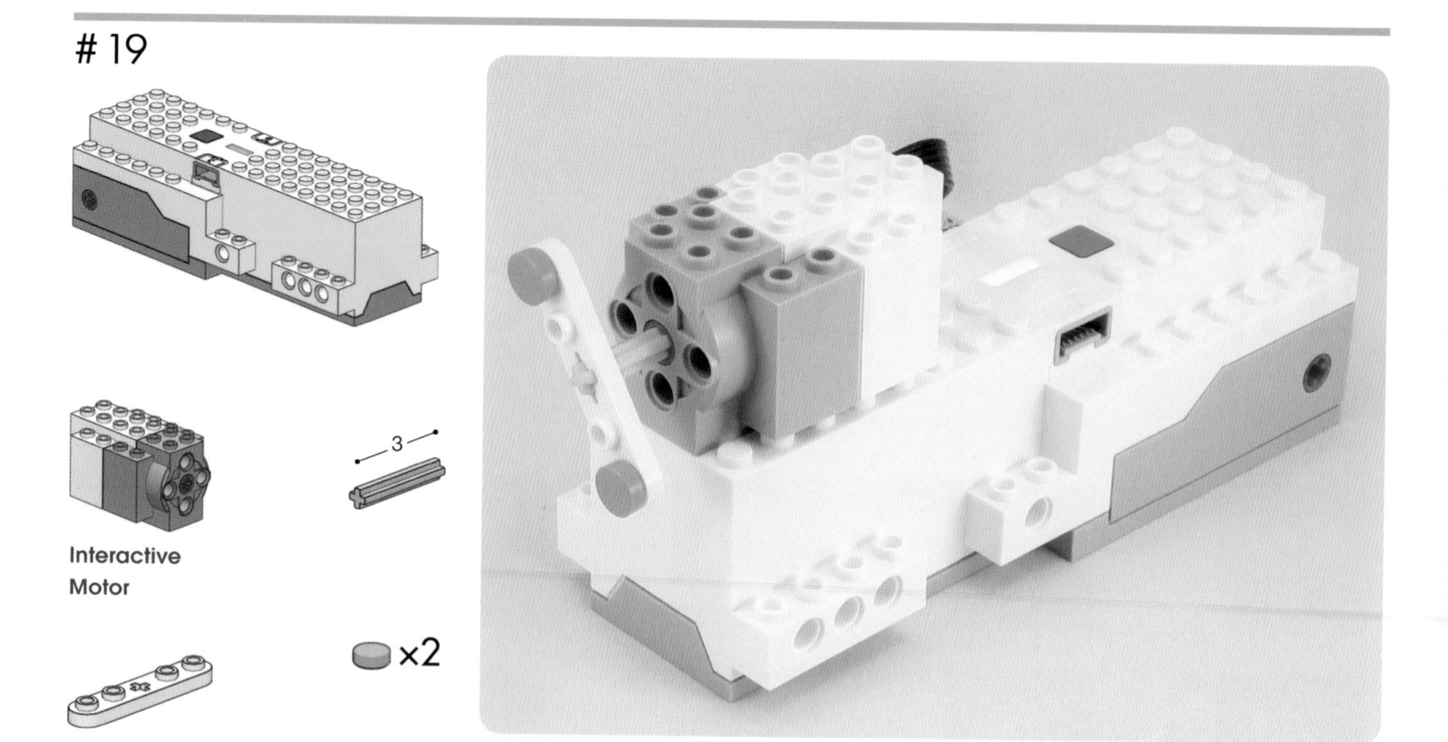

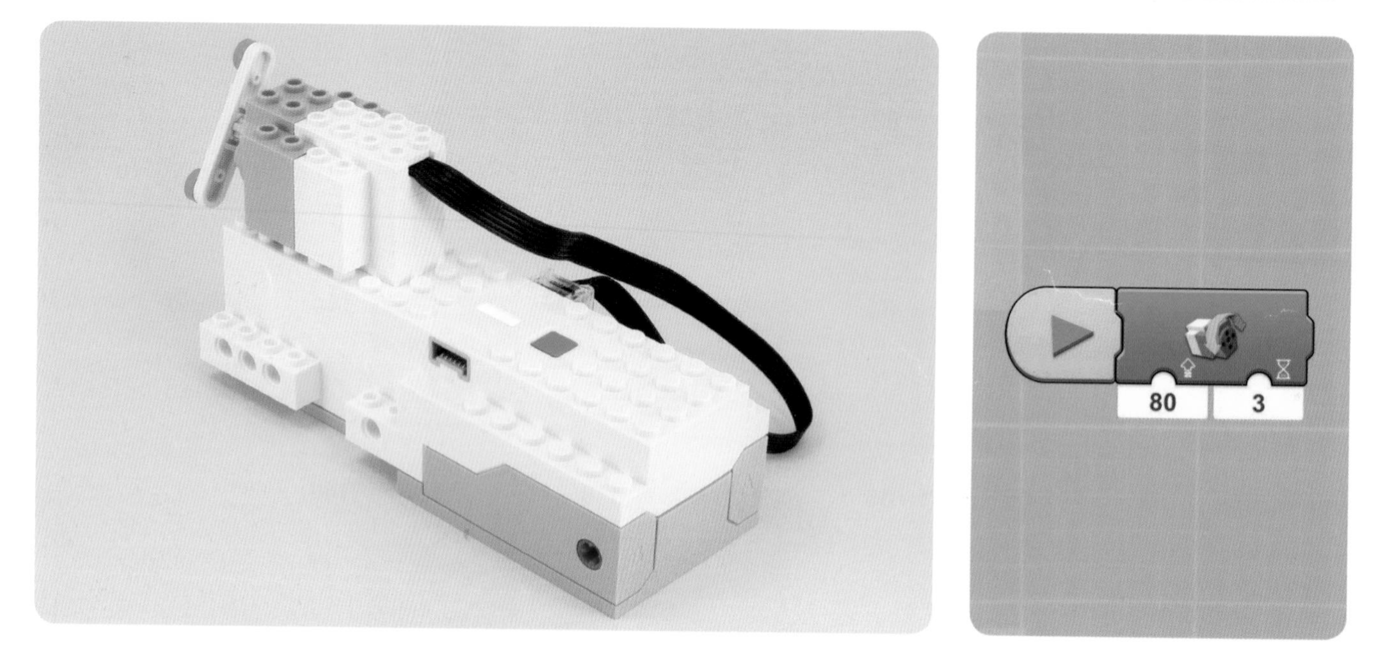

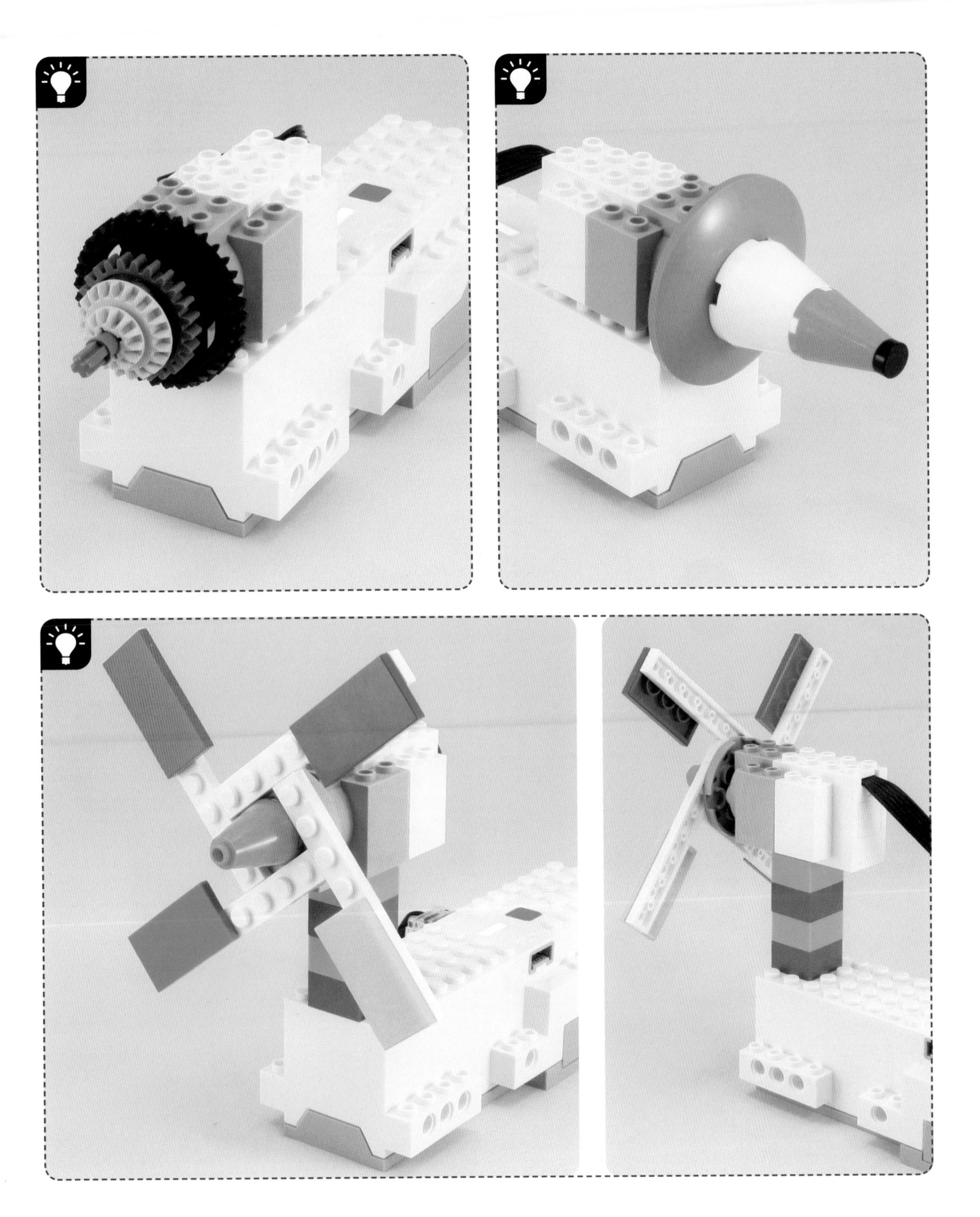

#20

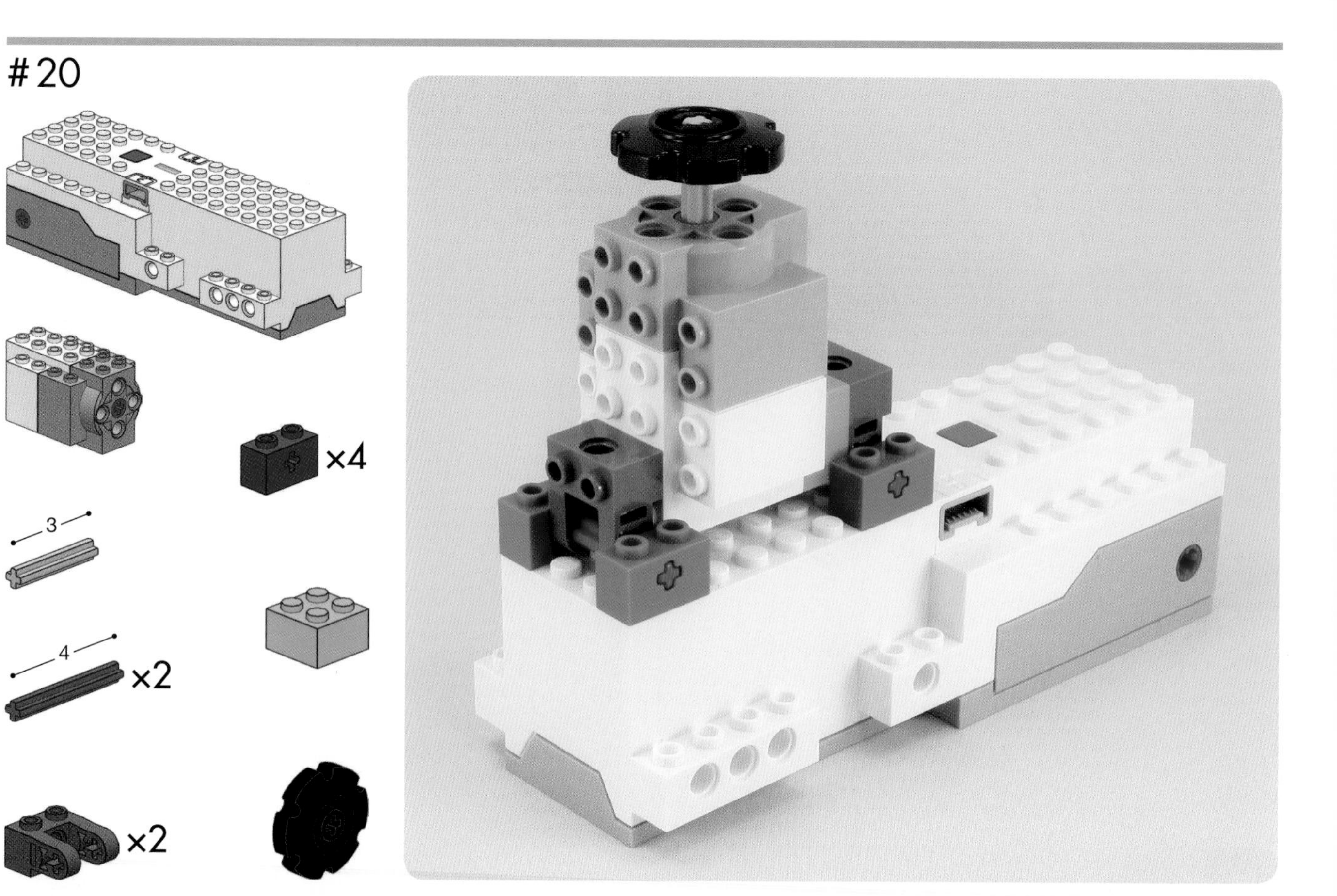

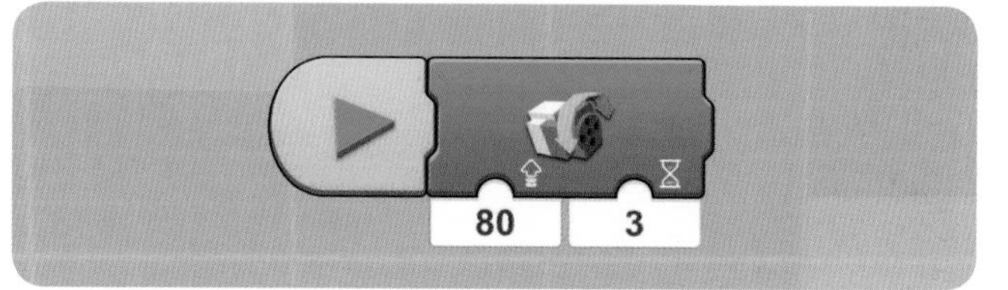

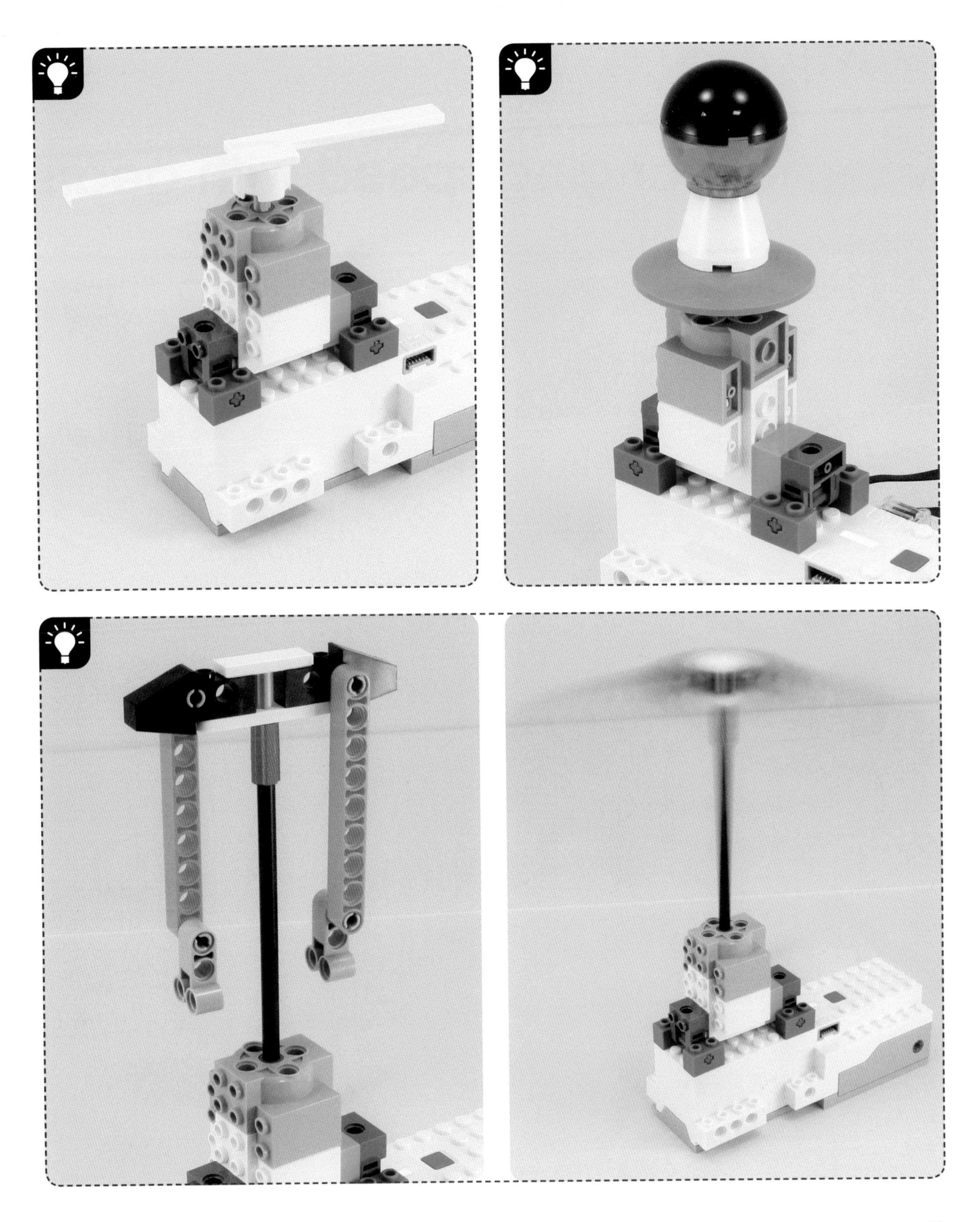

Changing speed with gears

#21

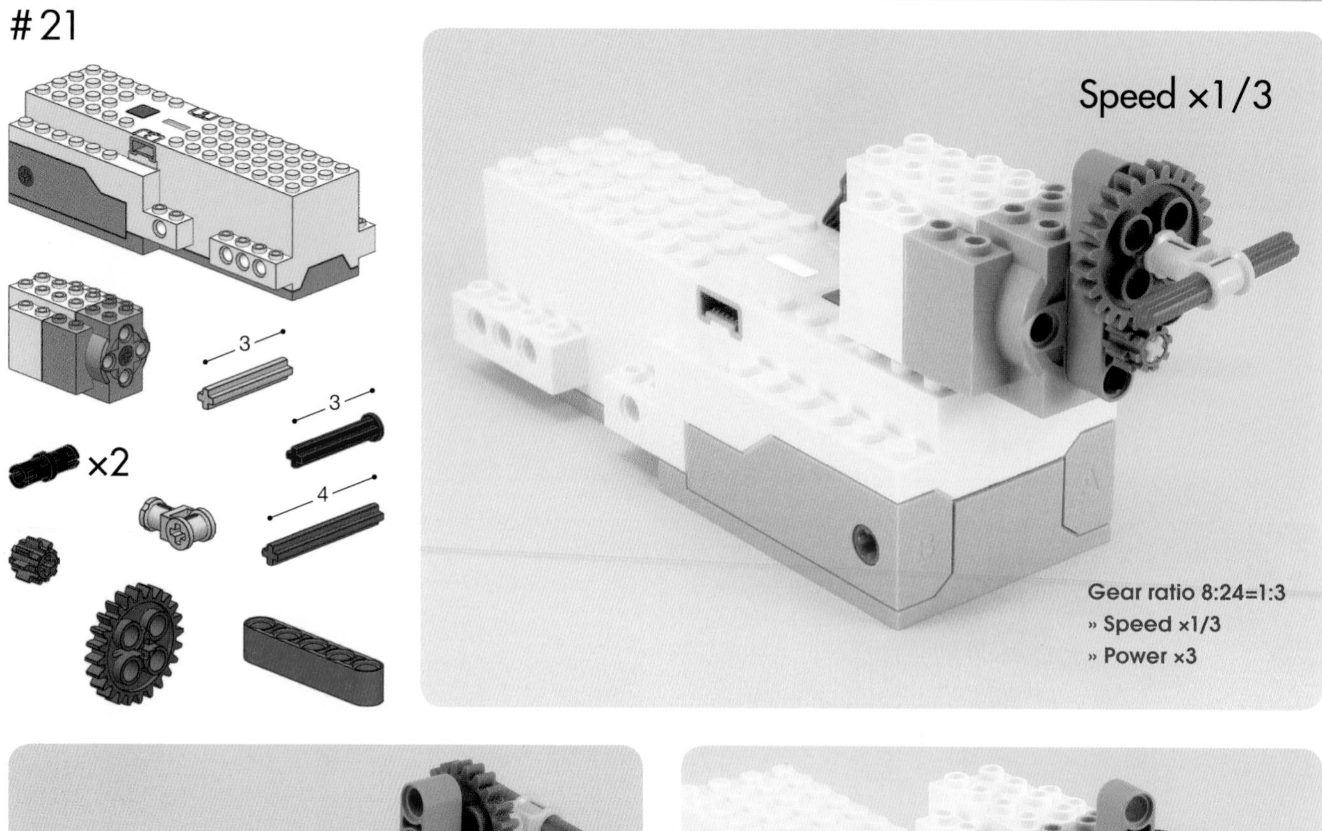

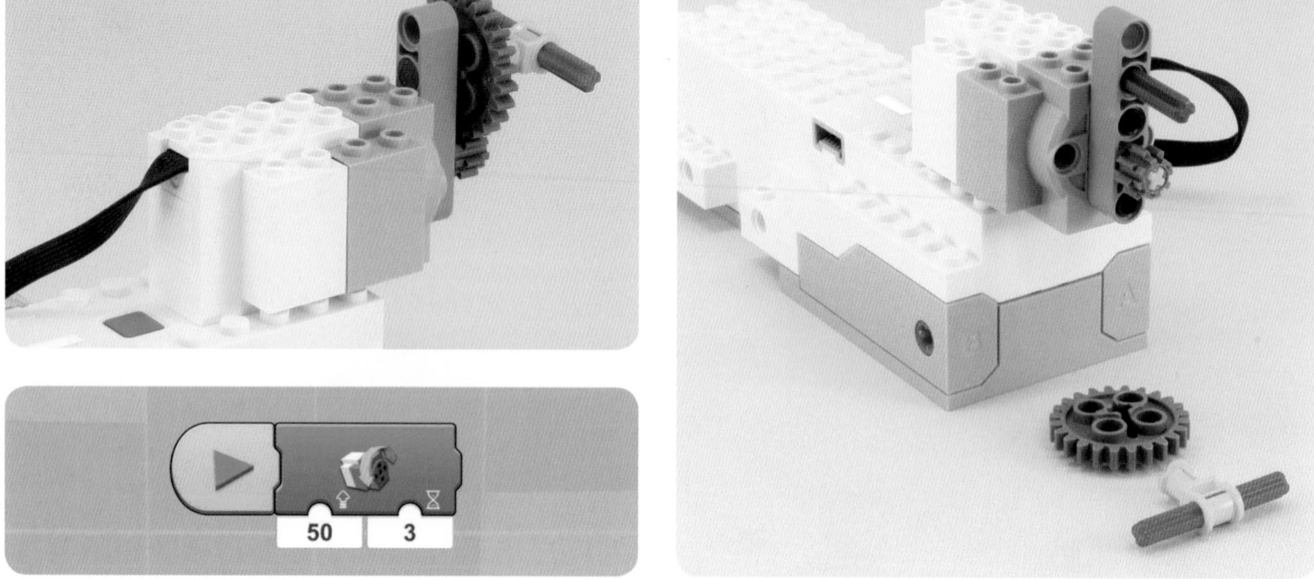

#22

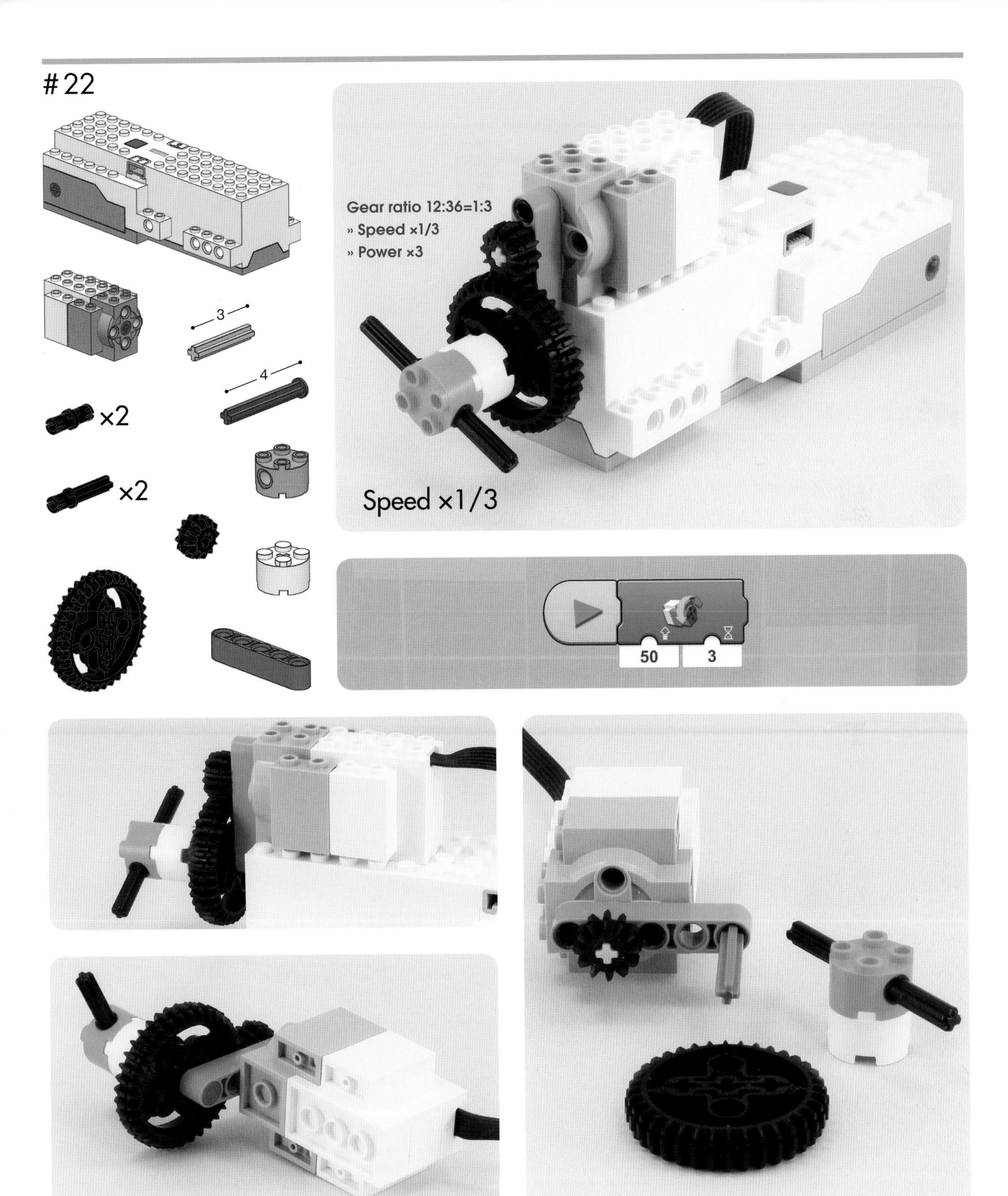

#23

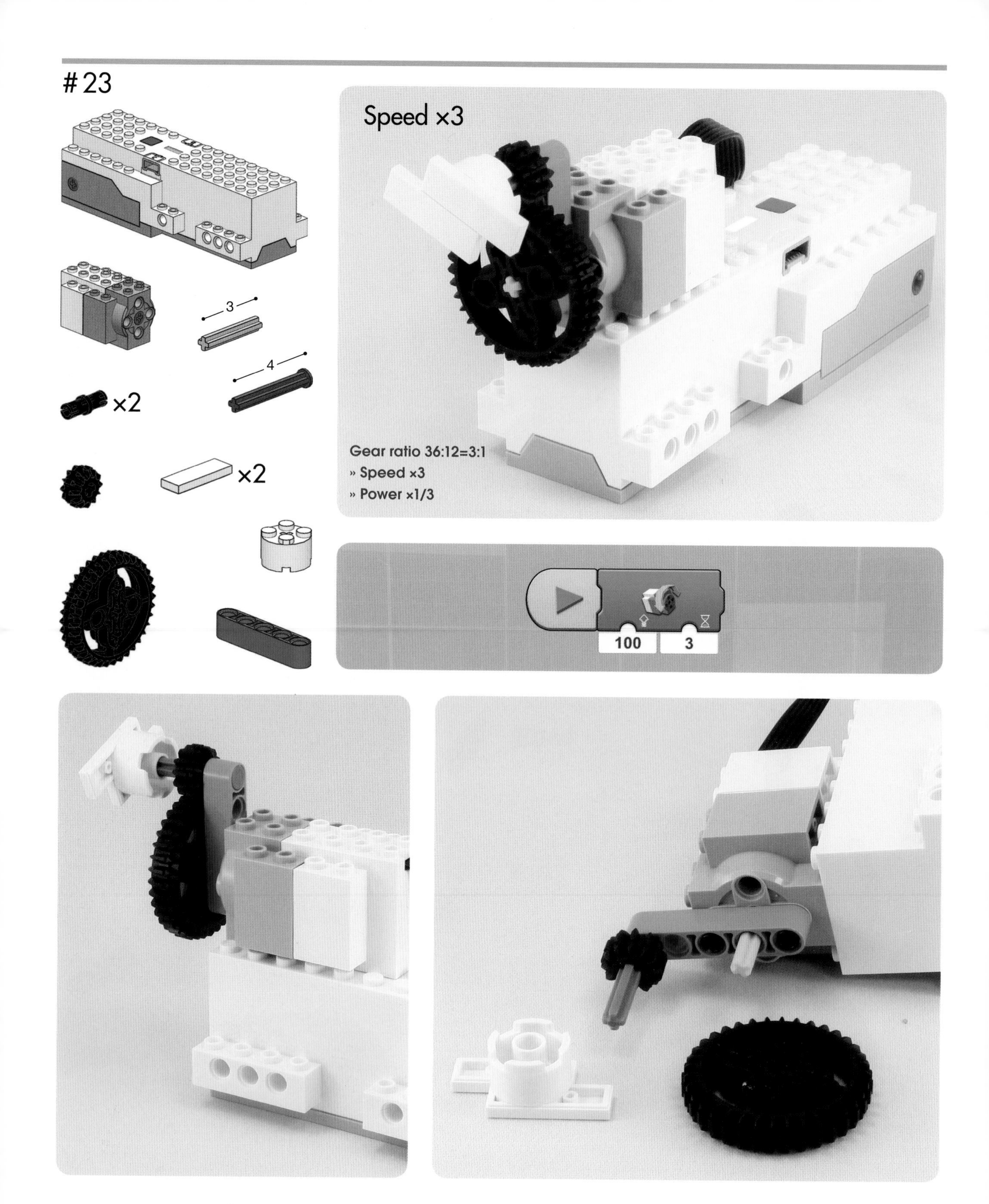

#24

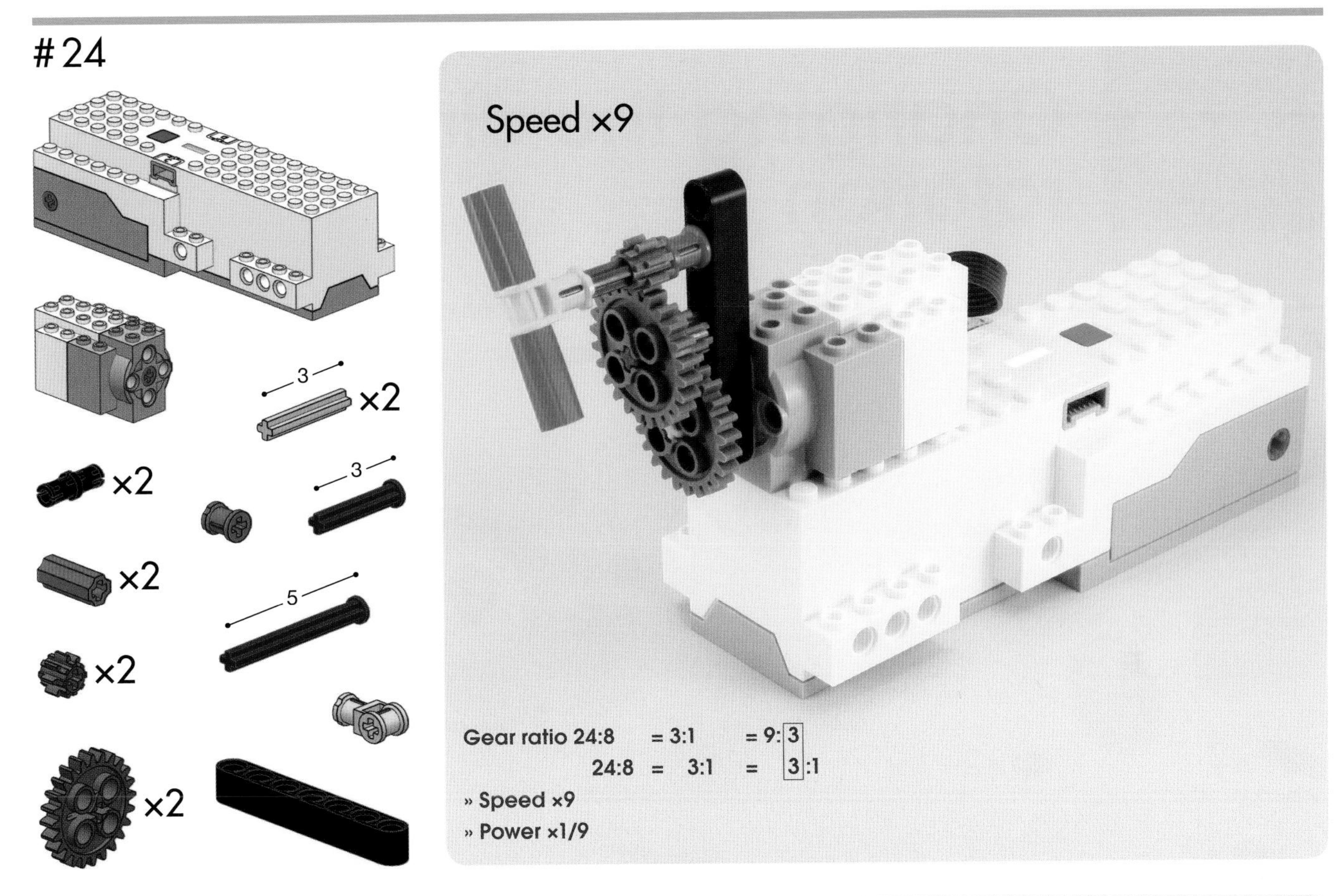
3
×2
×2
3
×2
5
×2
×2
Speed ×9
Gear ratio 24:8 = 3:1 = 9:3
24:8 = 3:1 = 3:1
» Speed ×9
» Power ×1/9

3

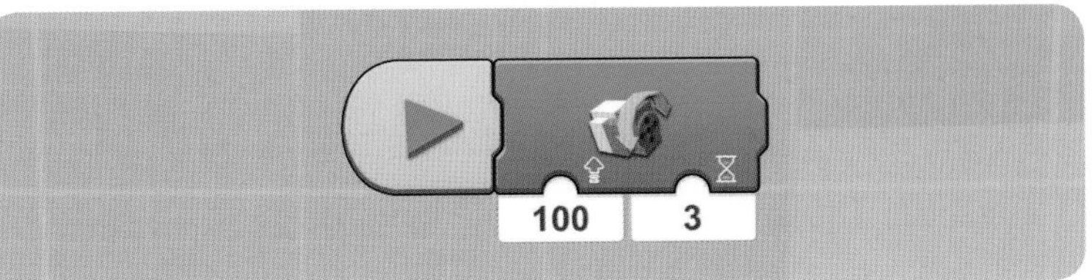
100
3

Changing the direction of rotation

#25

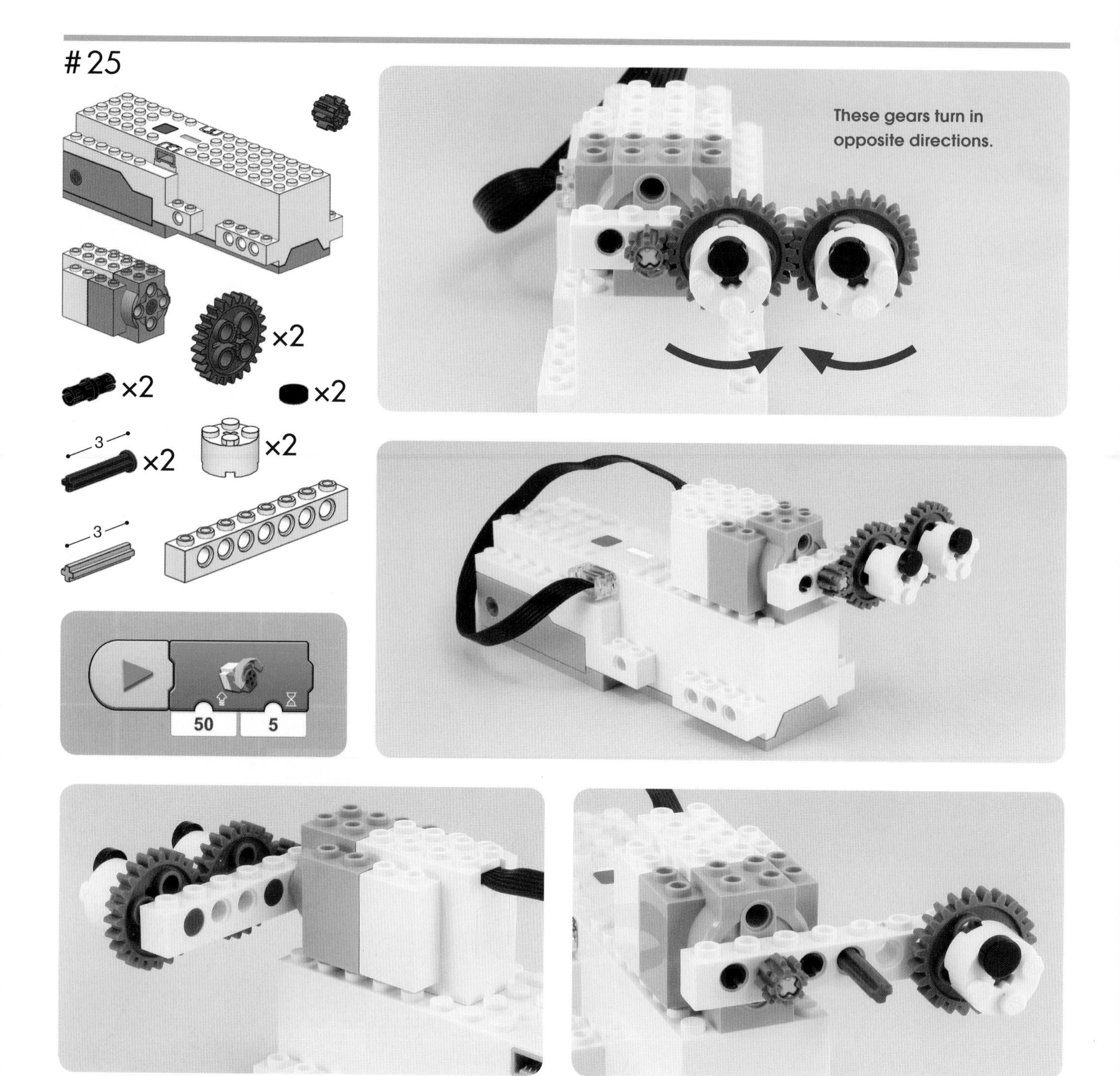

26

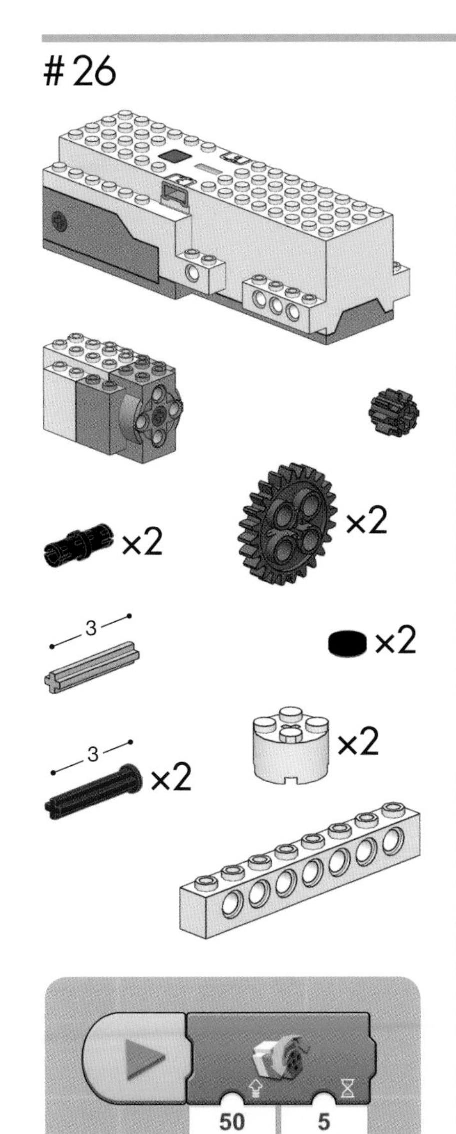

When you add a gear in the middle, two gears at each end turn in the same direction.

Changing the orientation of rotation

27

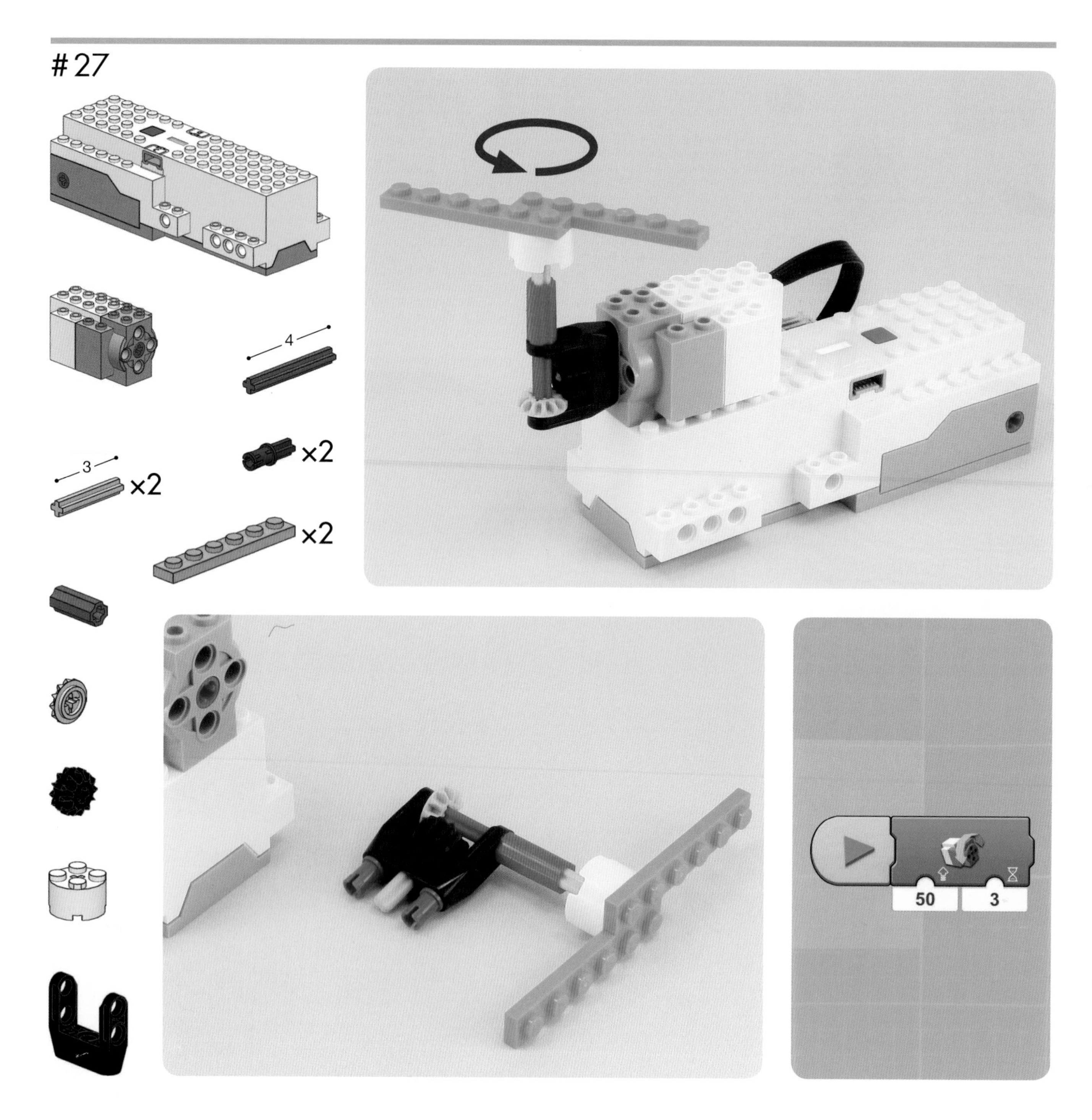

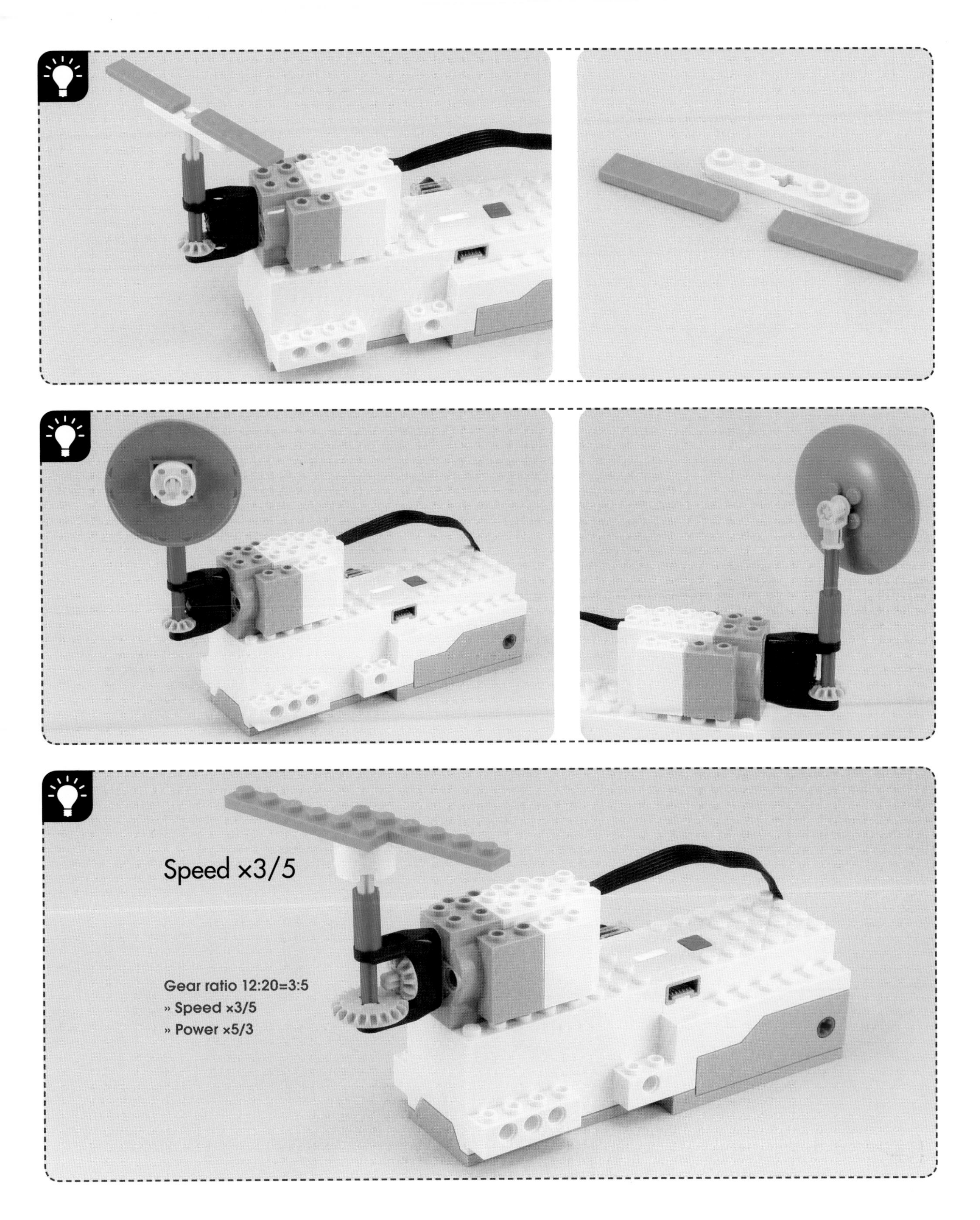
Speed ×3/5
Gear ratio 12:20=3:5
» Speed ×3/5
» Power ×5/3

#28

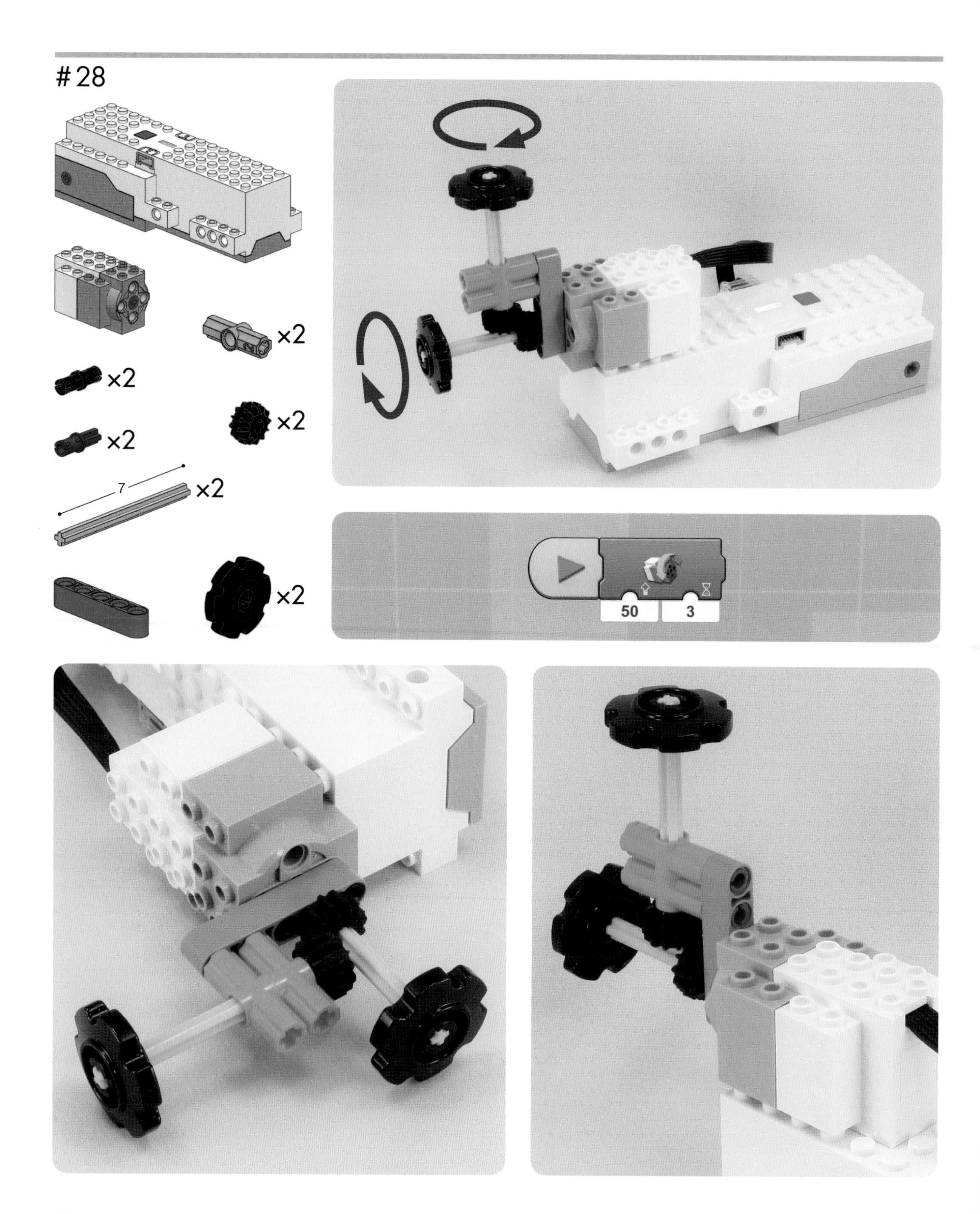

#29

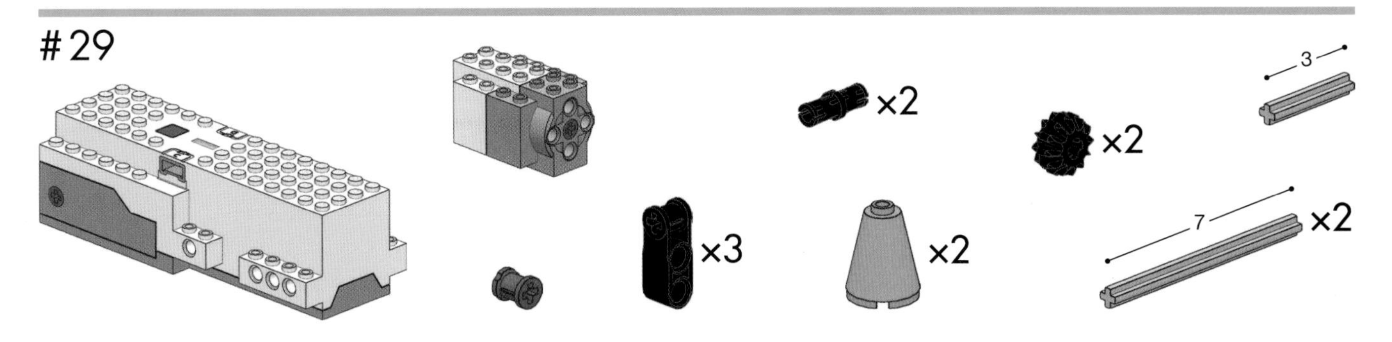

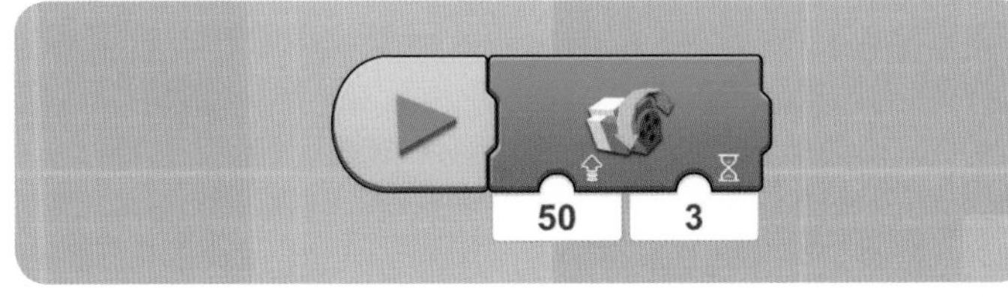

#30

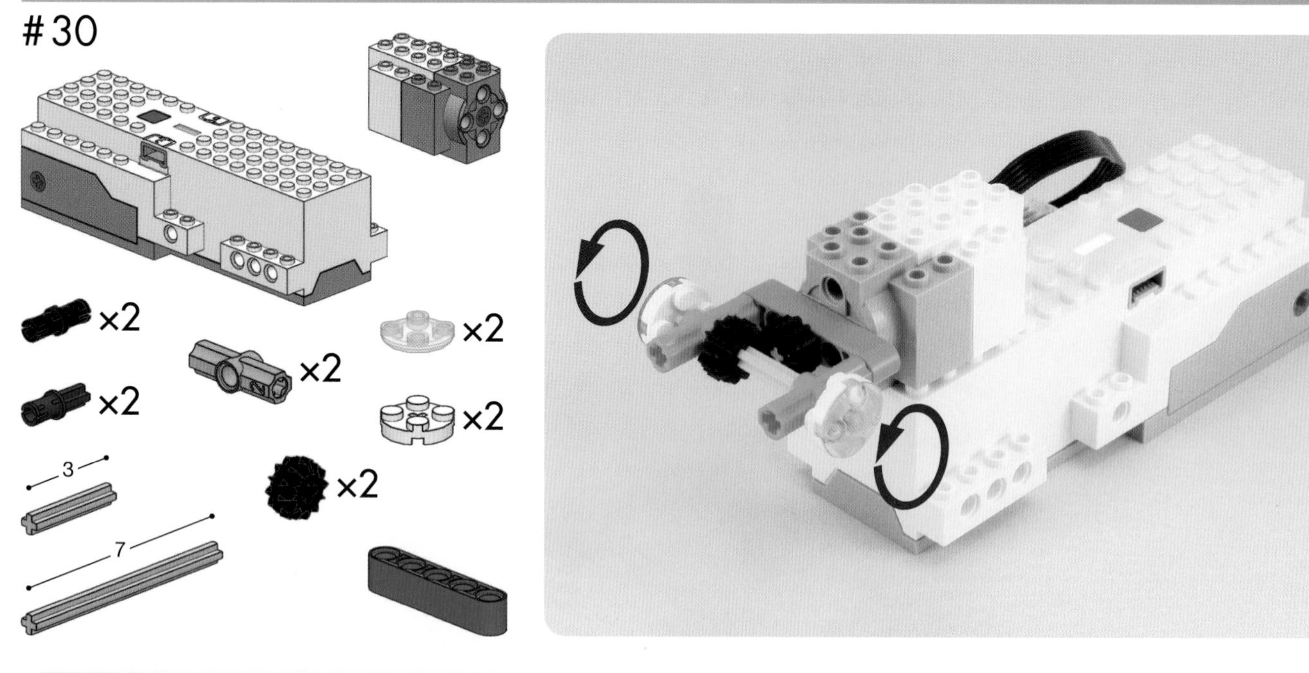

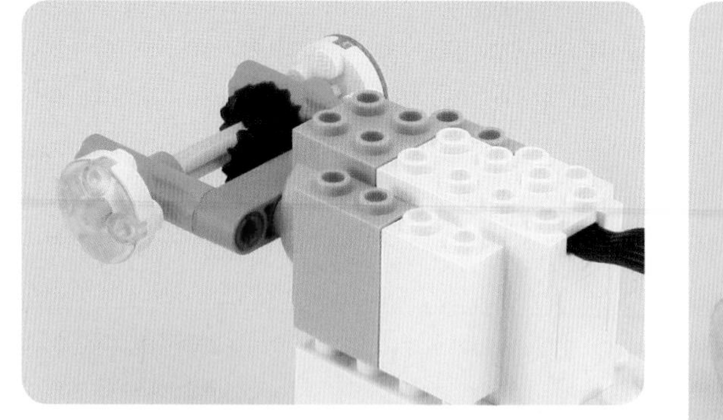

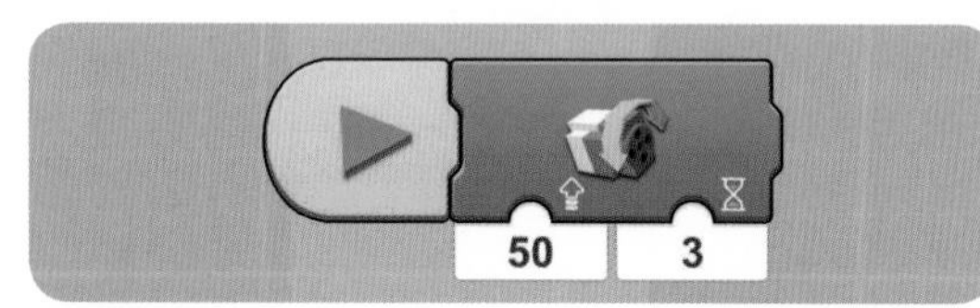

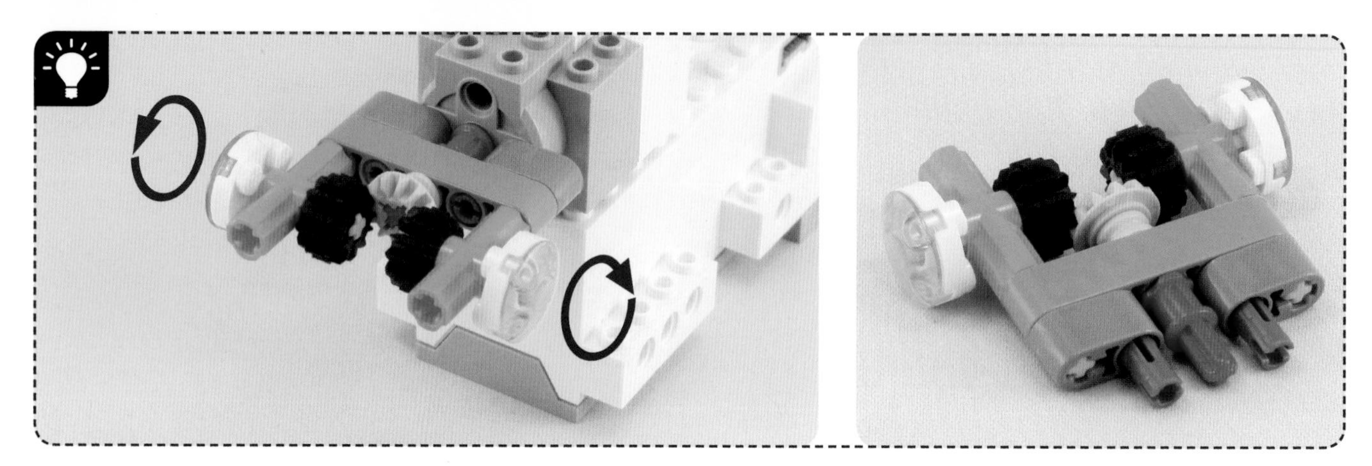

31

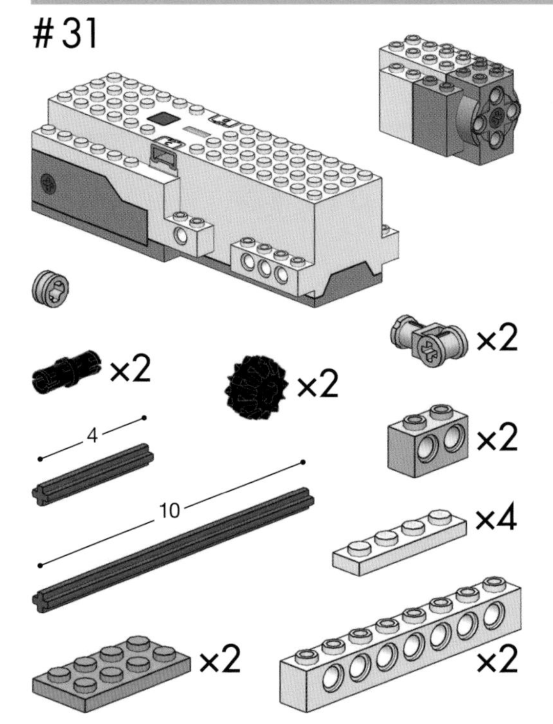

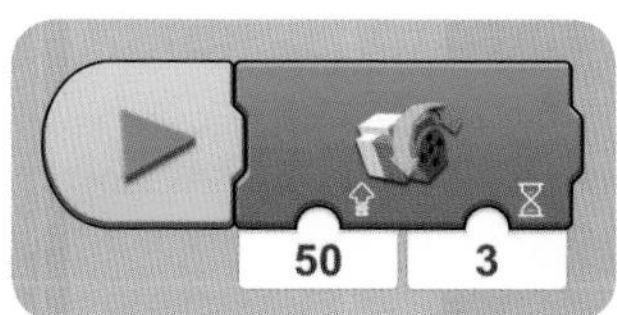

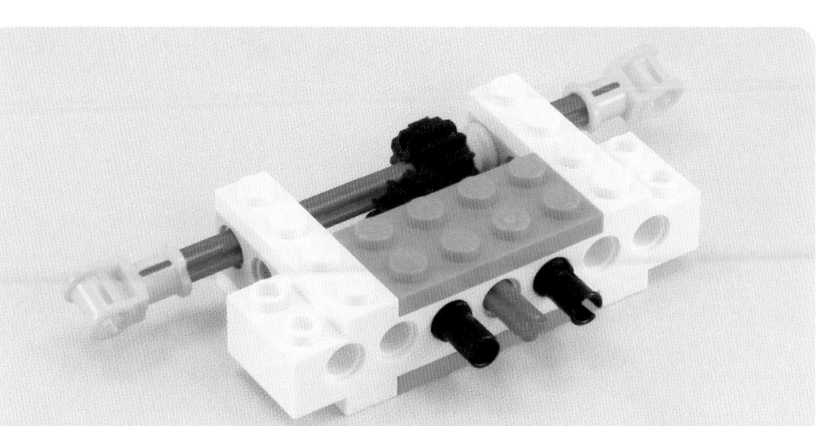

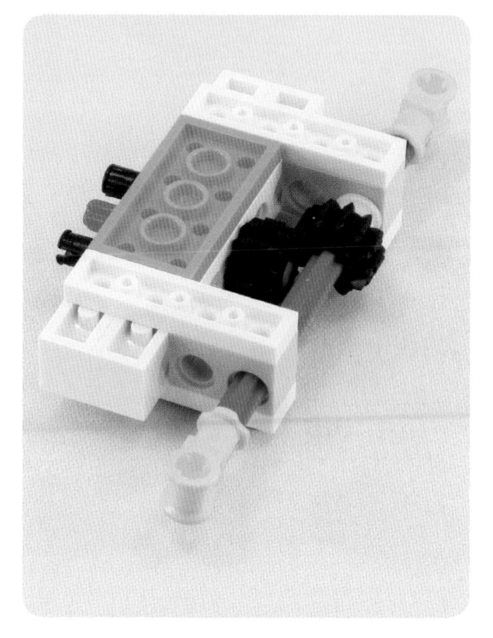

Gear ratio 20:12=5:3
» Speed ×5/3
» Power ×3/5

Gear ratio 12:20=3:5
» Speed ×3/5
» Power ×5/3

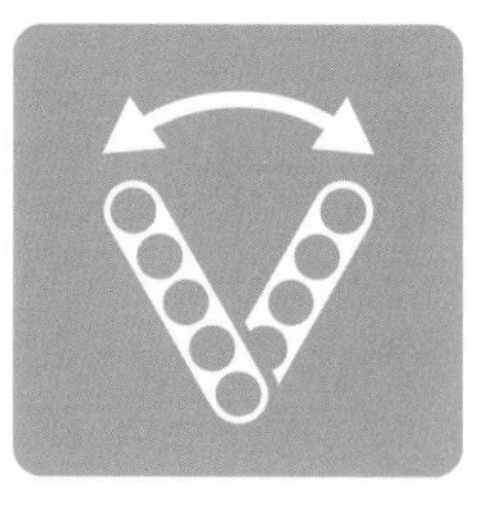

Swinging mechanisms

32

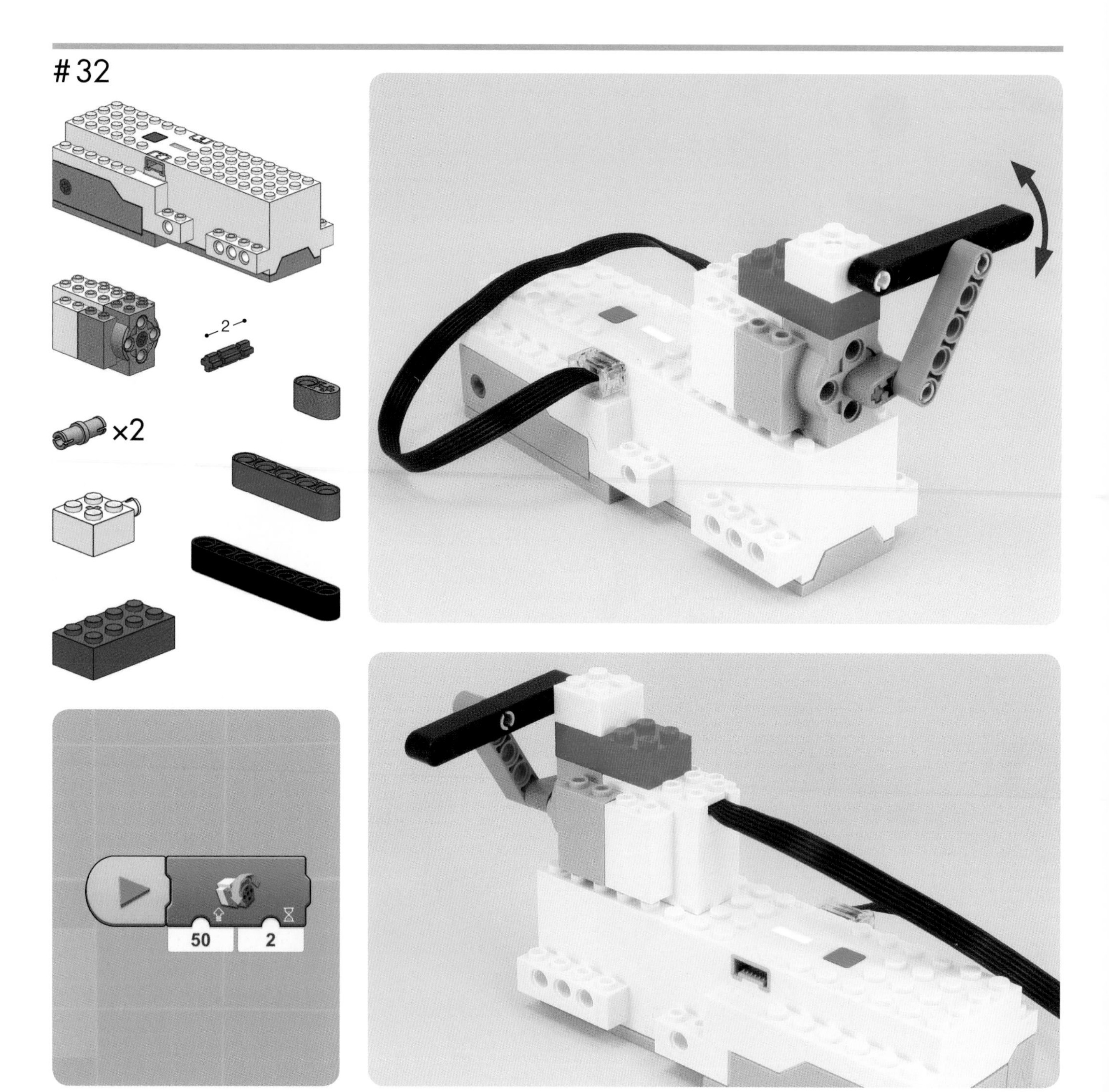

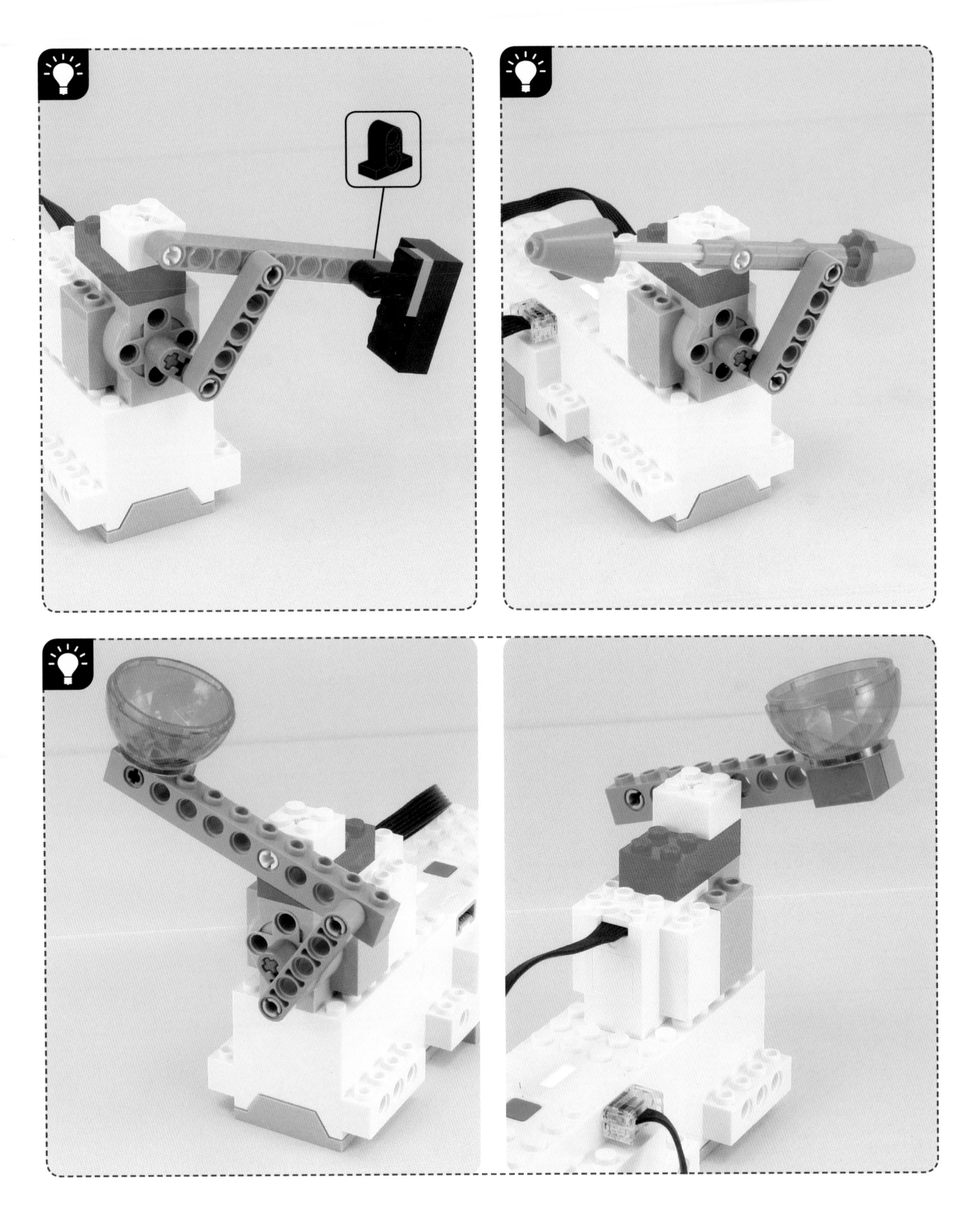

#33

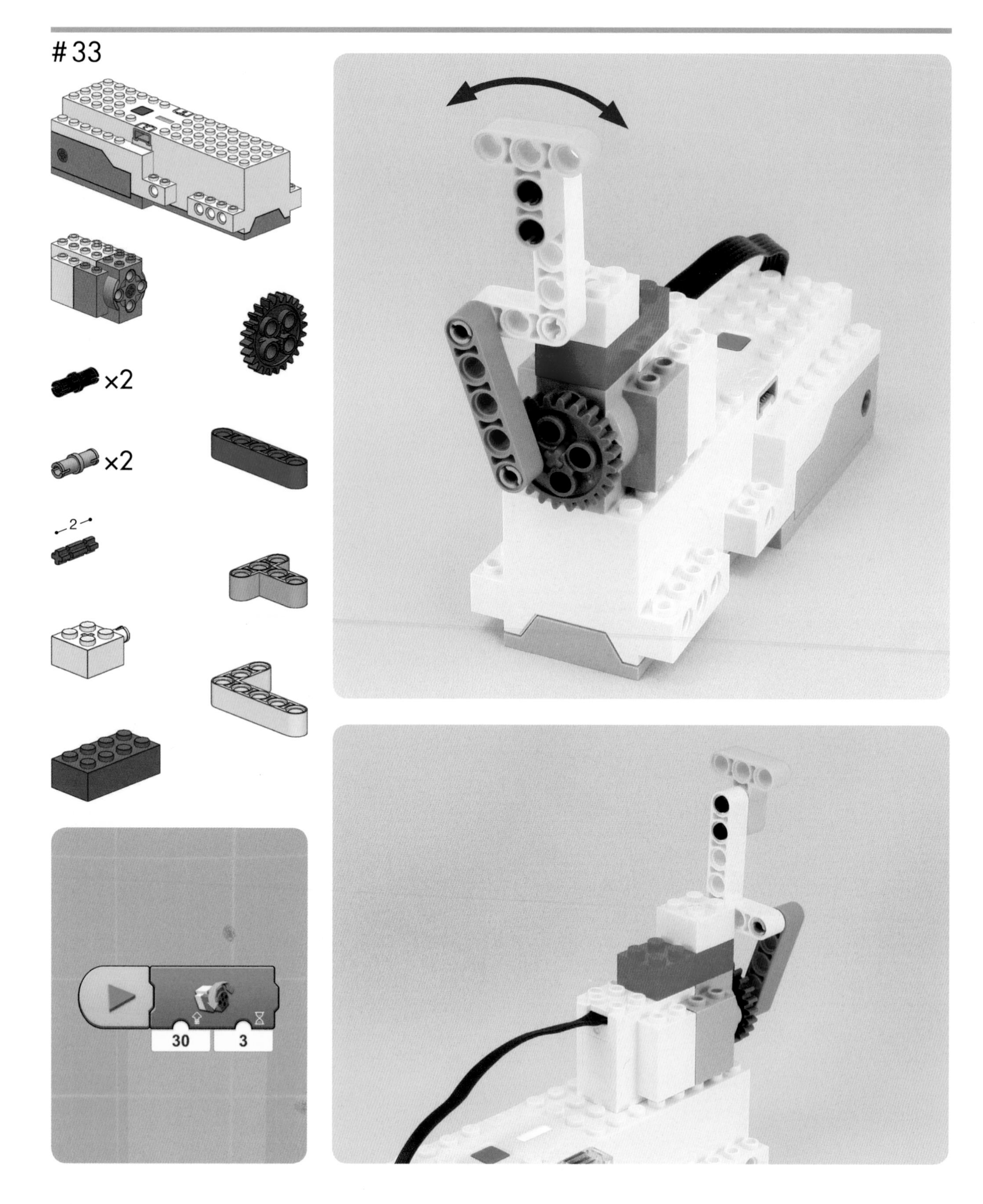

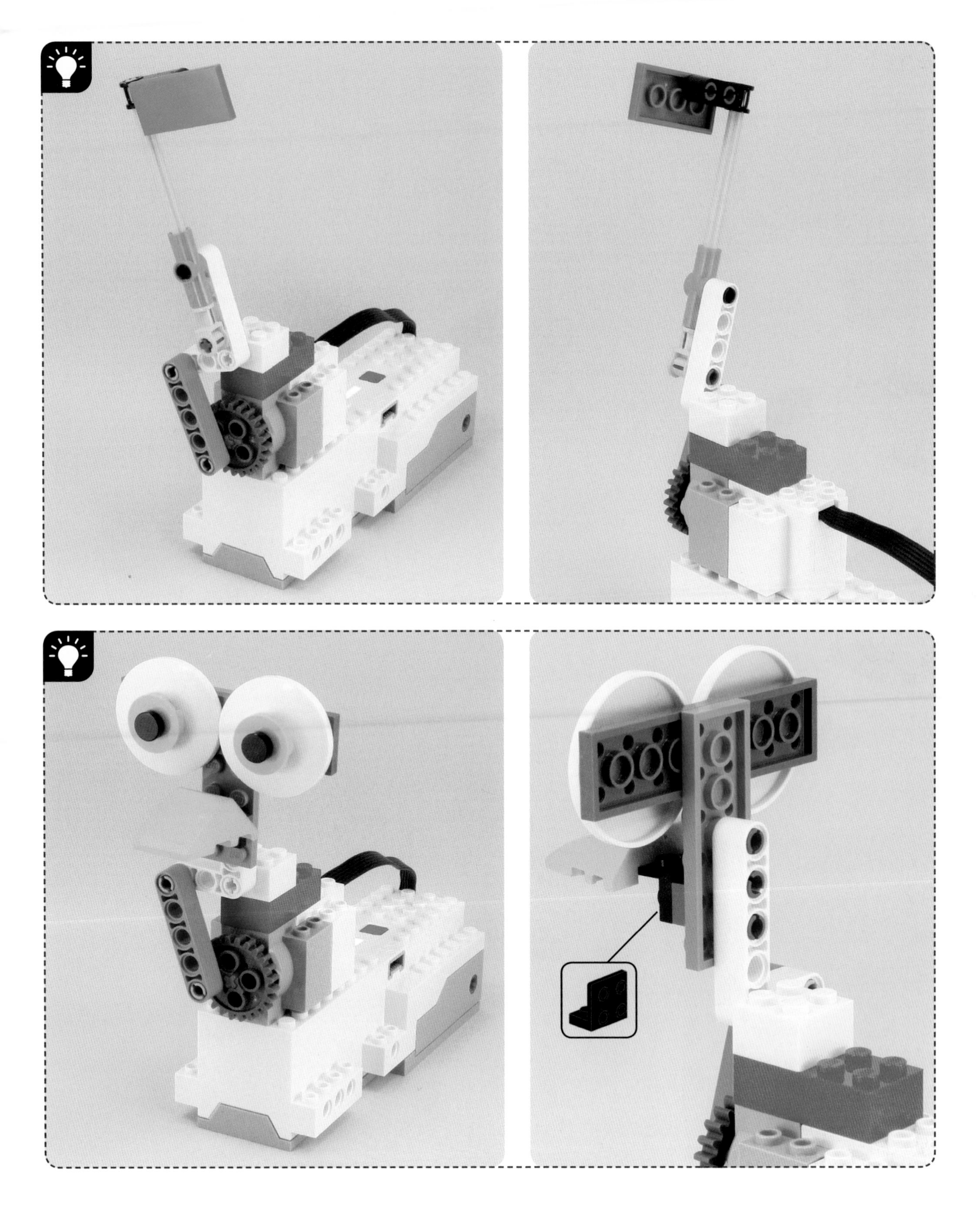

#34

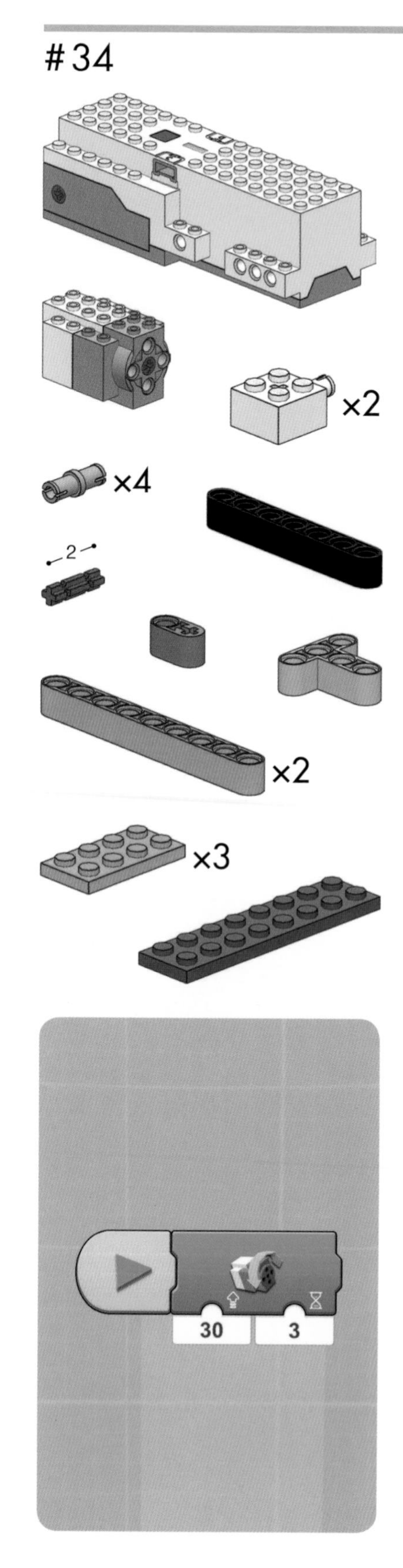

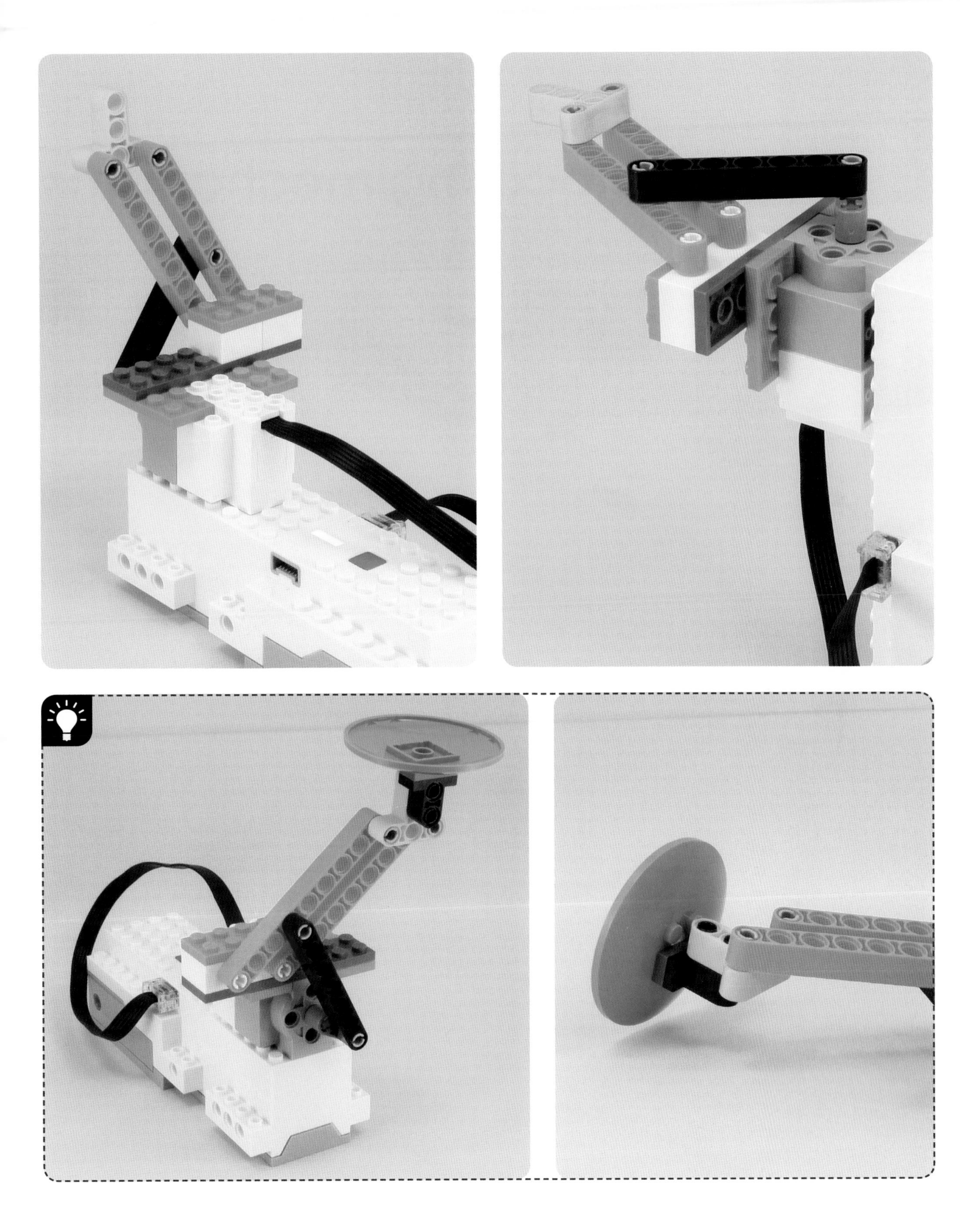

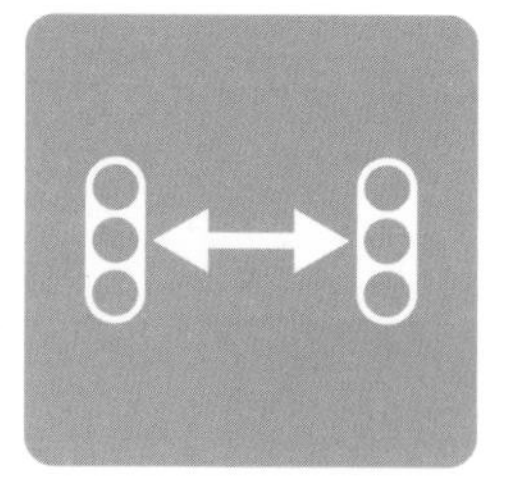

Reciprocating mechanisms

#35

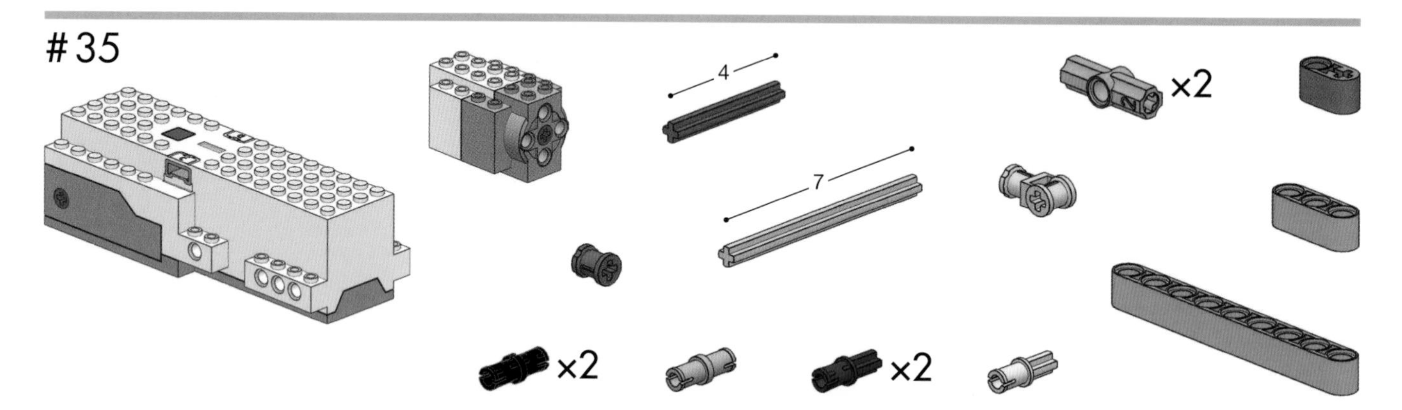

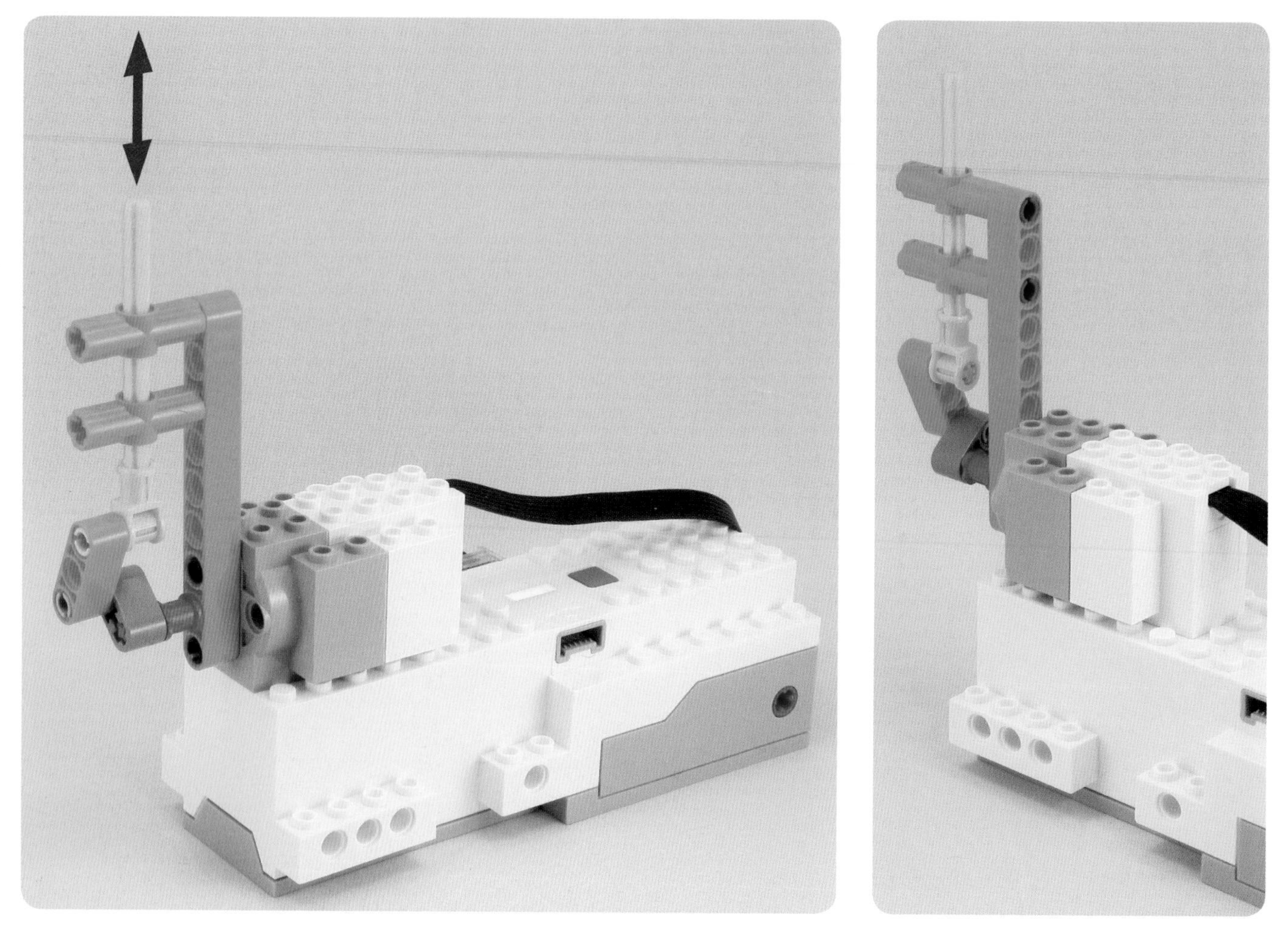

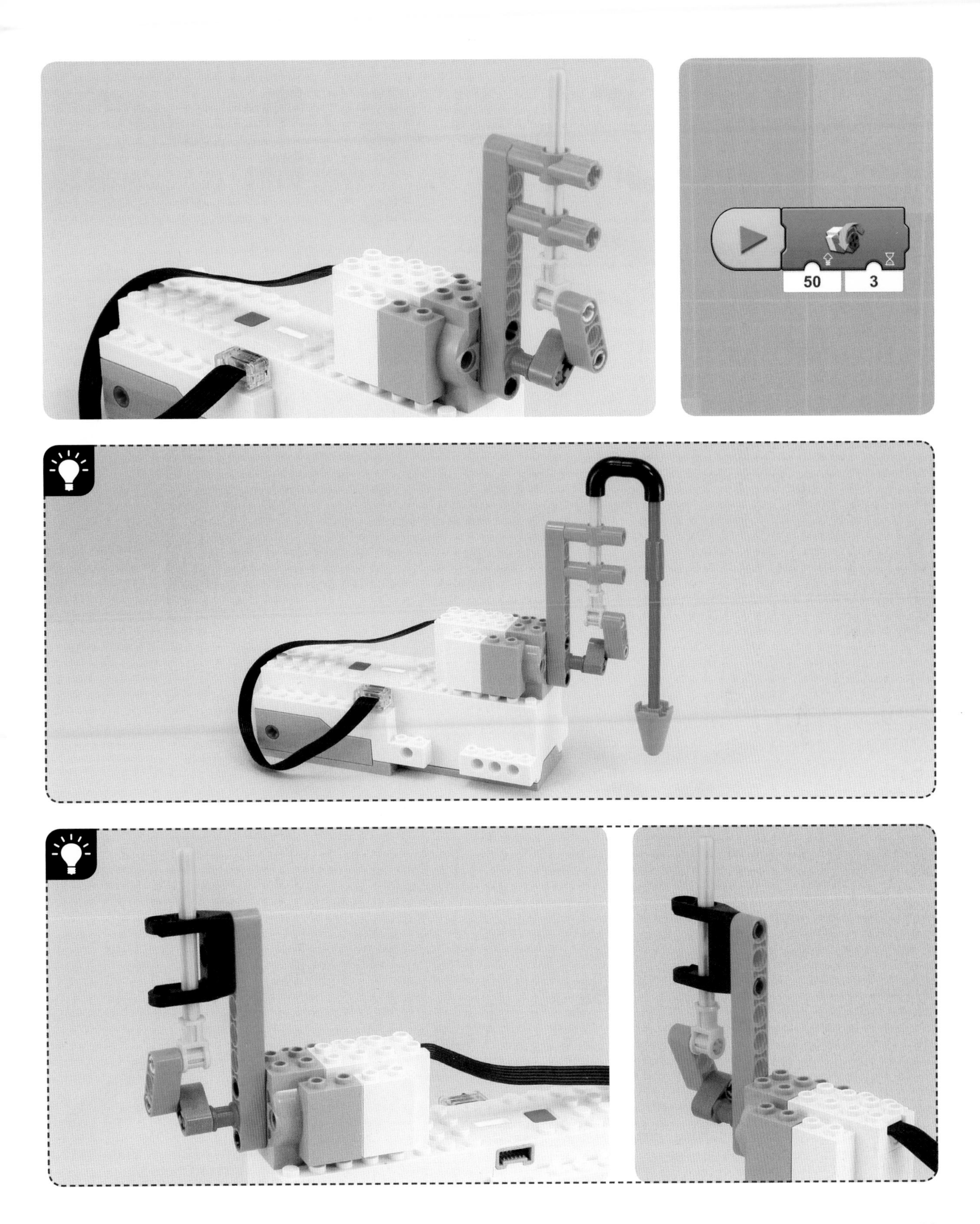
50
3

#36

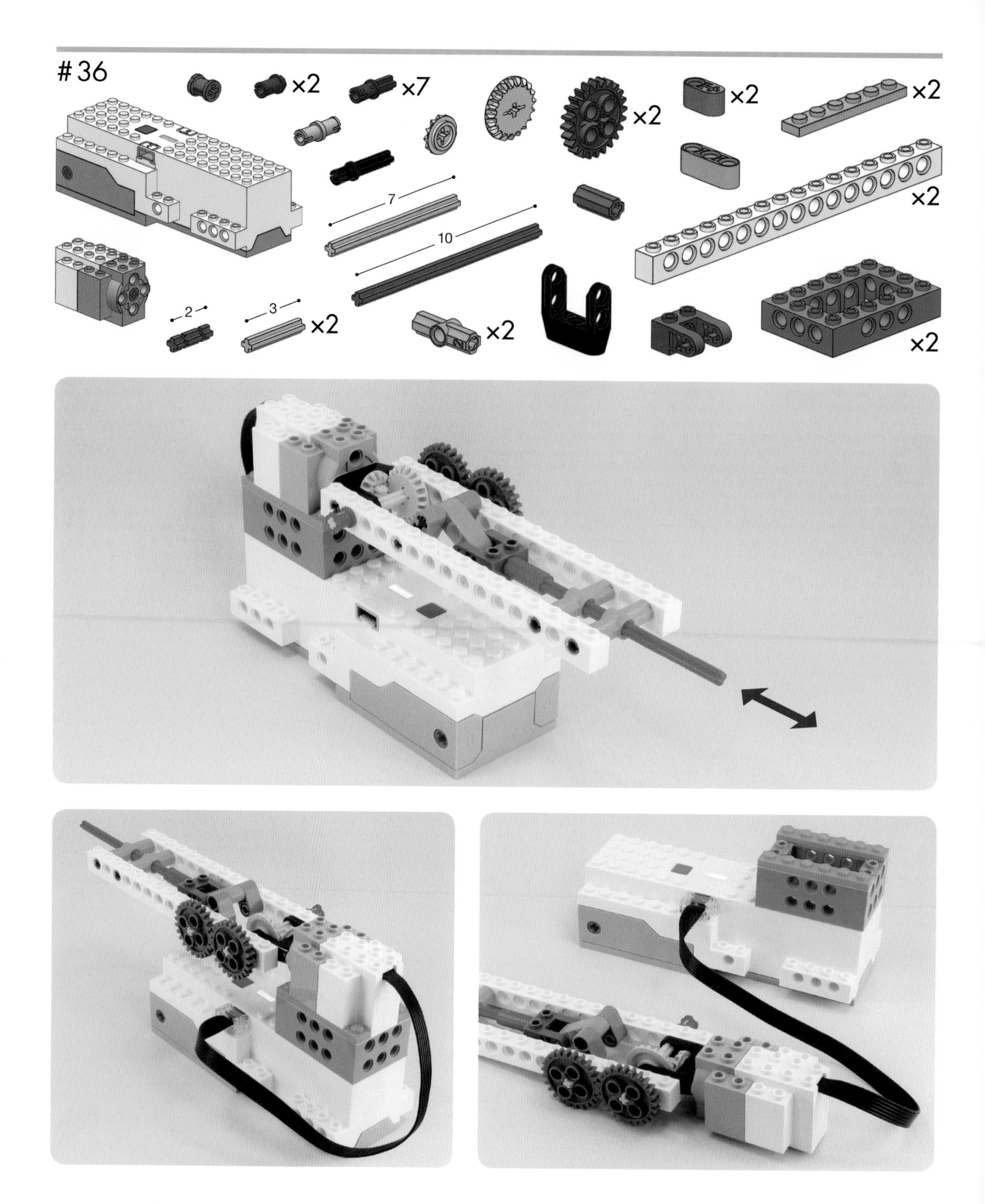

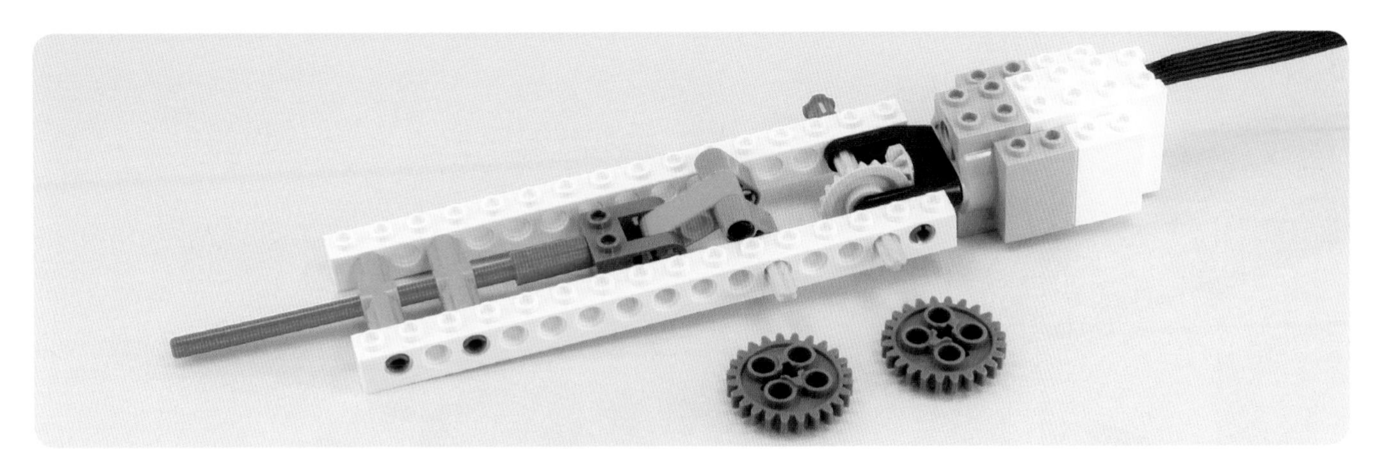

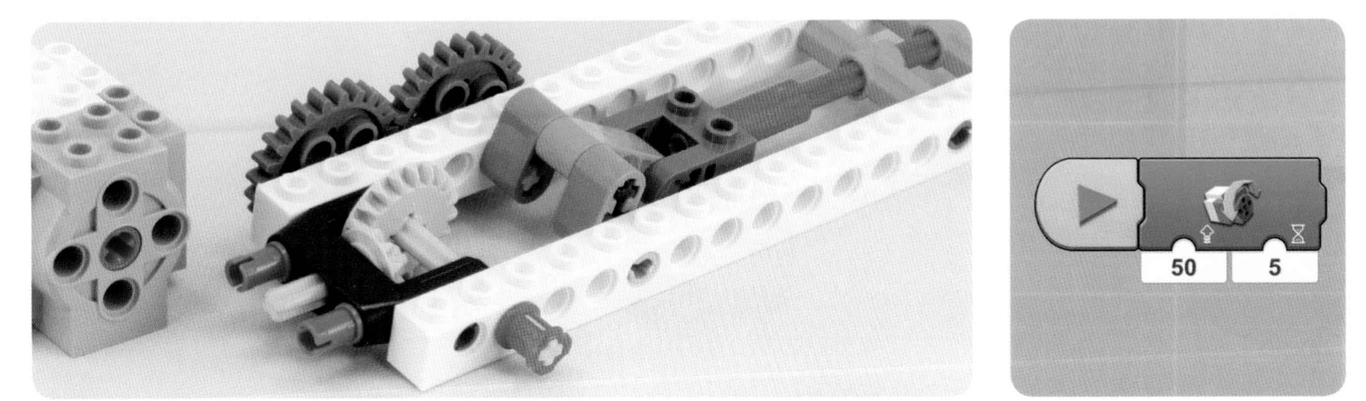
50
5

37

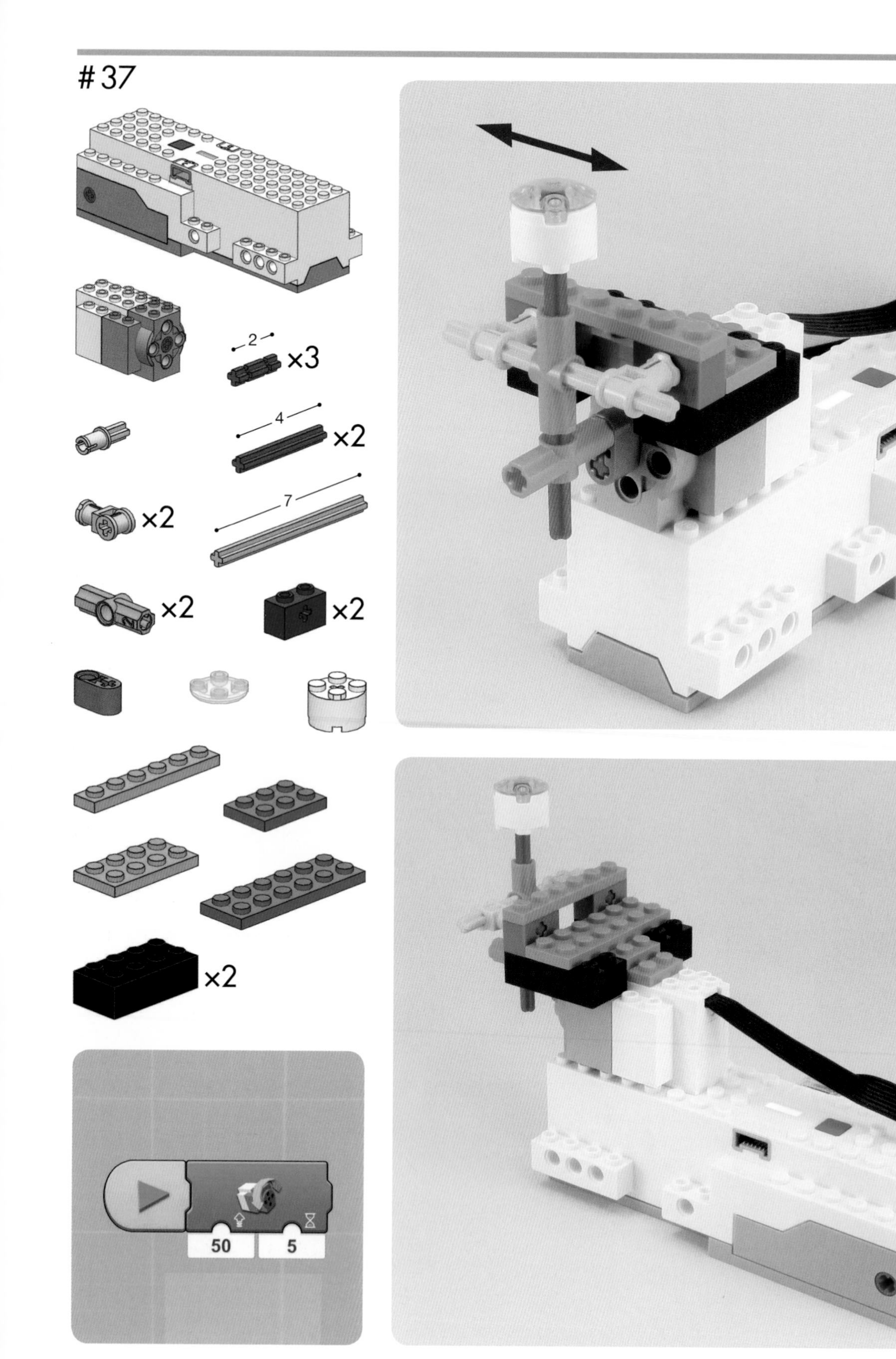

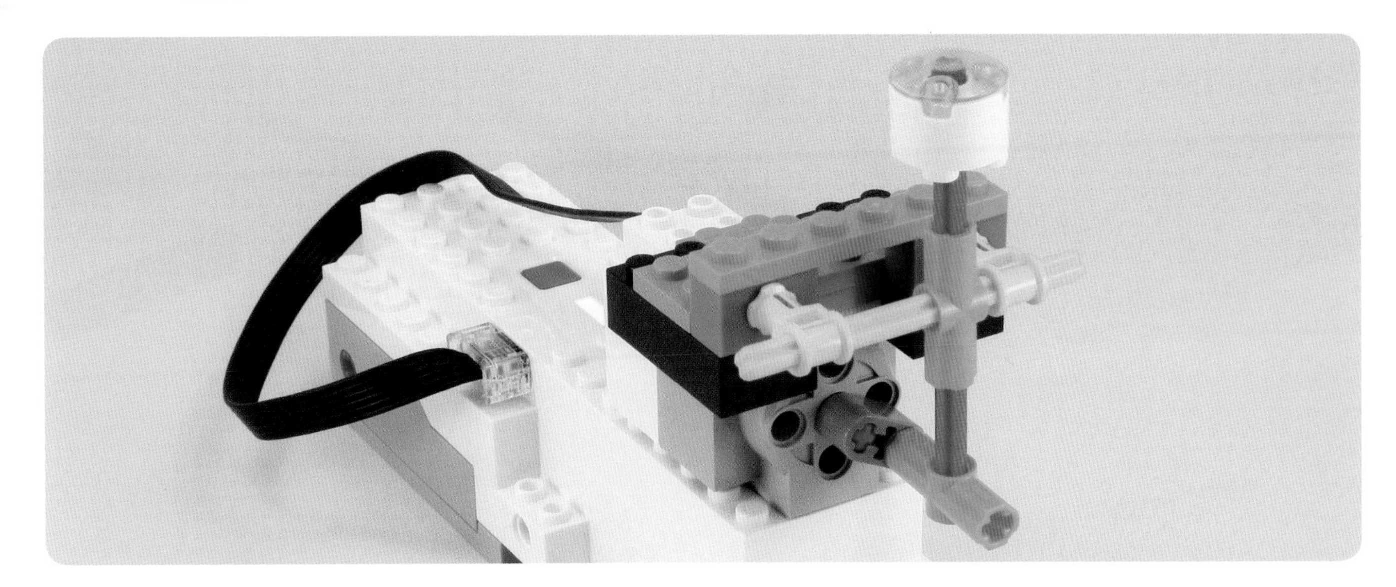

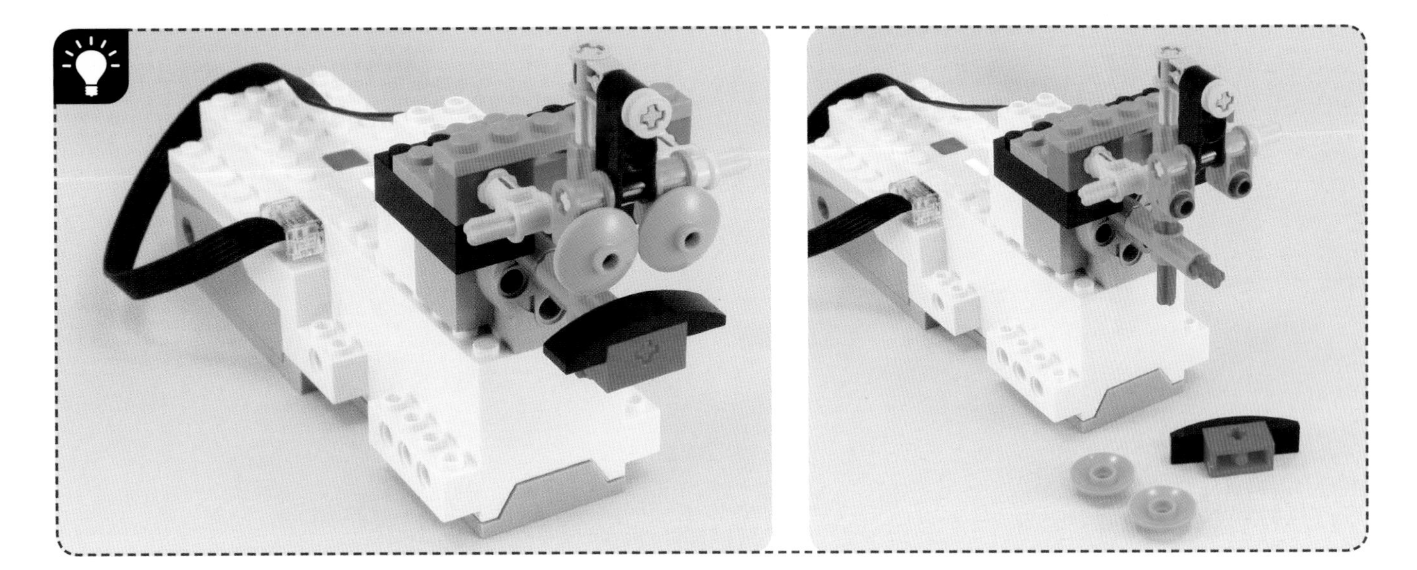

#38

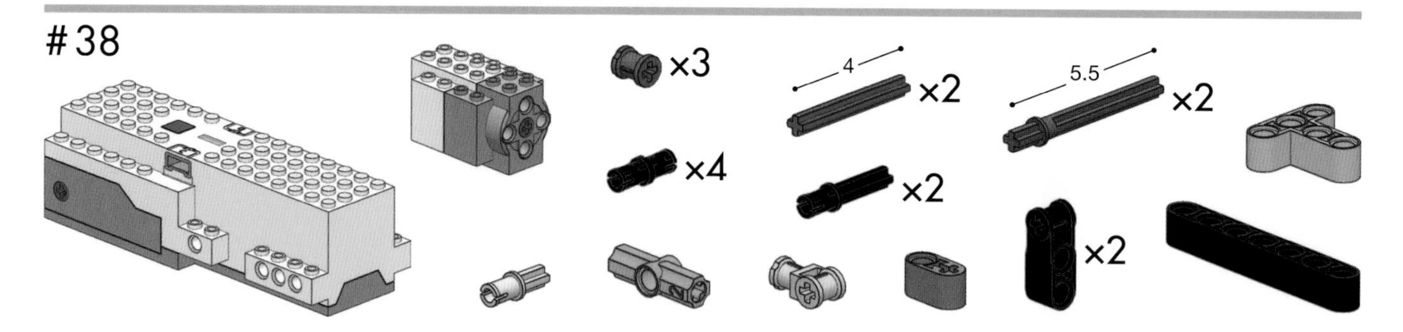

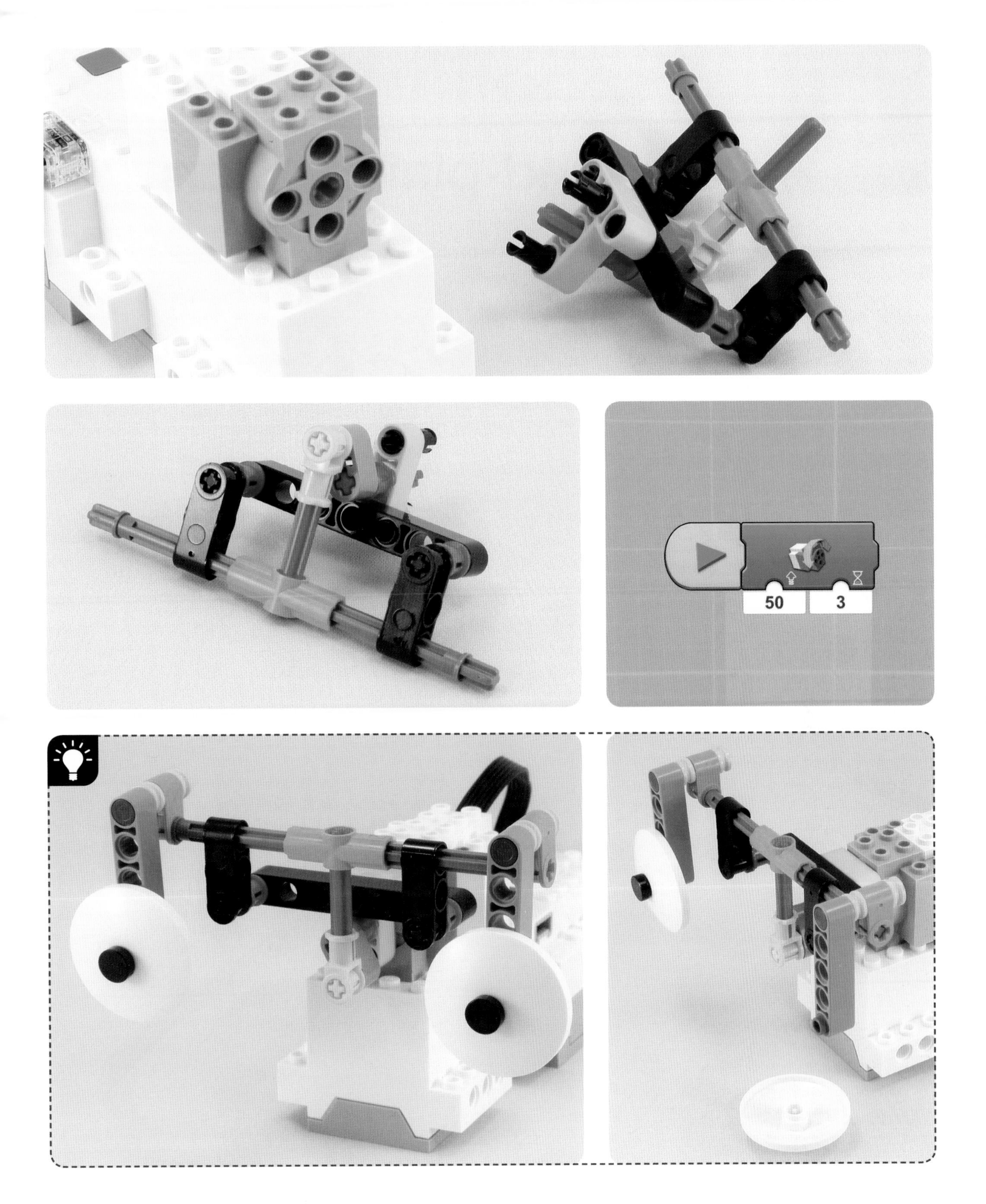
50
3

Rack-and-pinion gears

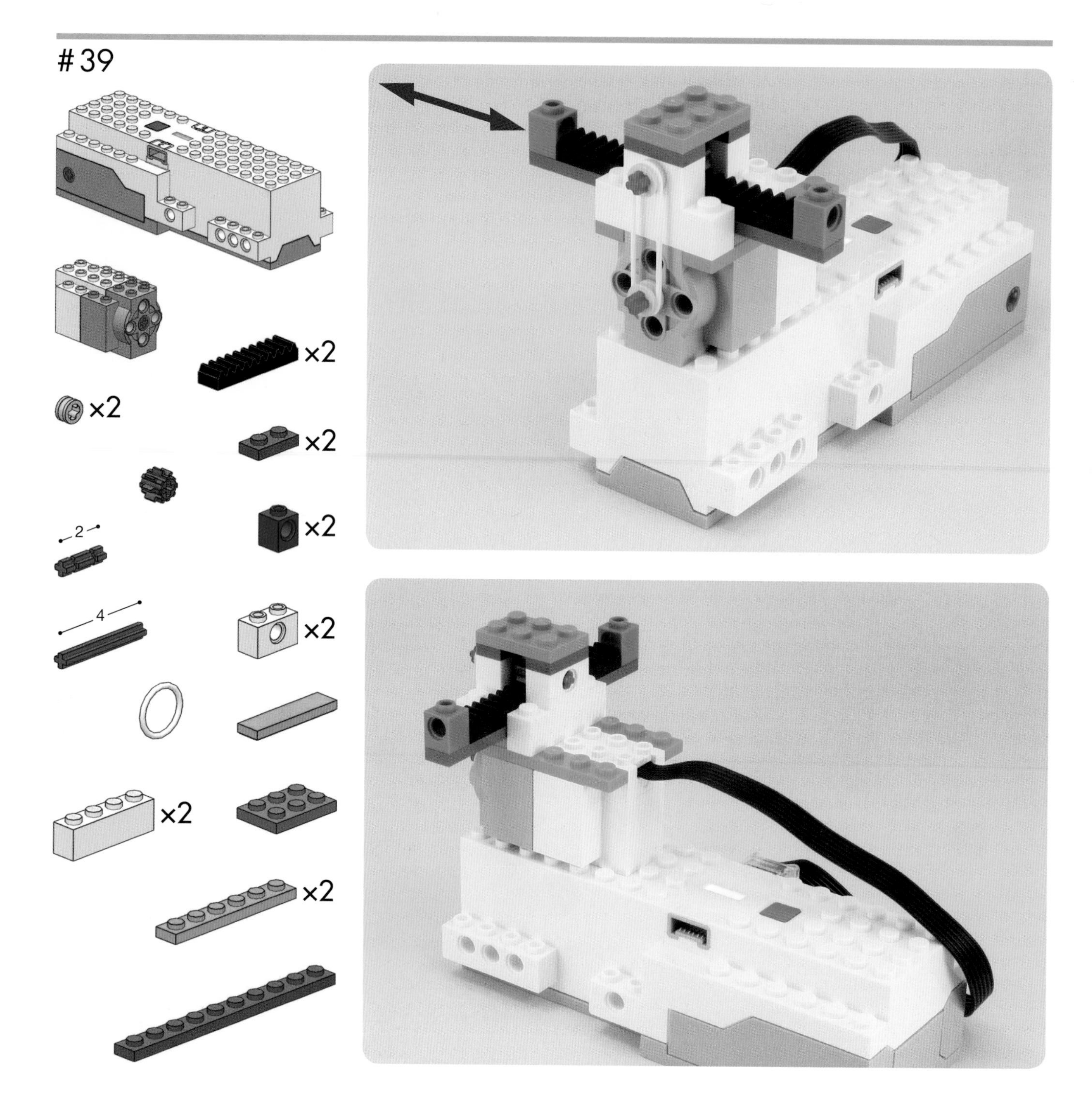

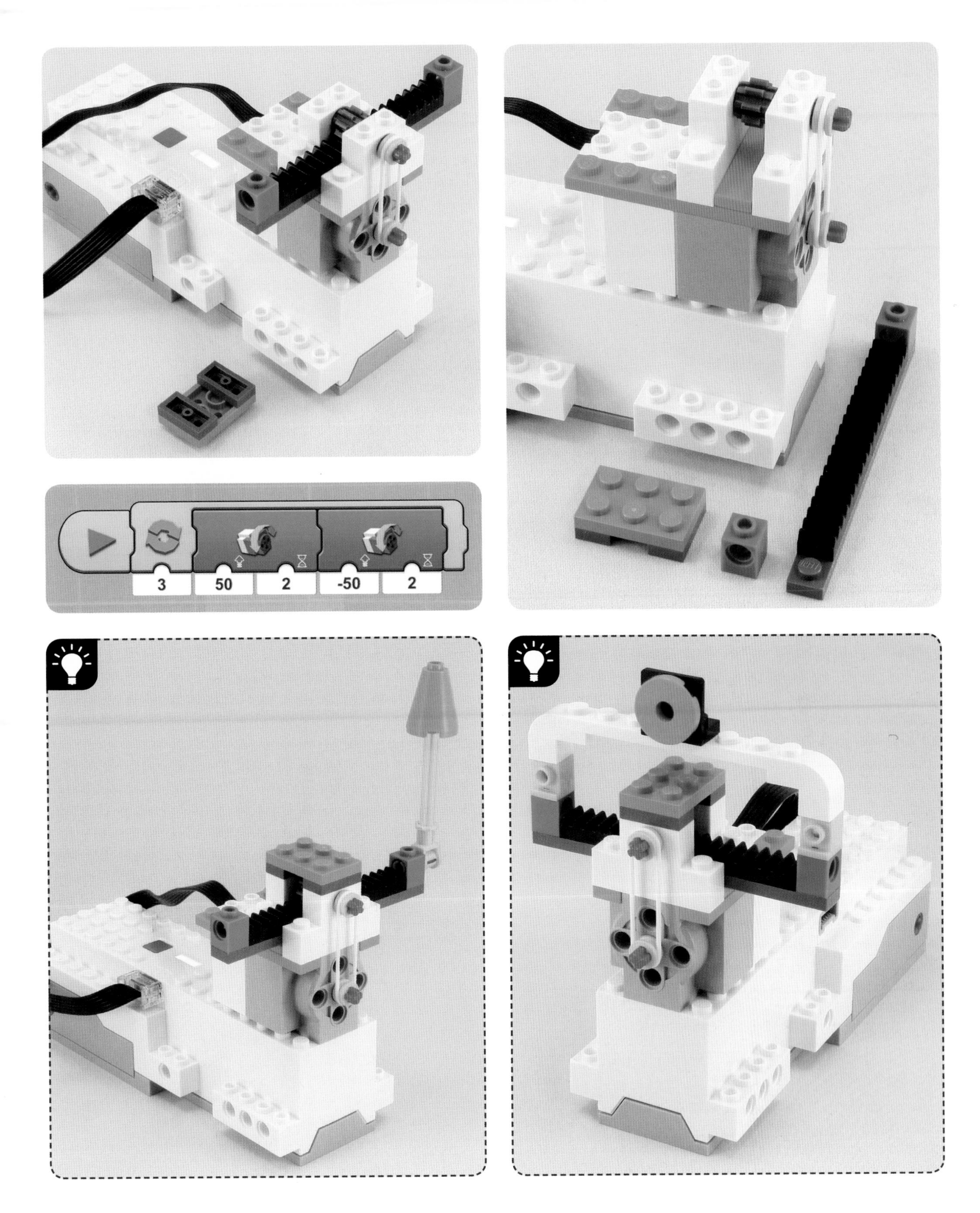
3
50
2
-50
2

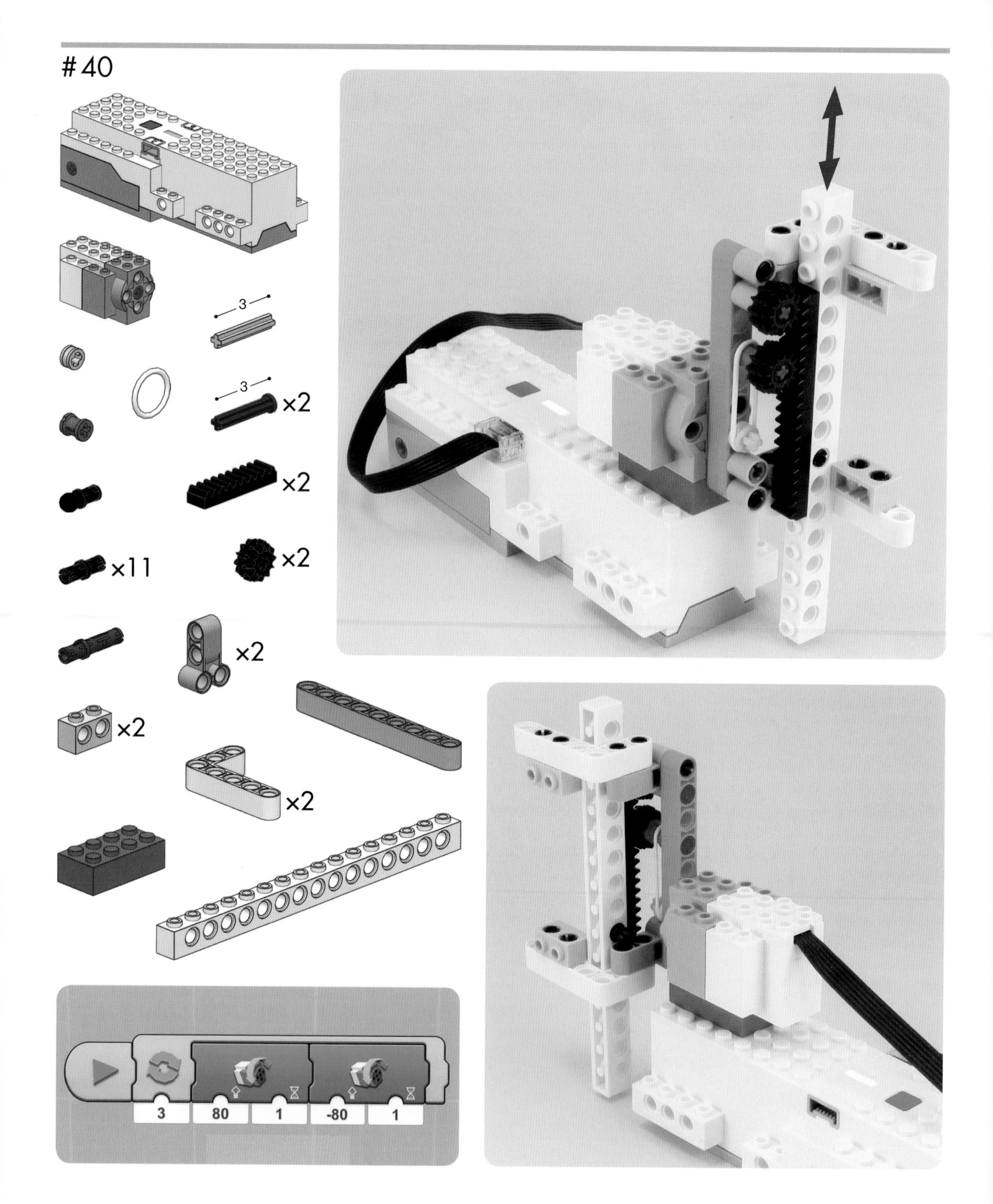

40
3
3
×2
×2
×11
×2
×2
×2
×2
3
80
1
-80
1

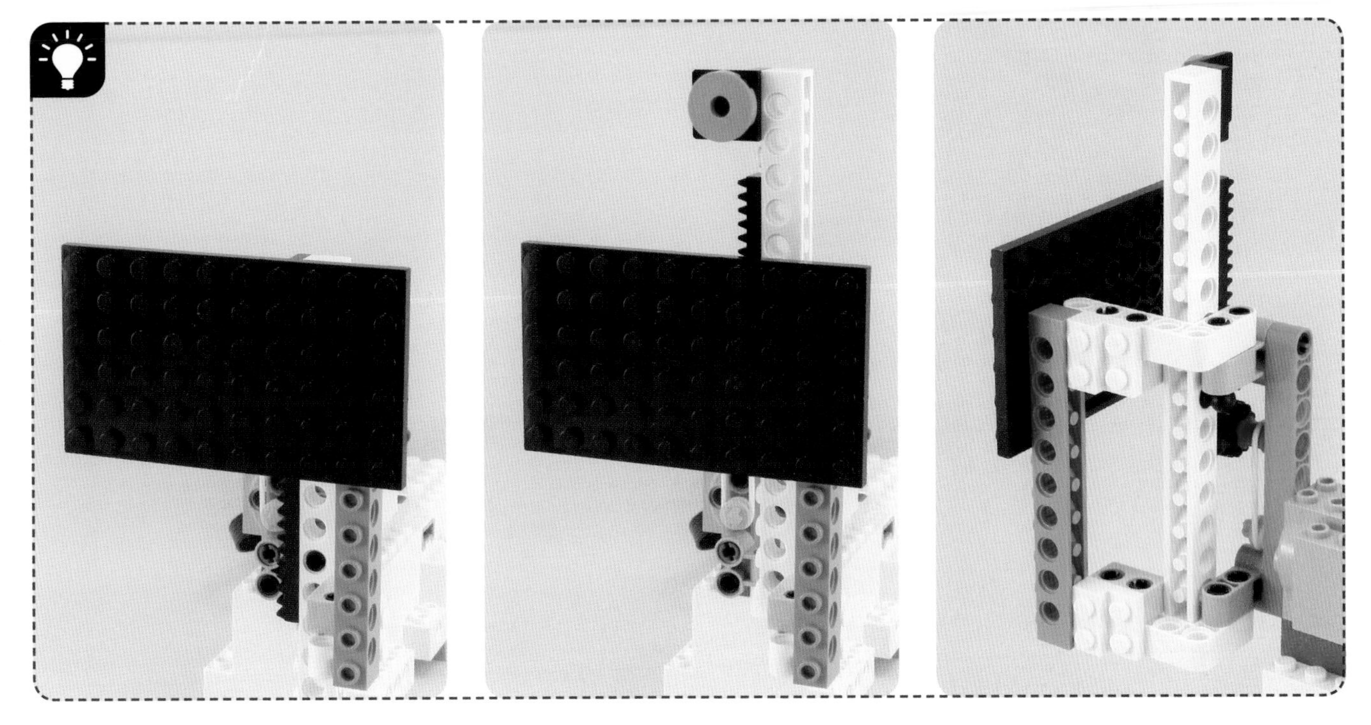

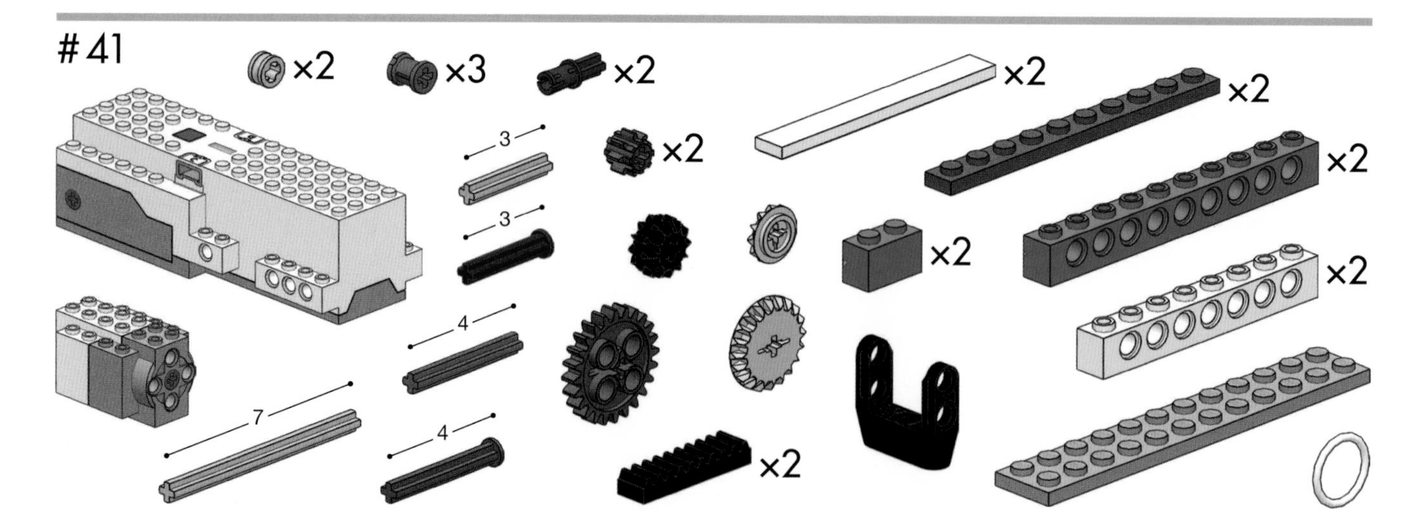
#41
×2
×3
×2
3
3
4
7
4
×2
×2
×2
×2
×2
×2
×2

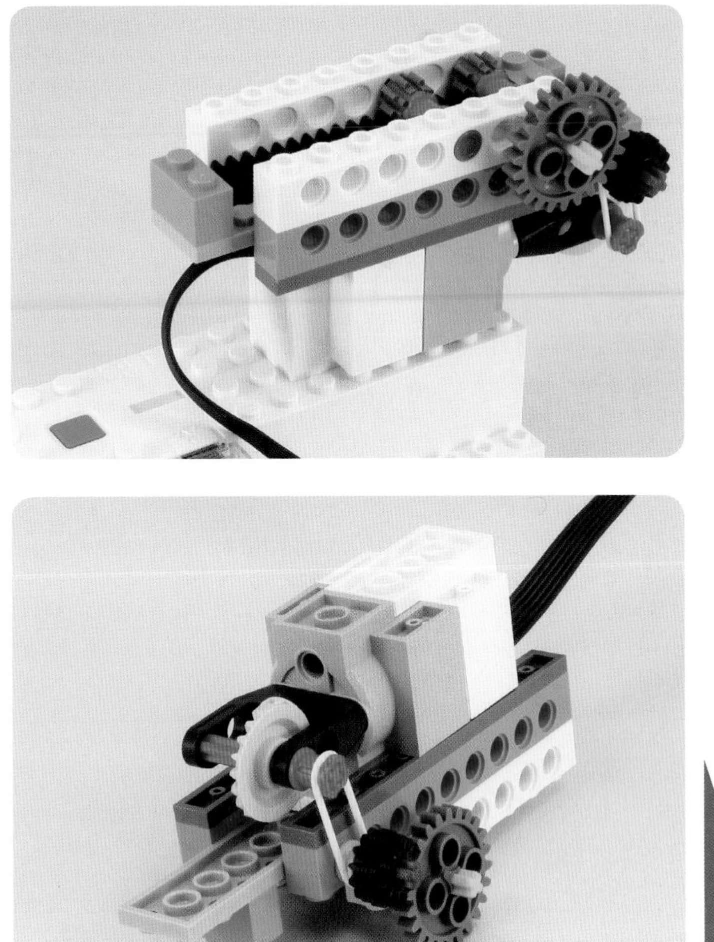

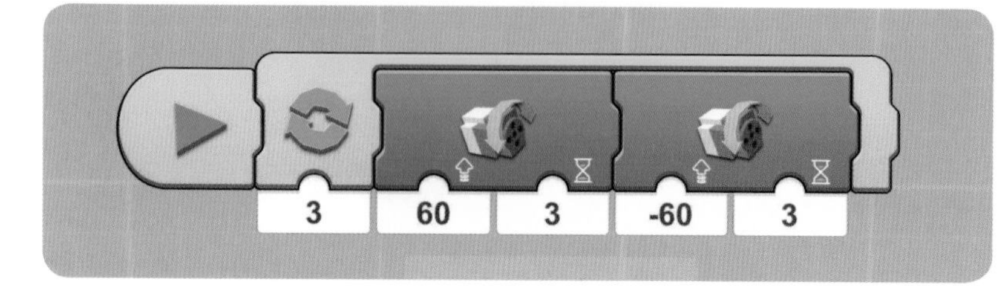
3
60
3
-60
3

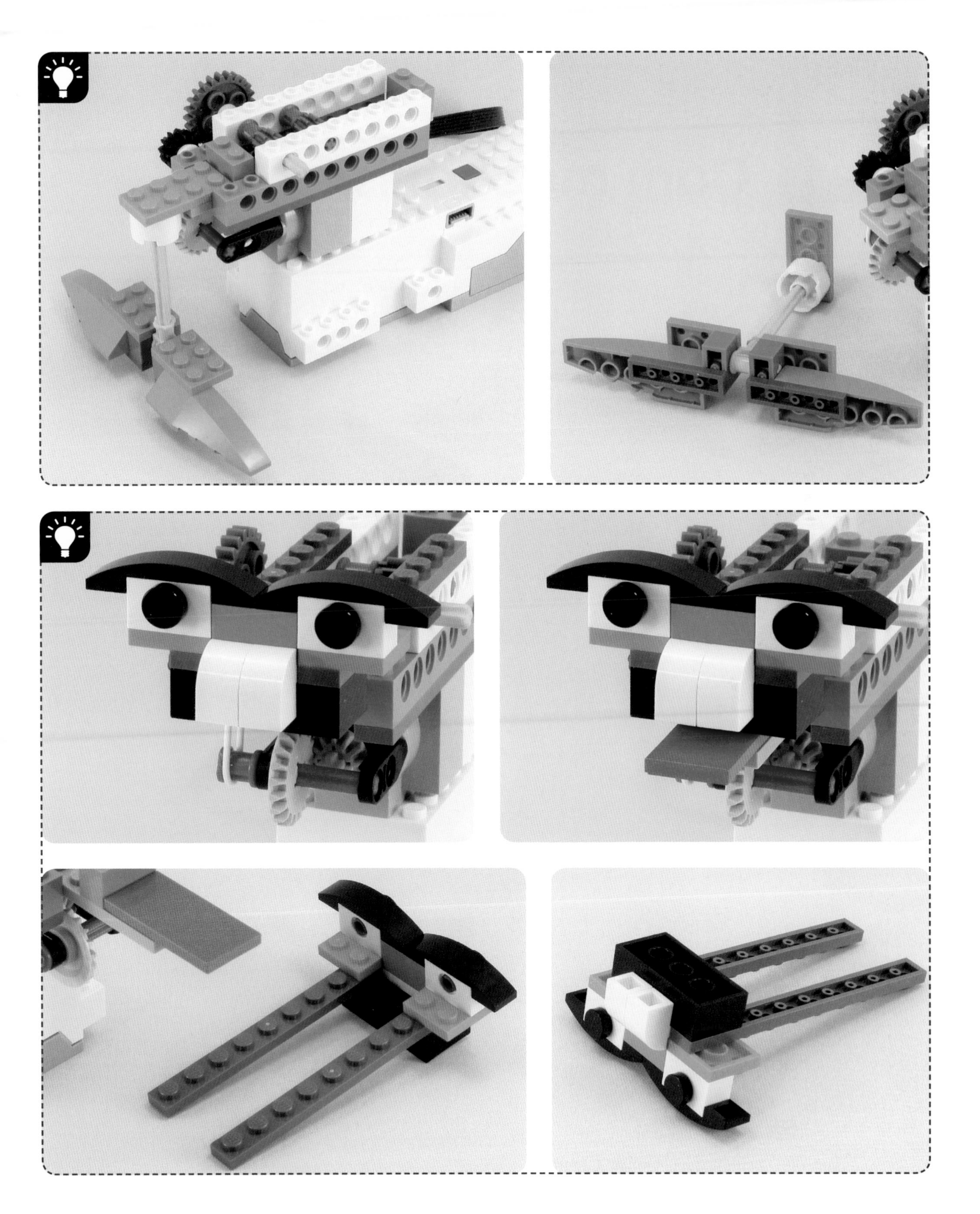

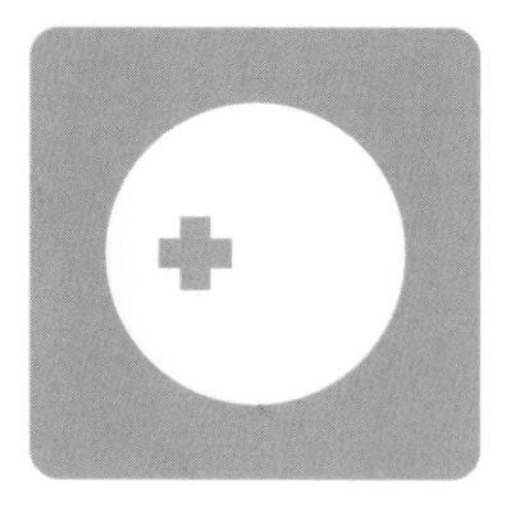

Cam mechanisms

42

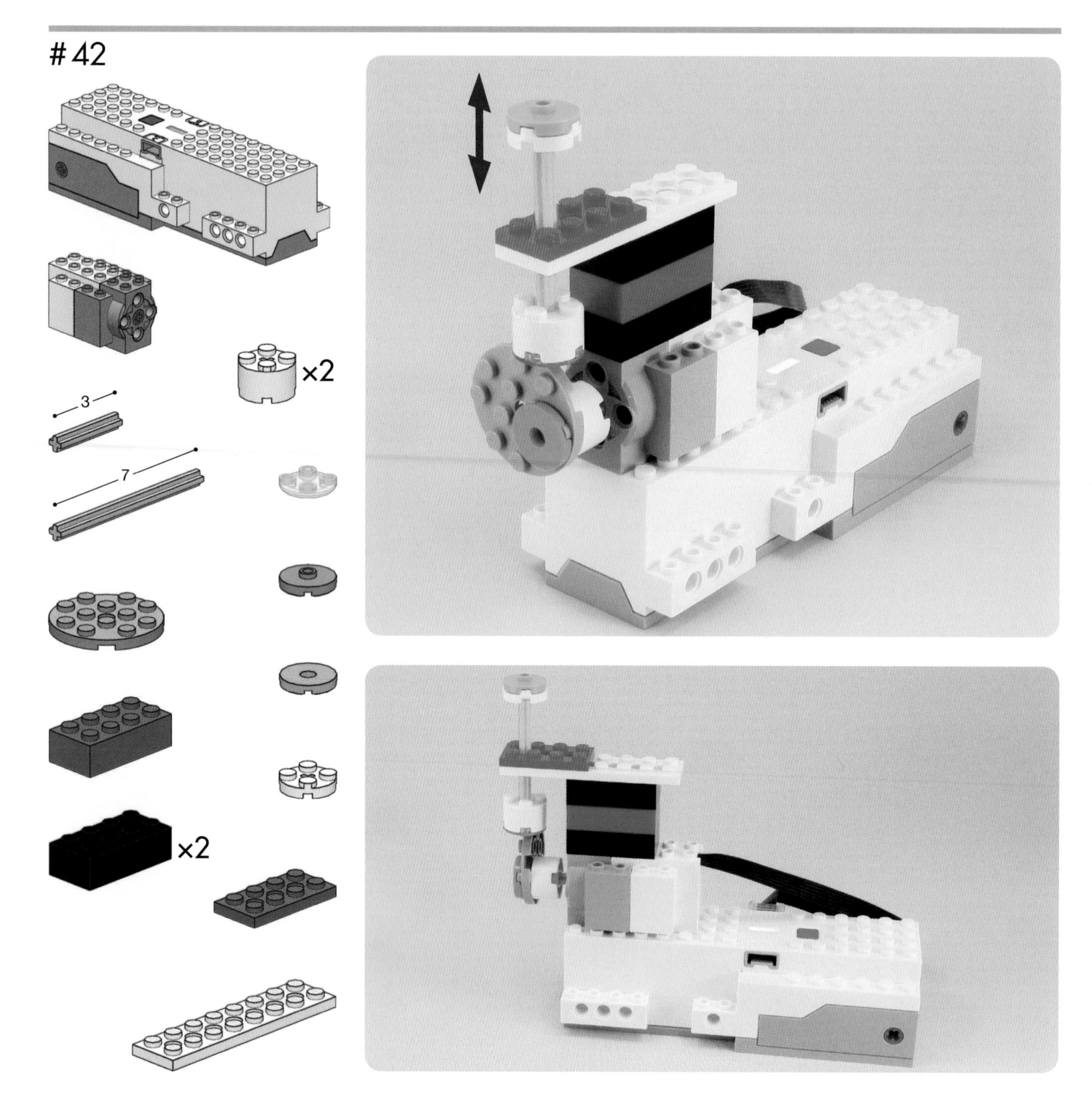

30
5

Off-center axes of rotation

43

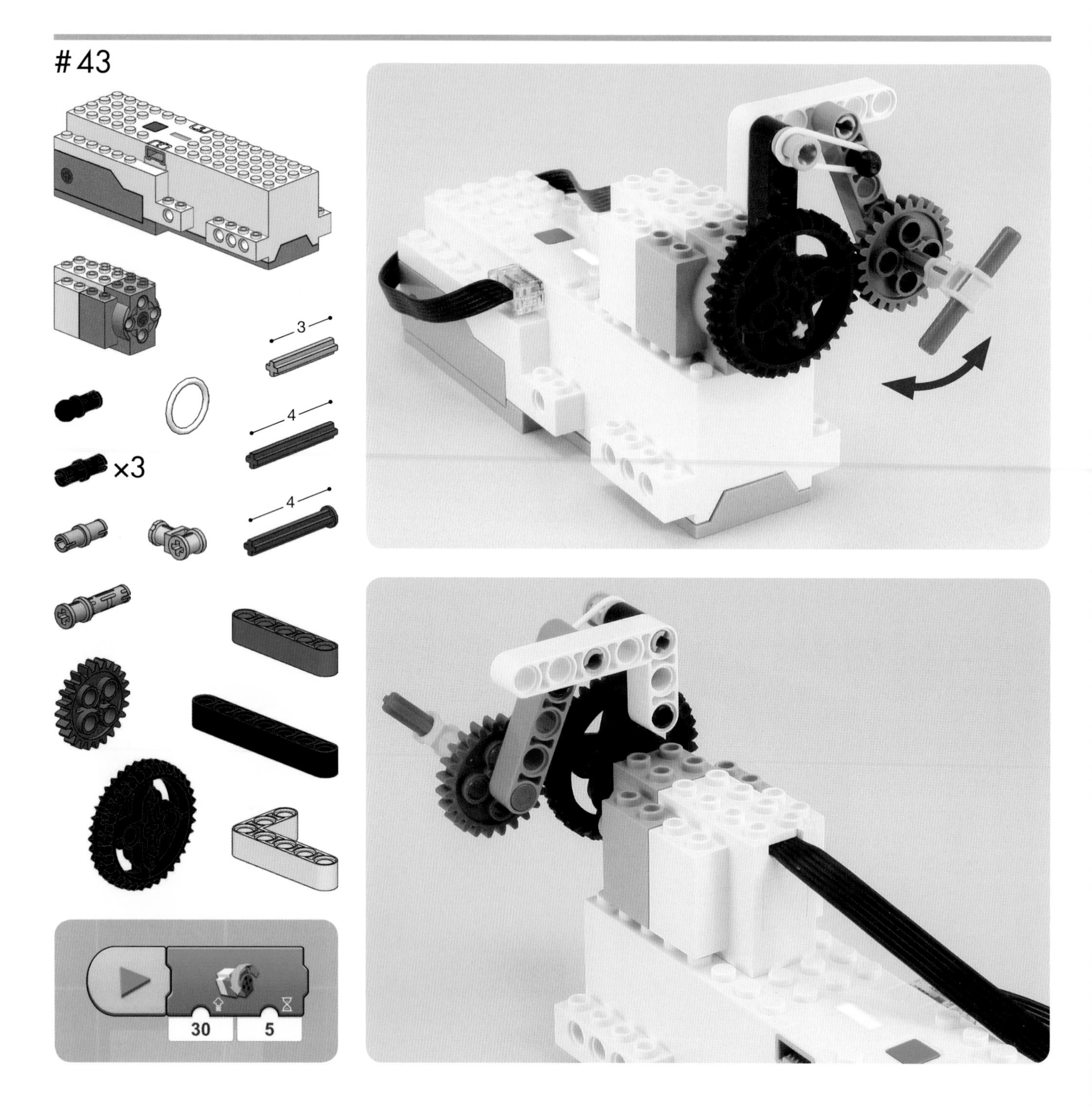

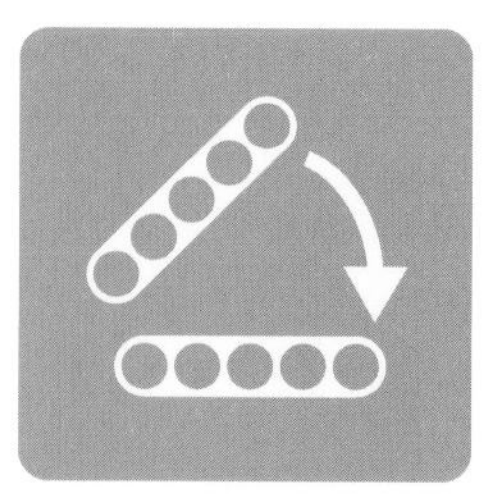

Chomping bots

#44

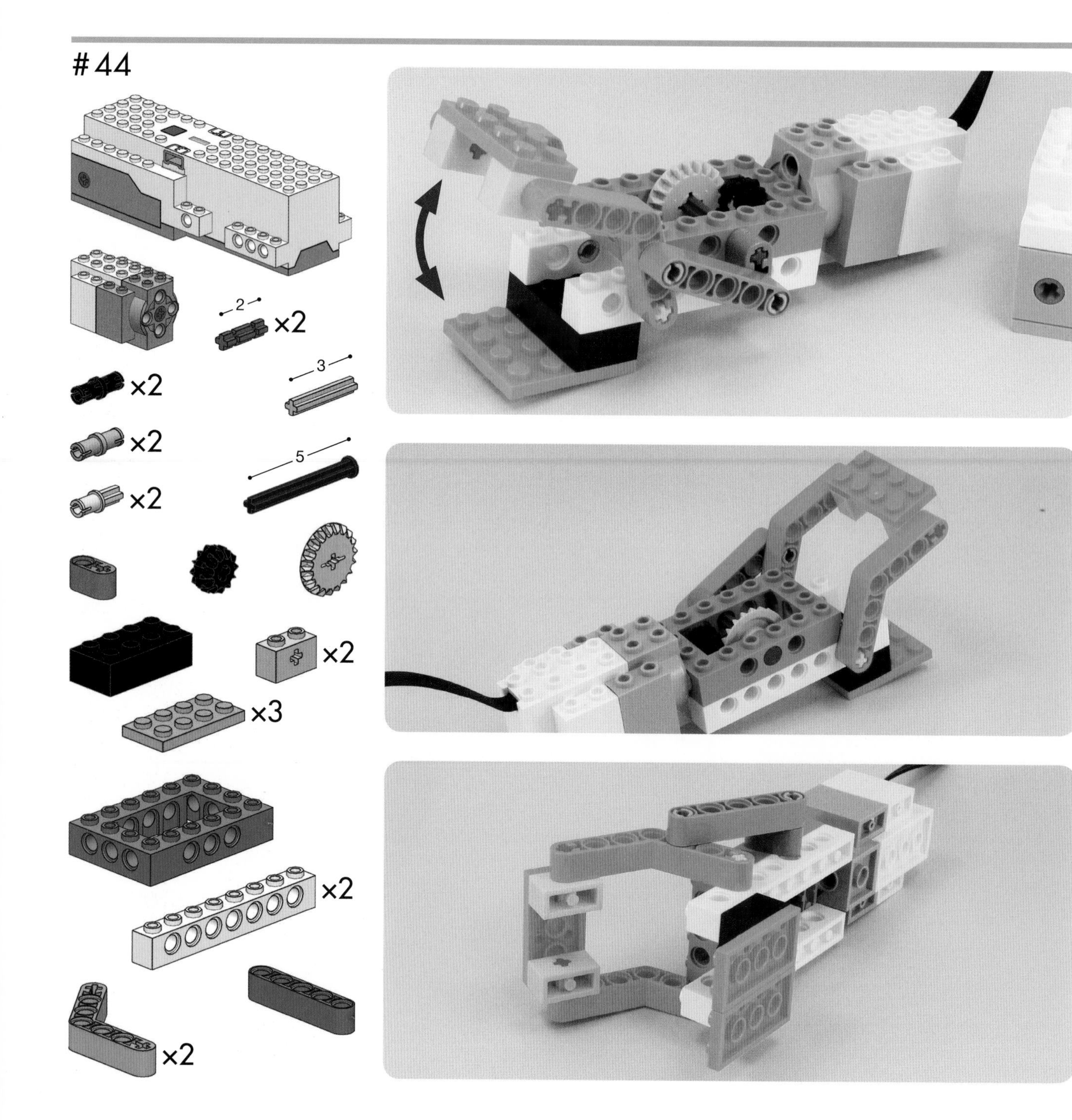

50
5

#45

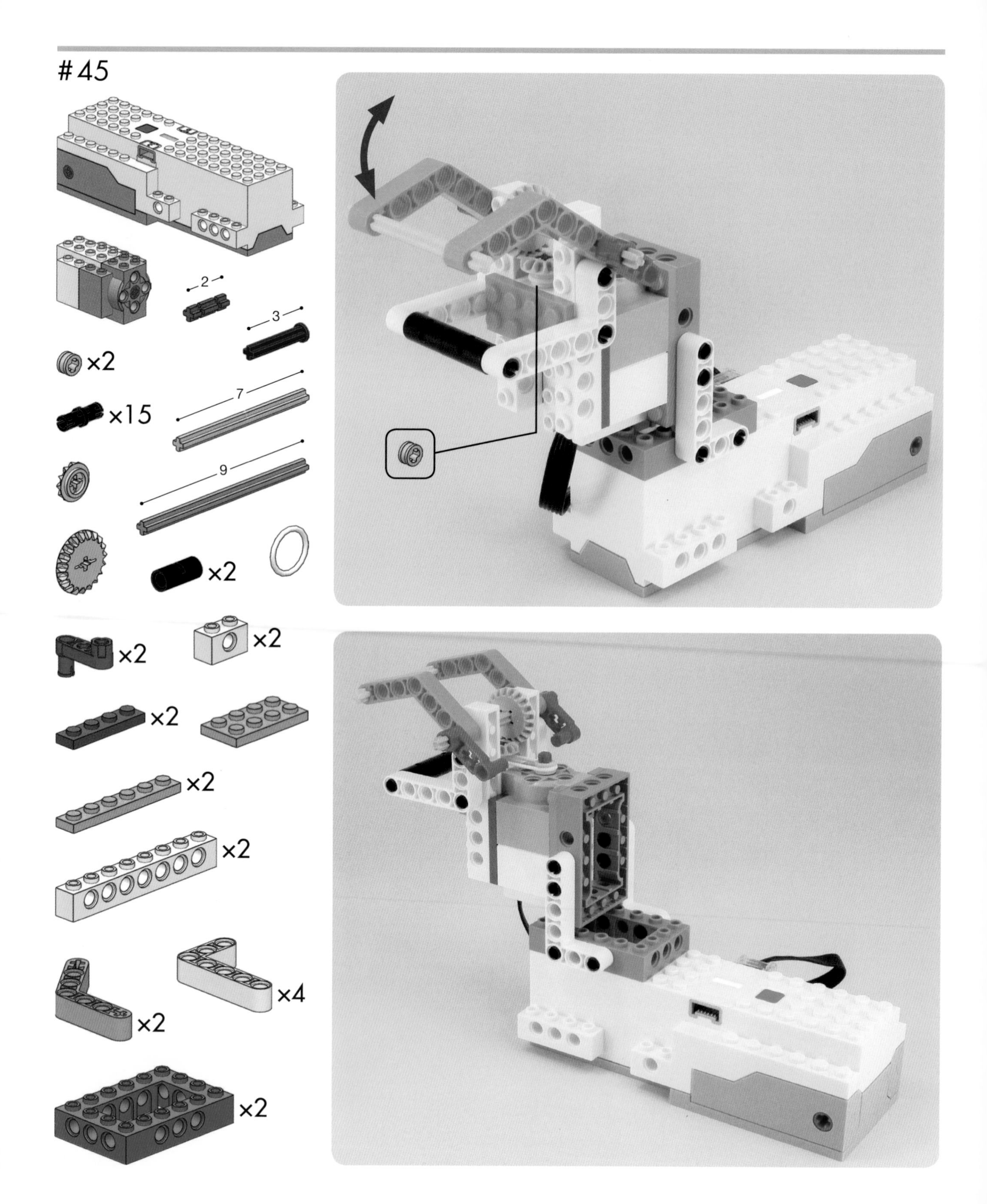

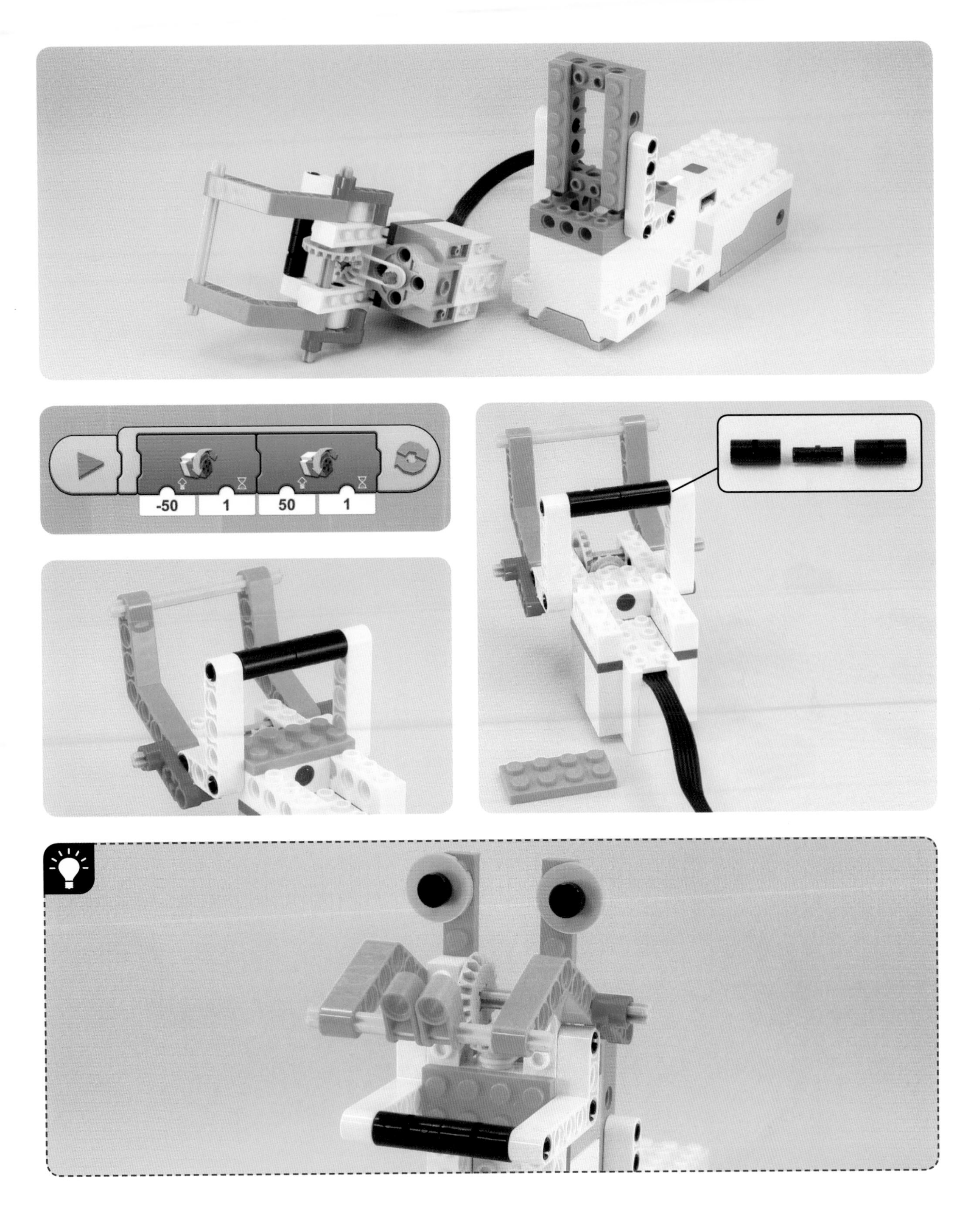
-50
1
50
1

Gripping fingers

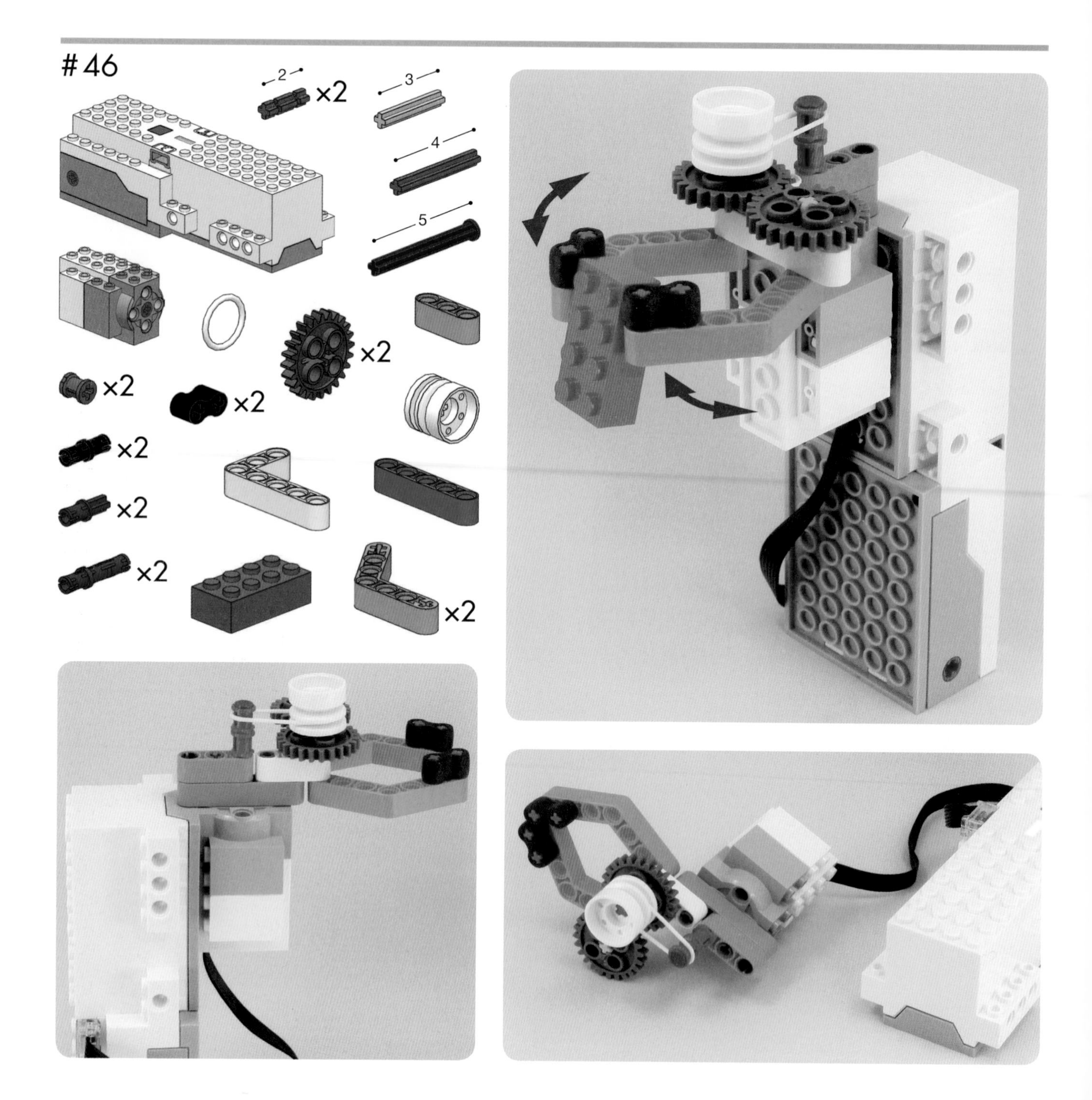

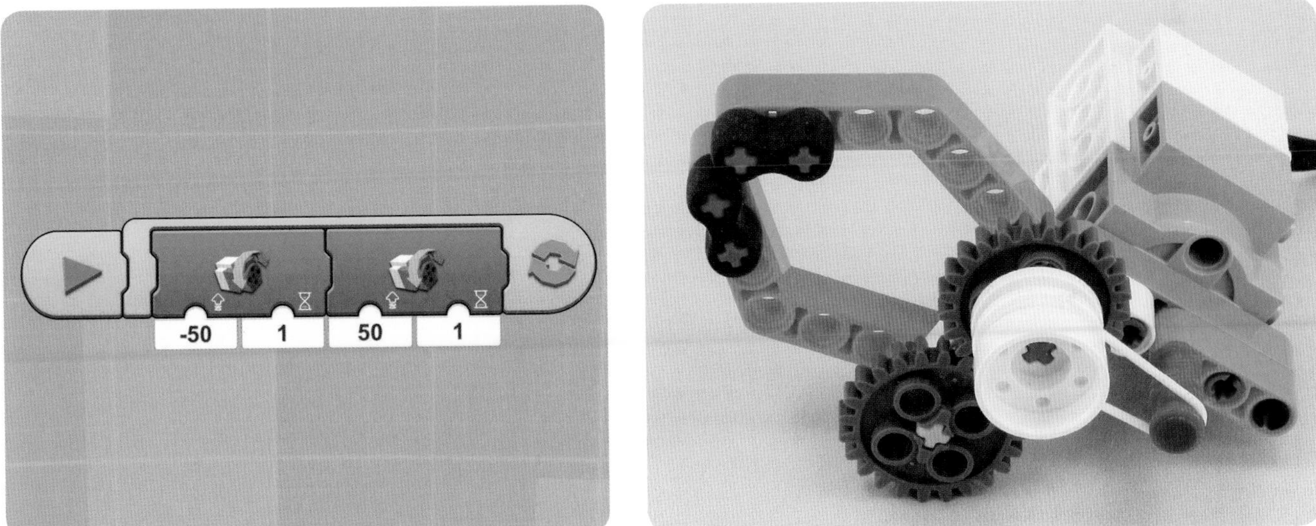
-50
1
50
1

#47

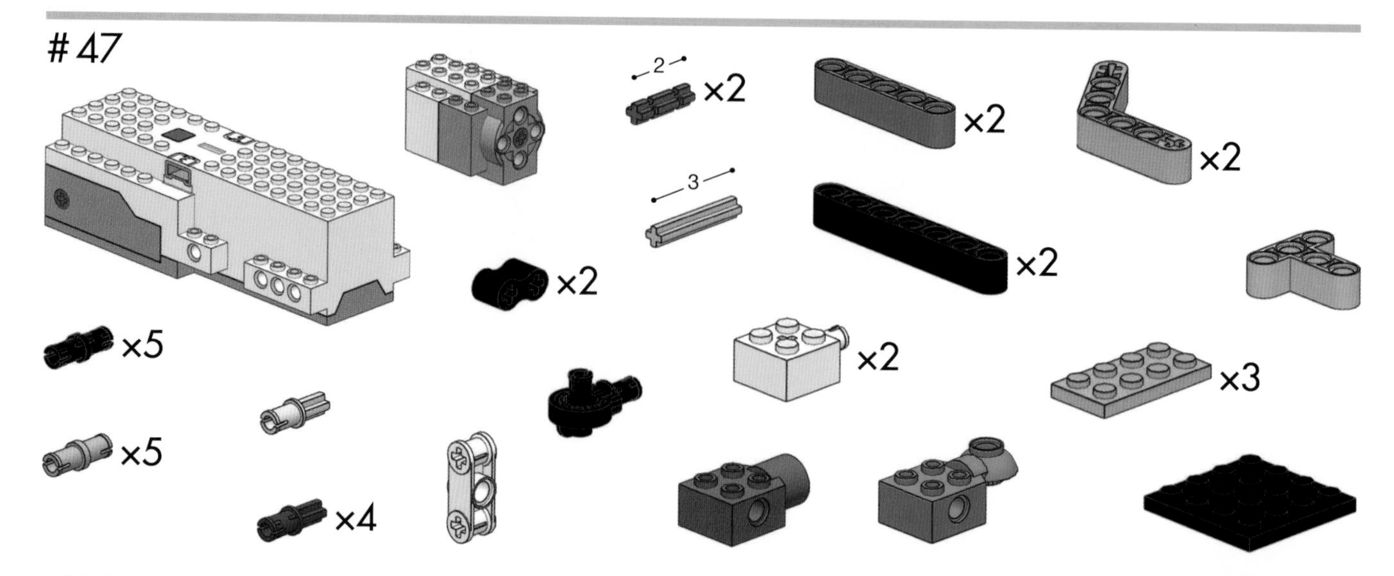

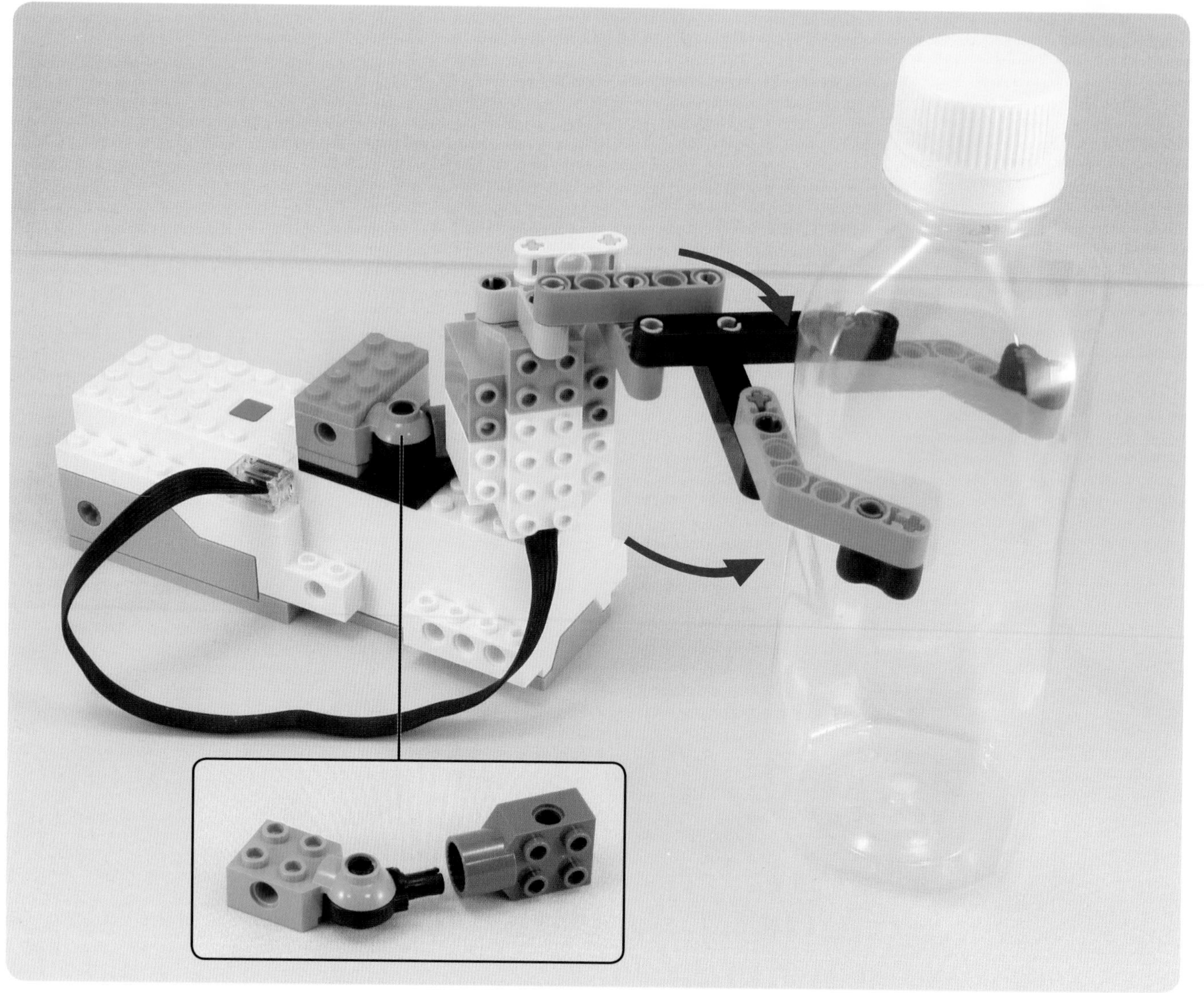

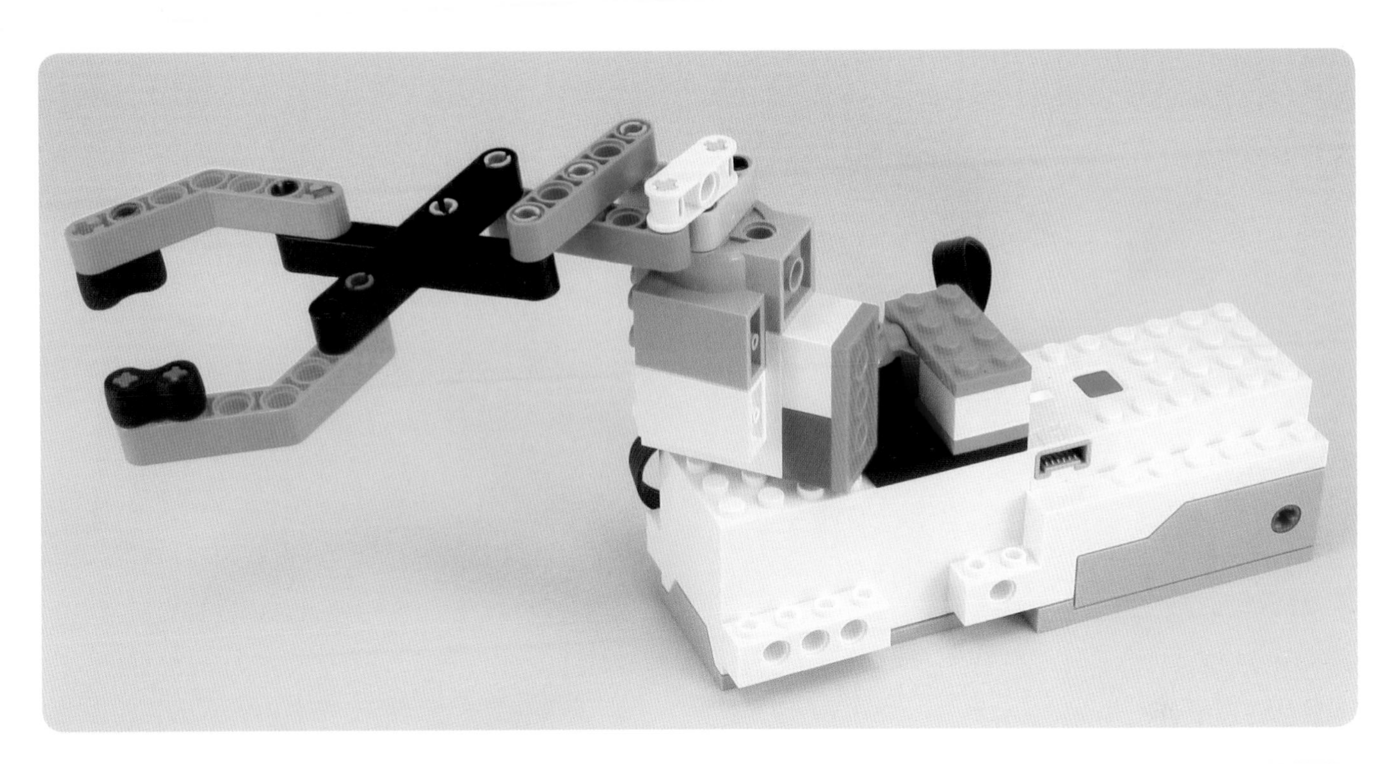

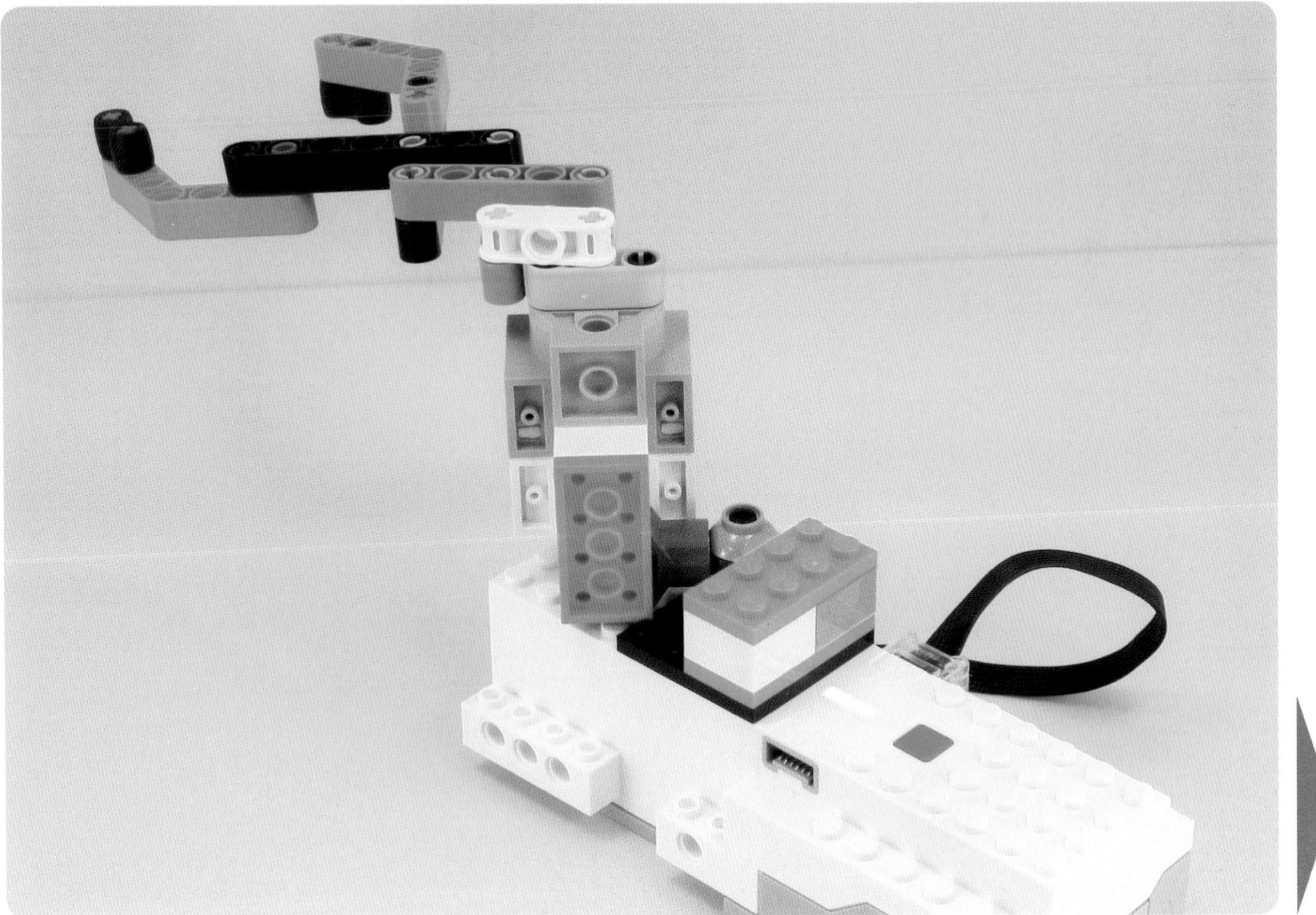

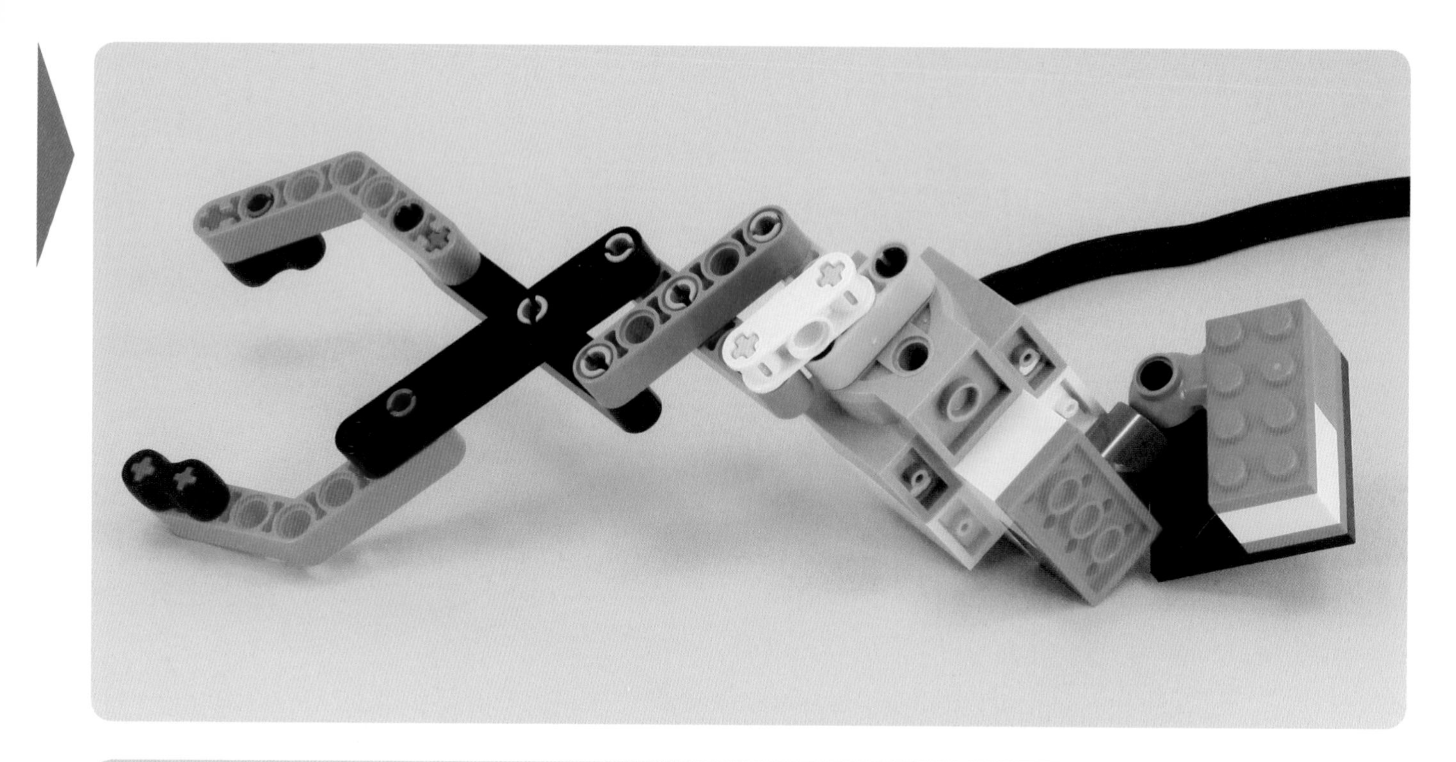

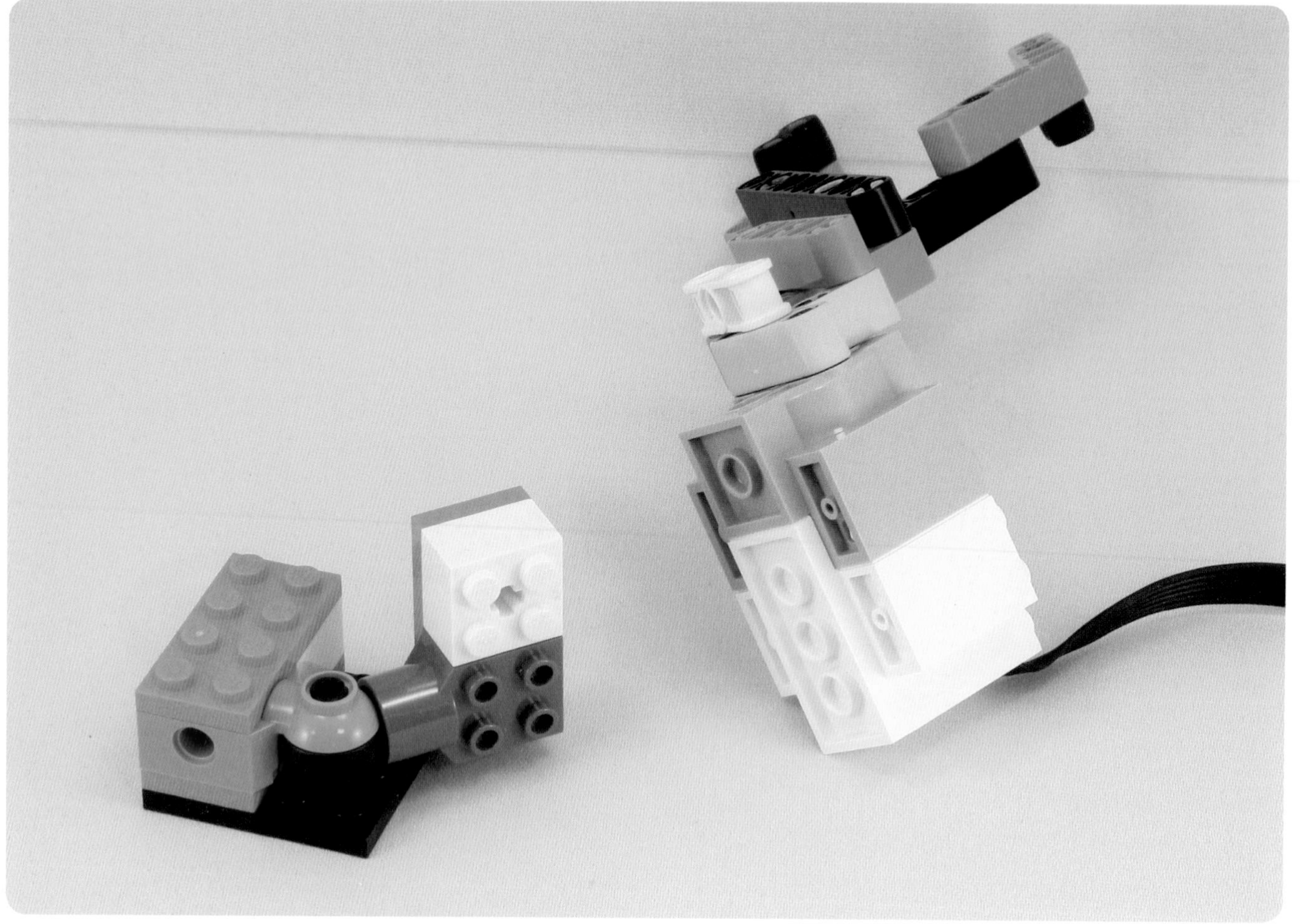

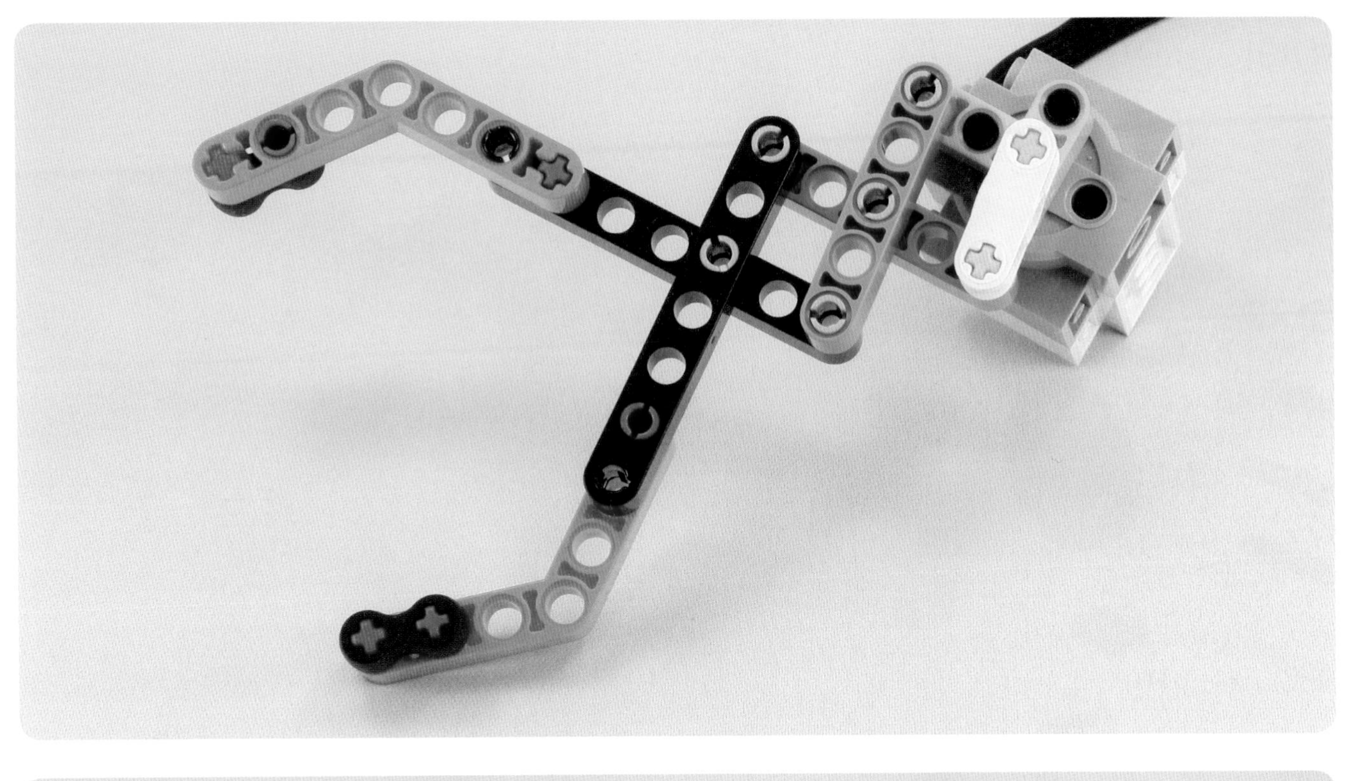

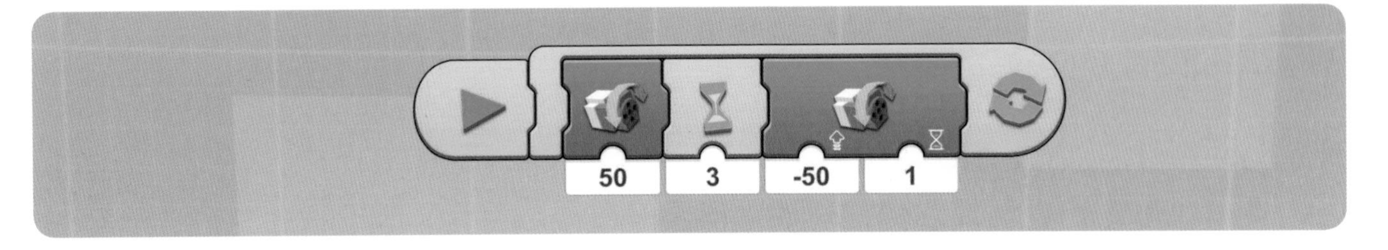
50
3
-50
1

#48

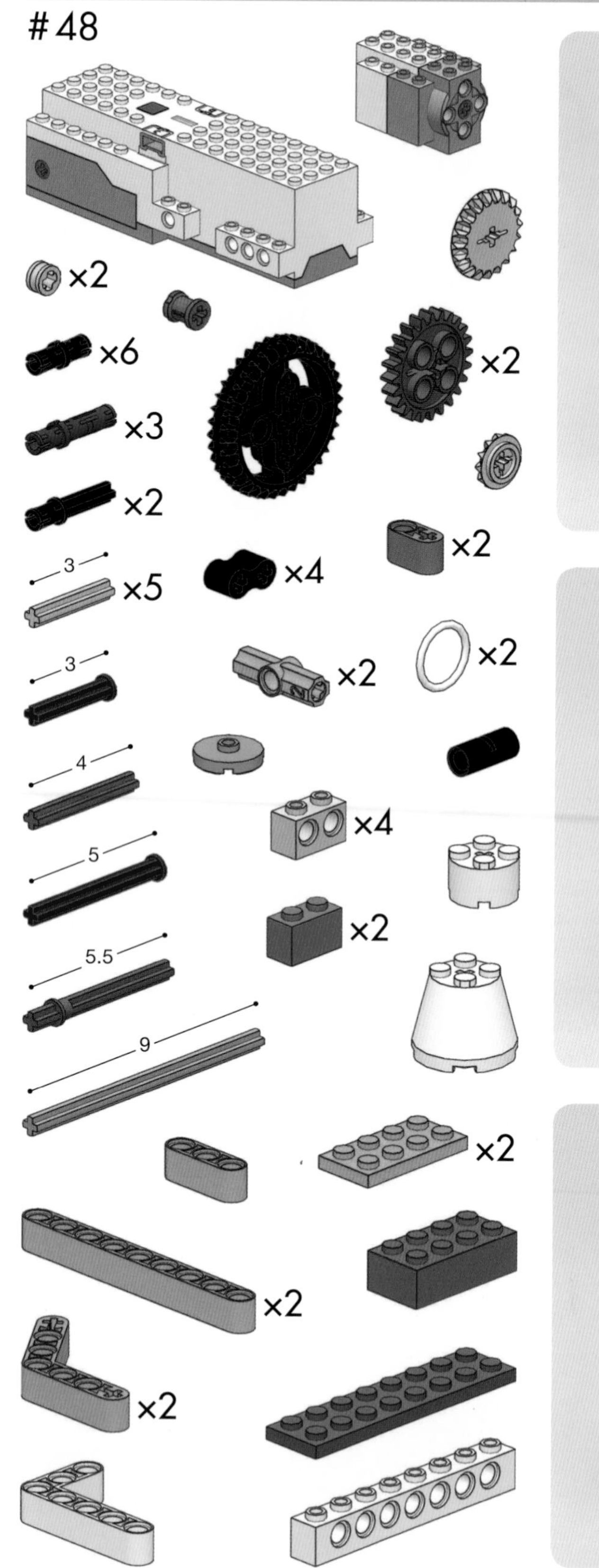

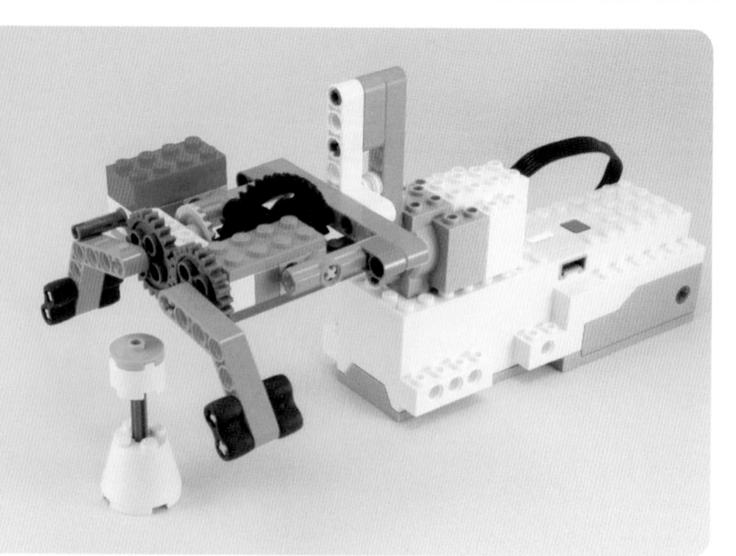

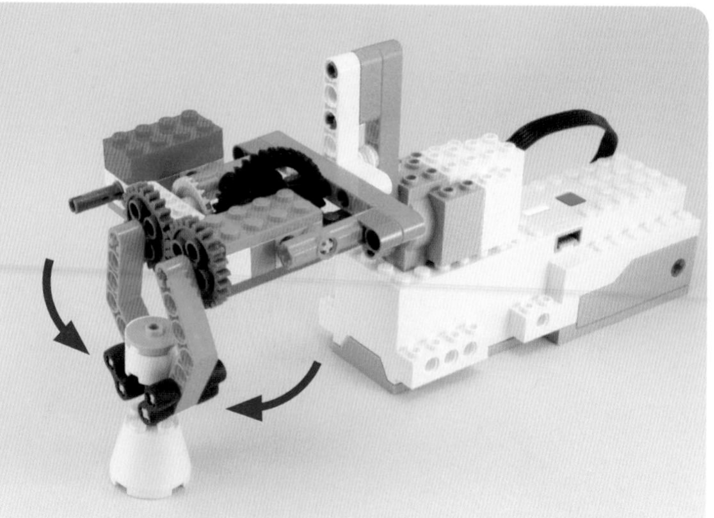

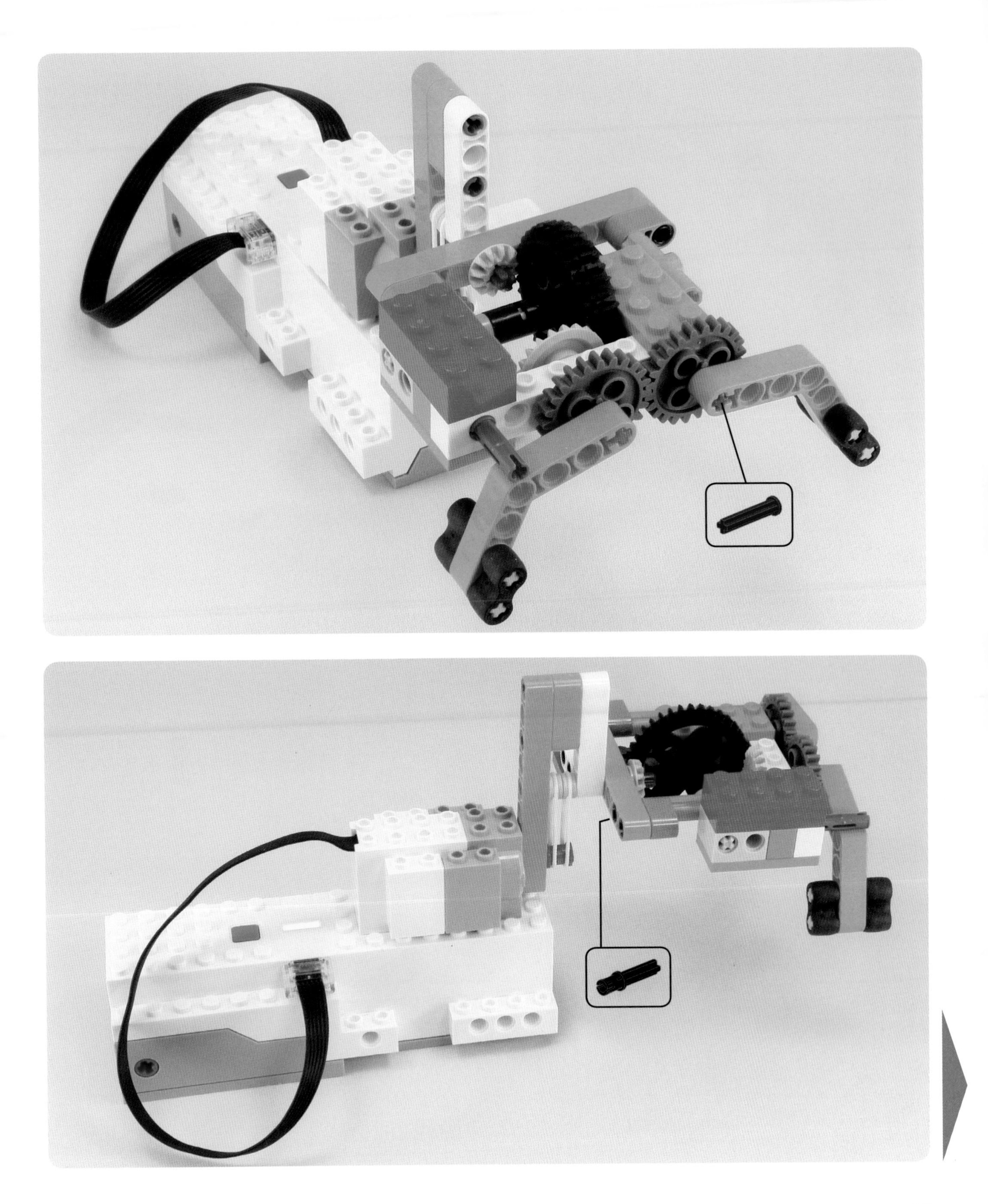

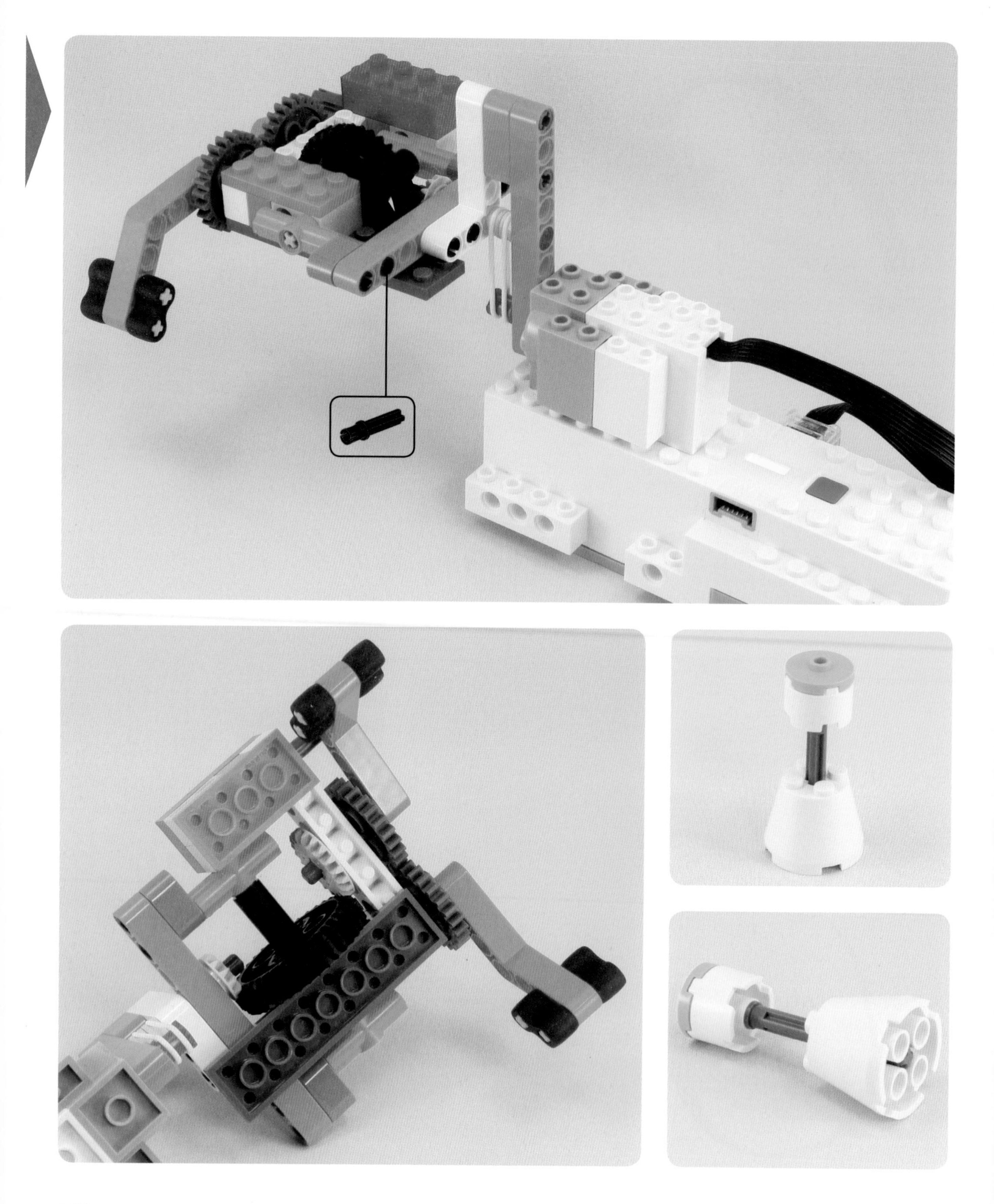

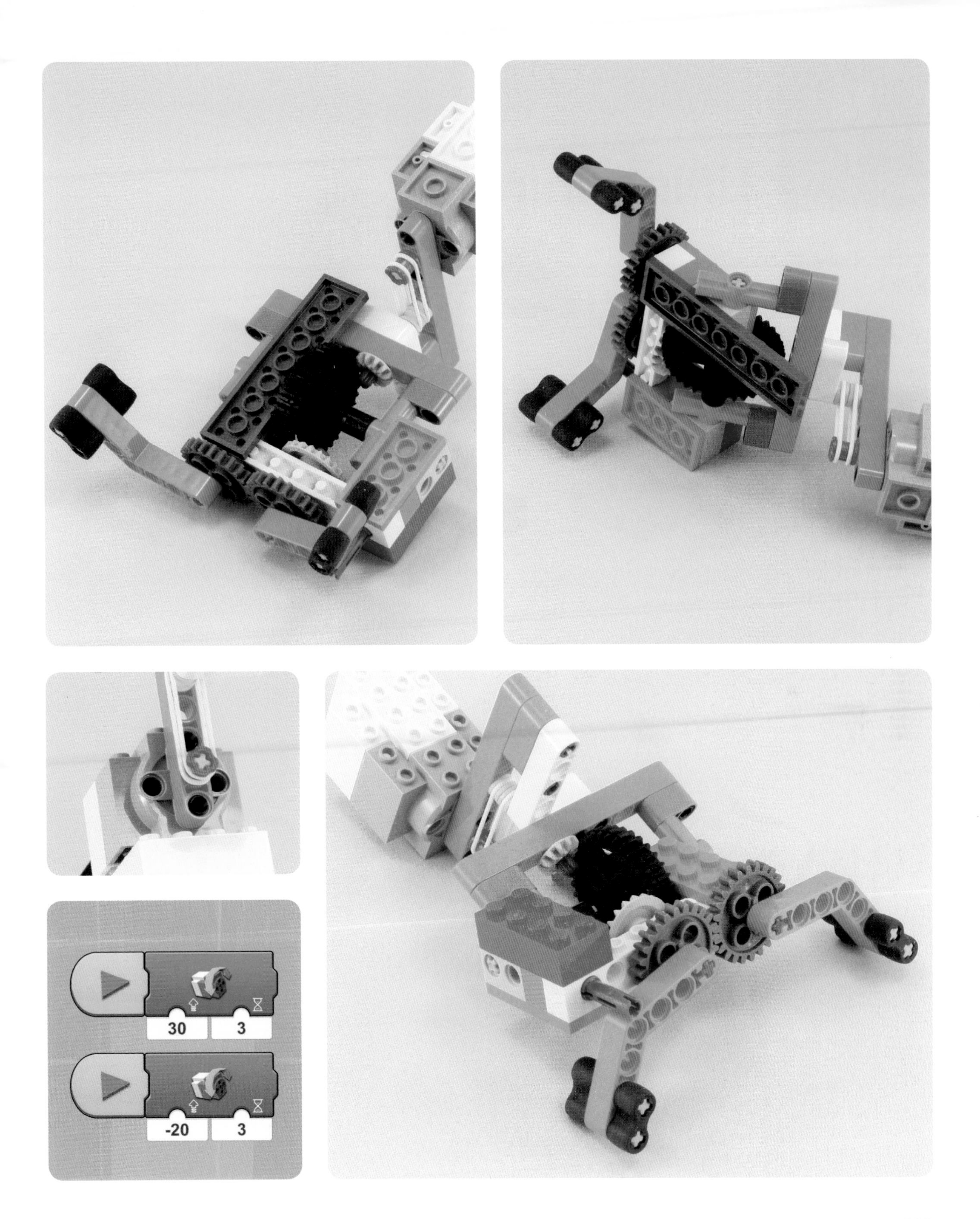
30
3
-20
3

Lifting things

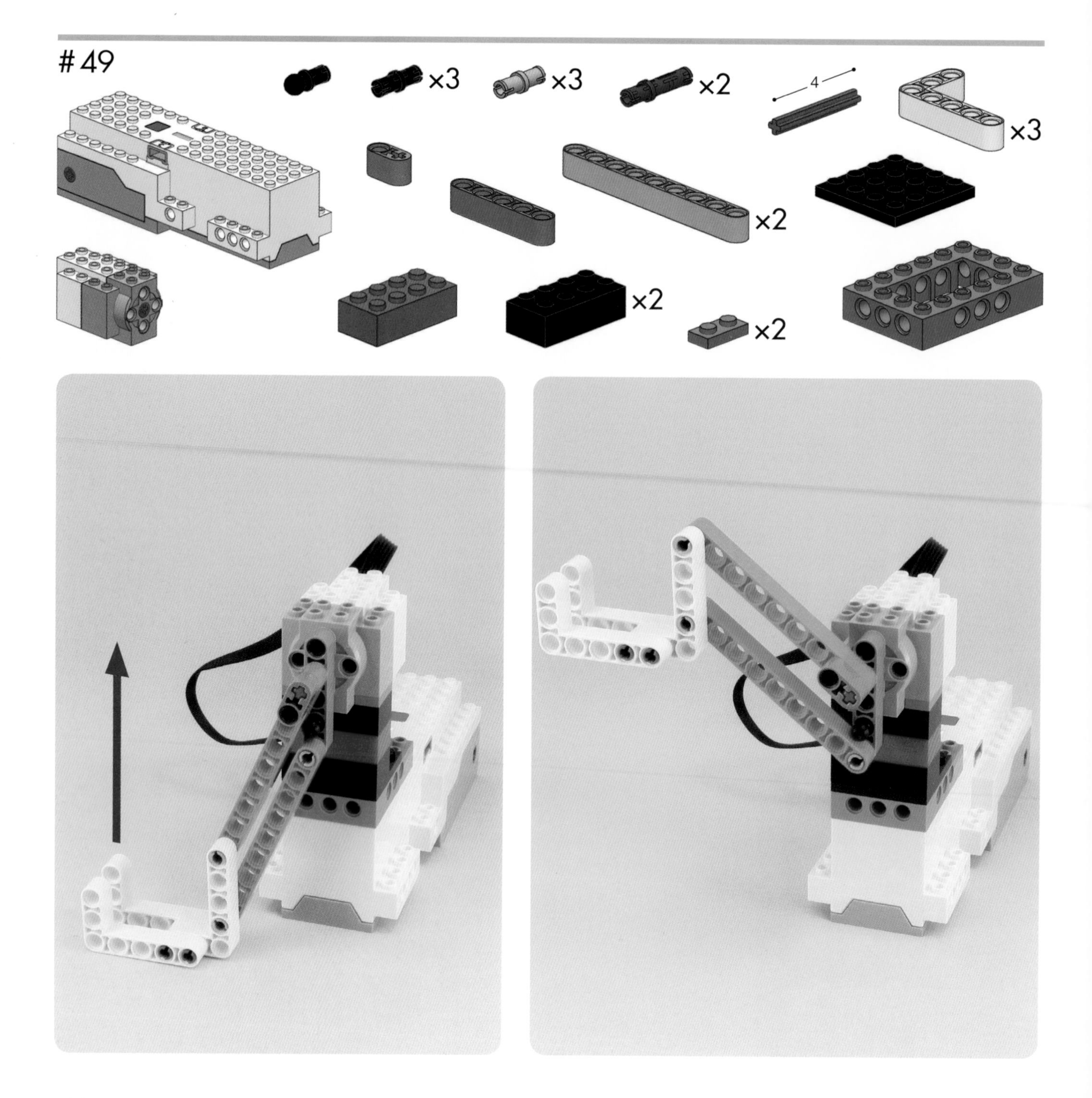

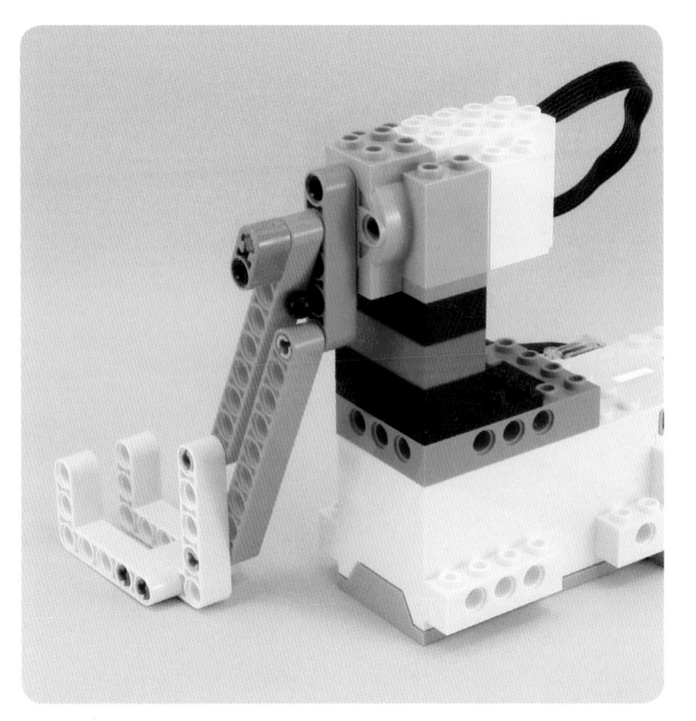

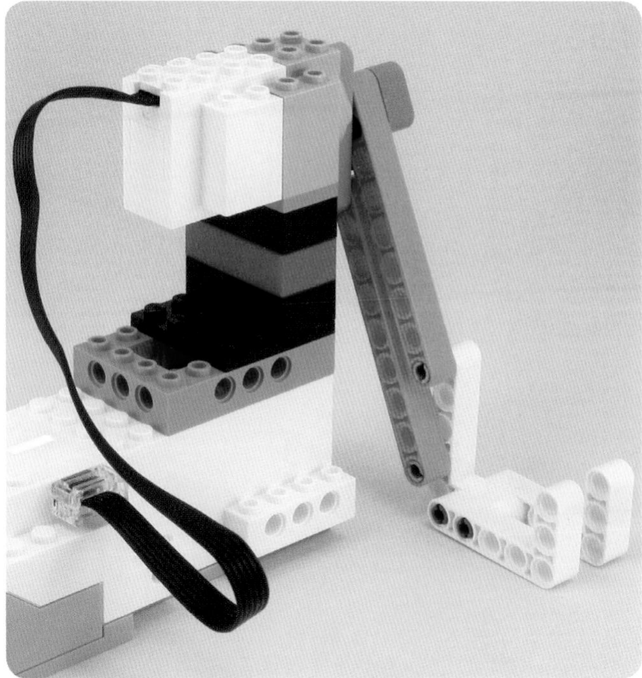

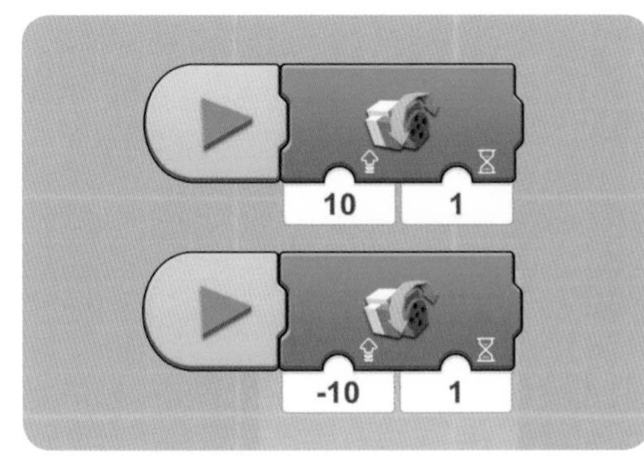
10
1
-10
1

50

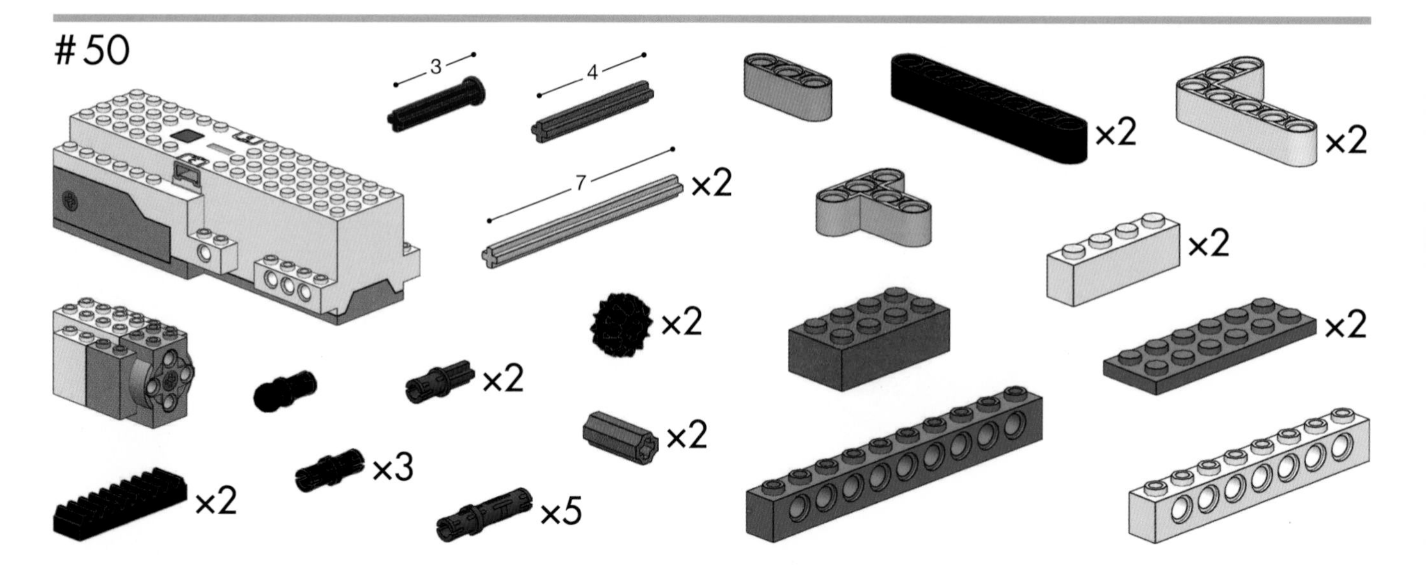

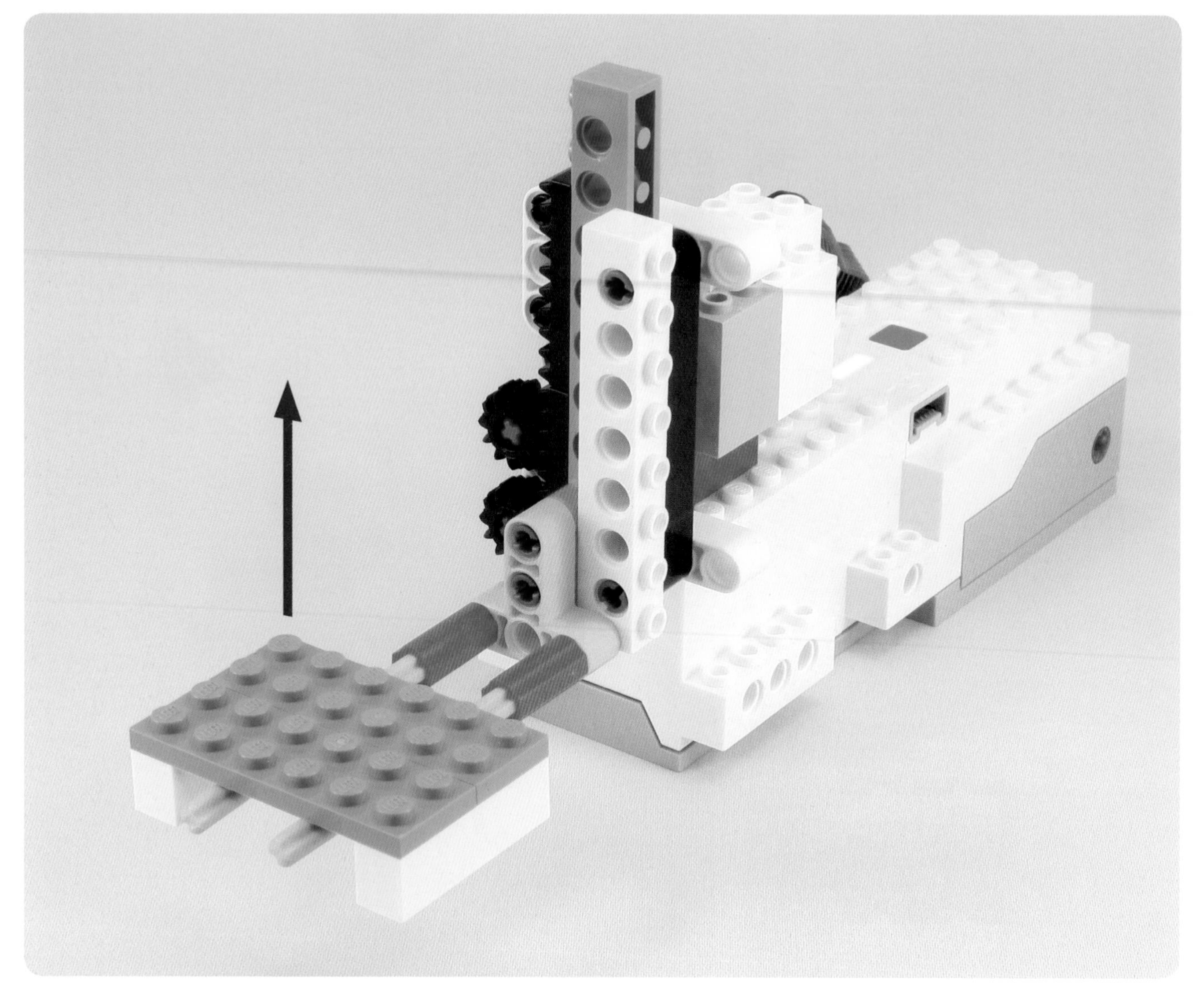

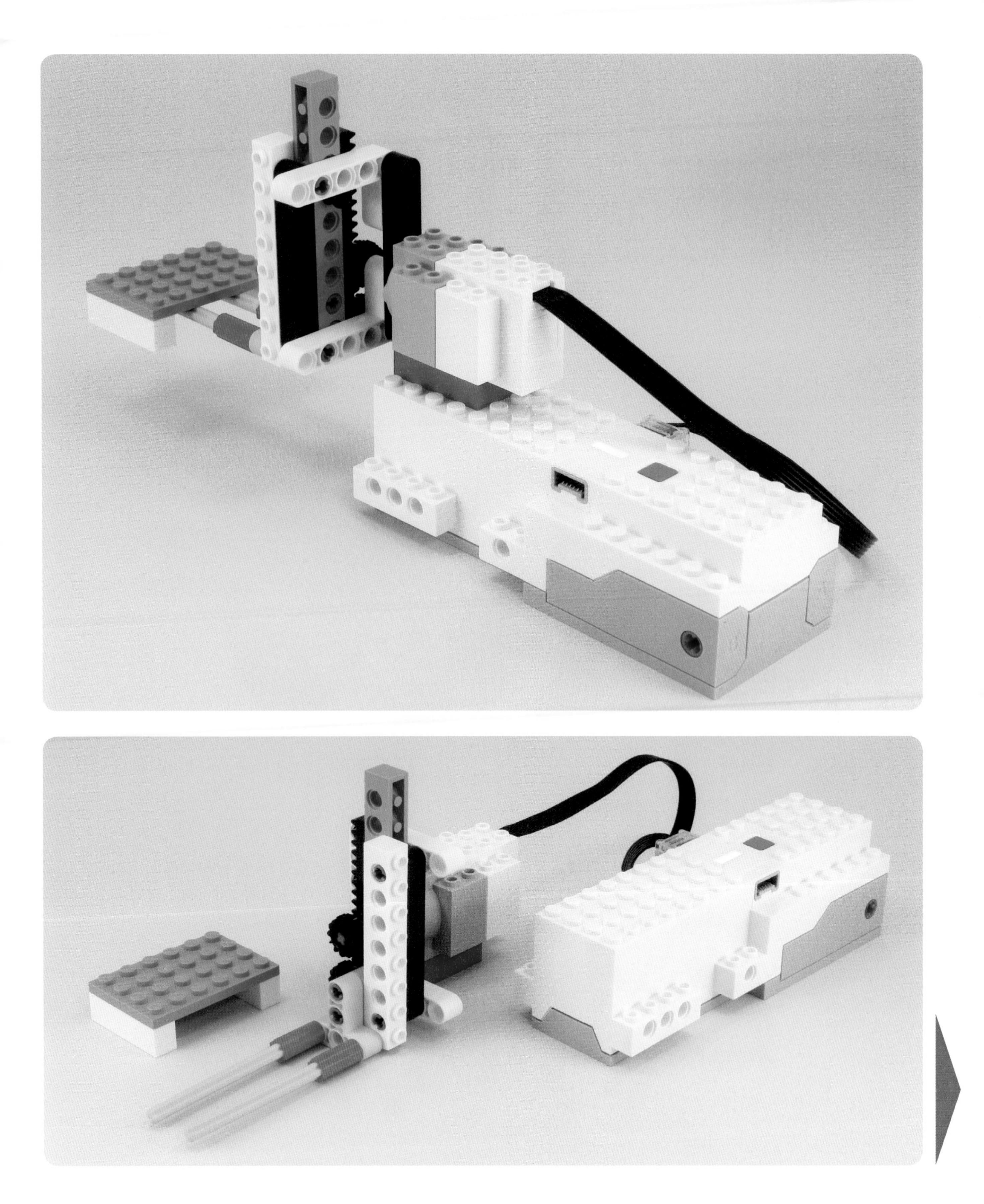

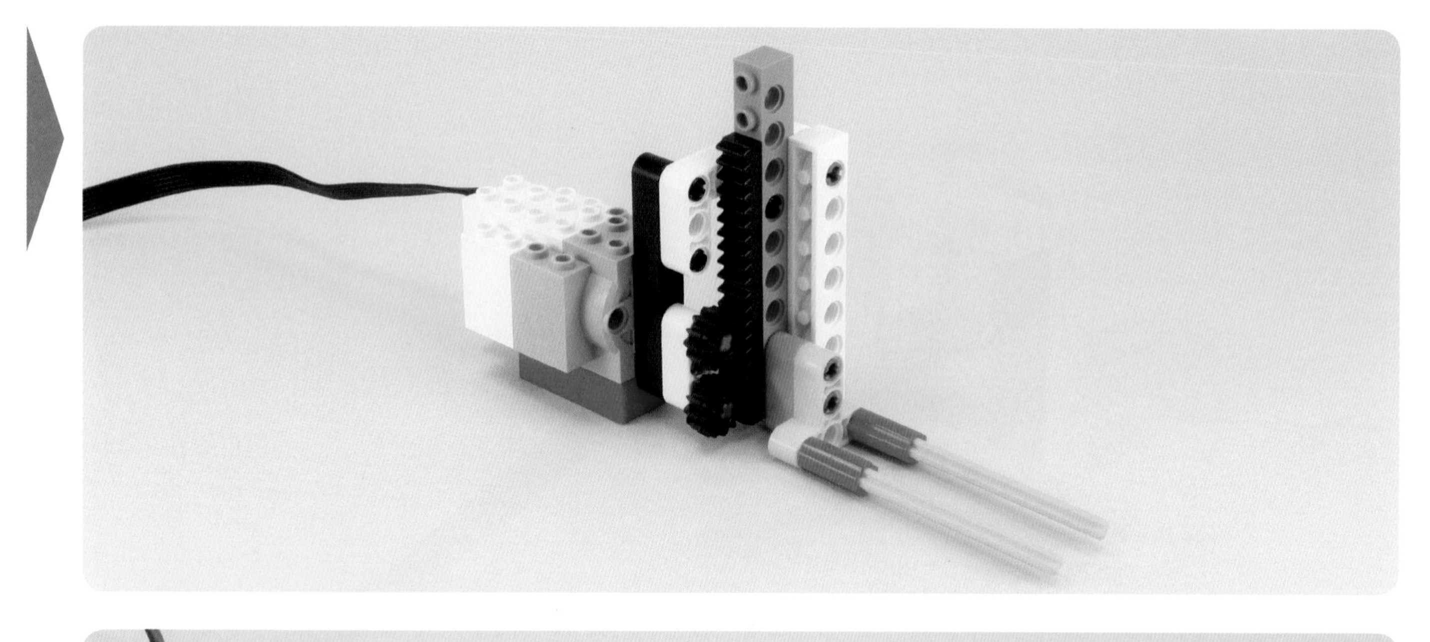

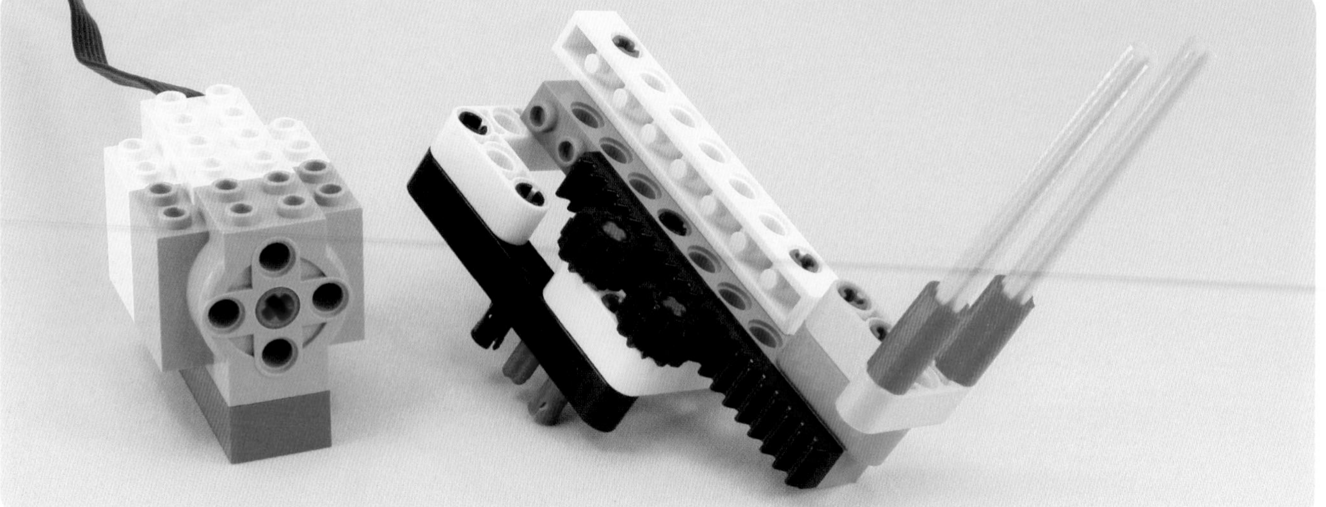

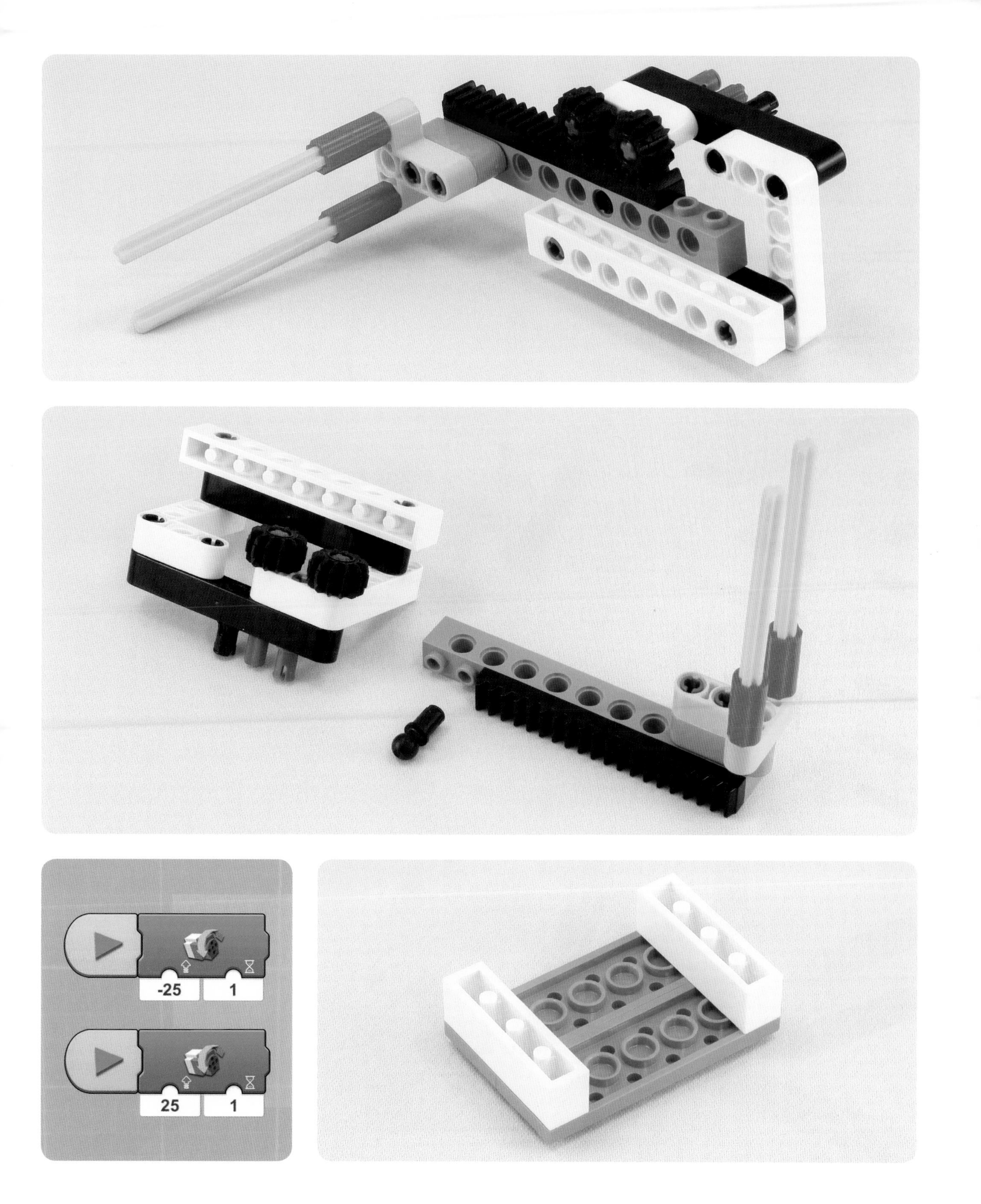
-25
1
25
1

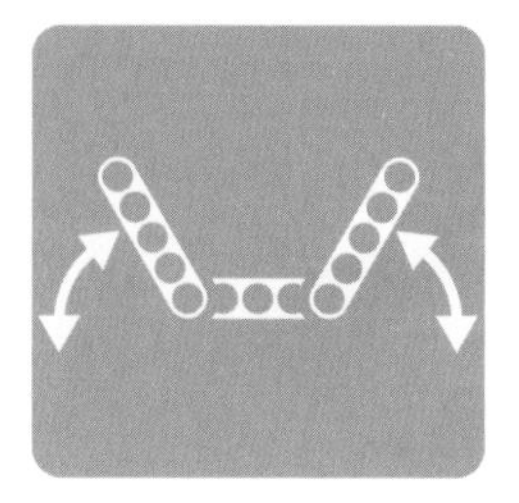

Flapping wings

#51

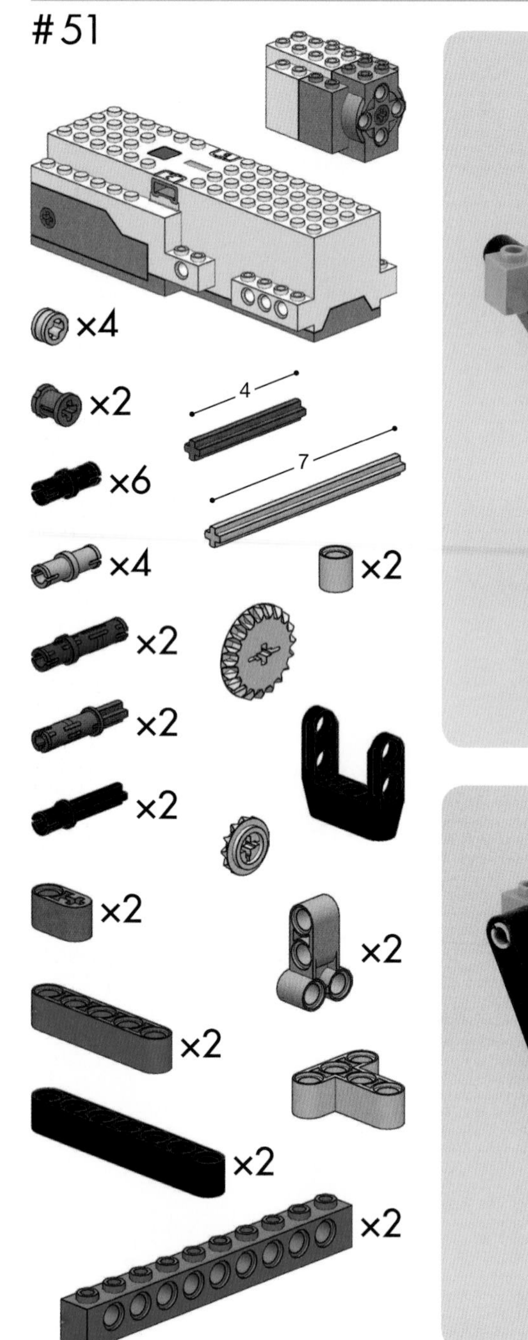

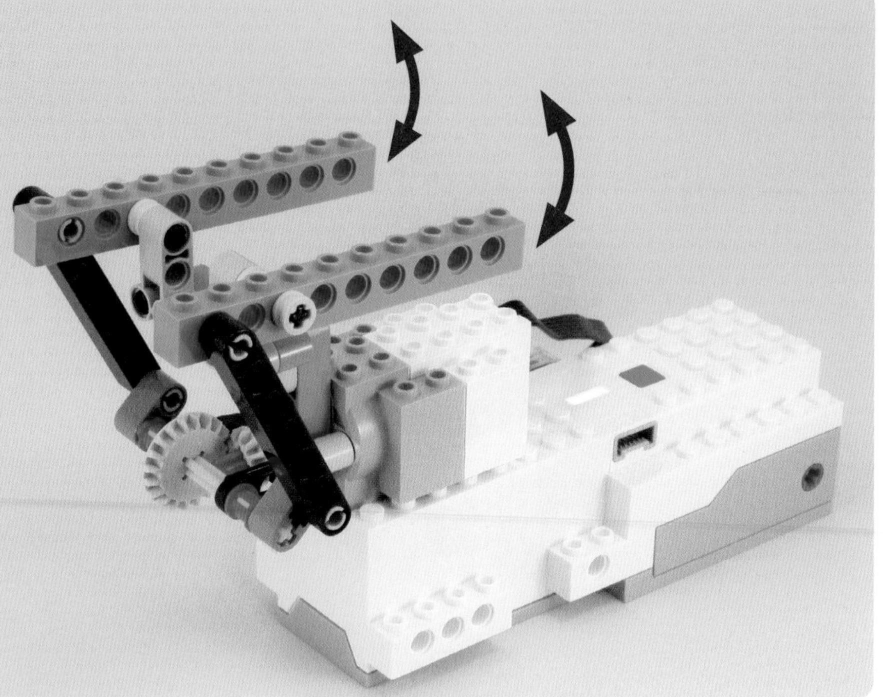

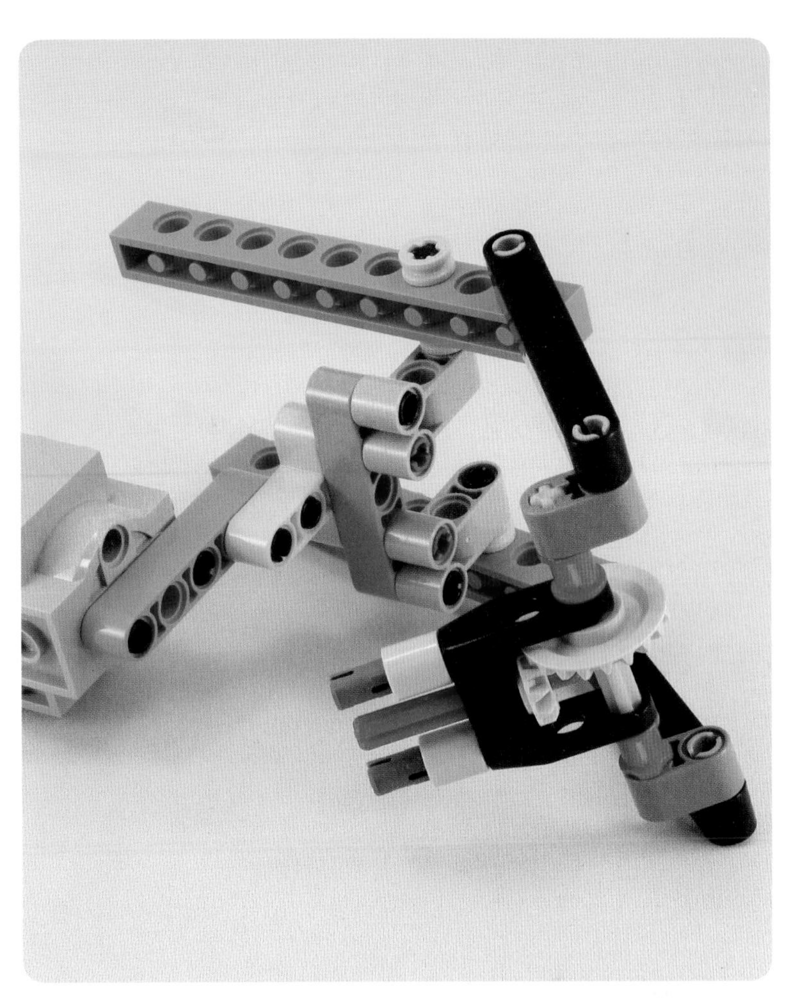

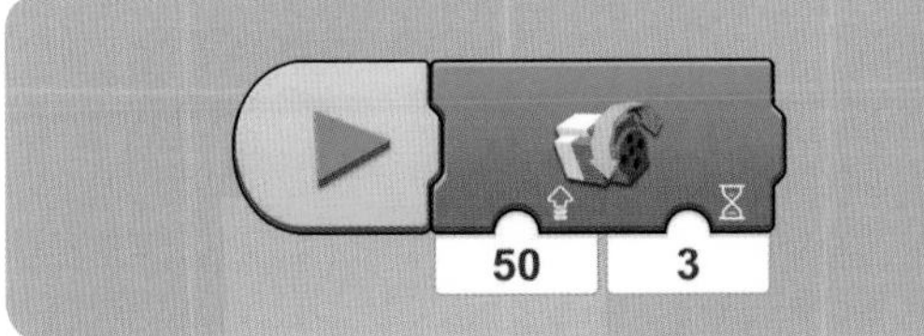
50
3

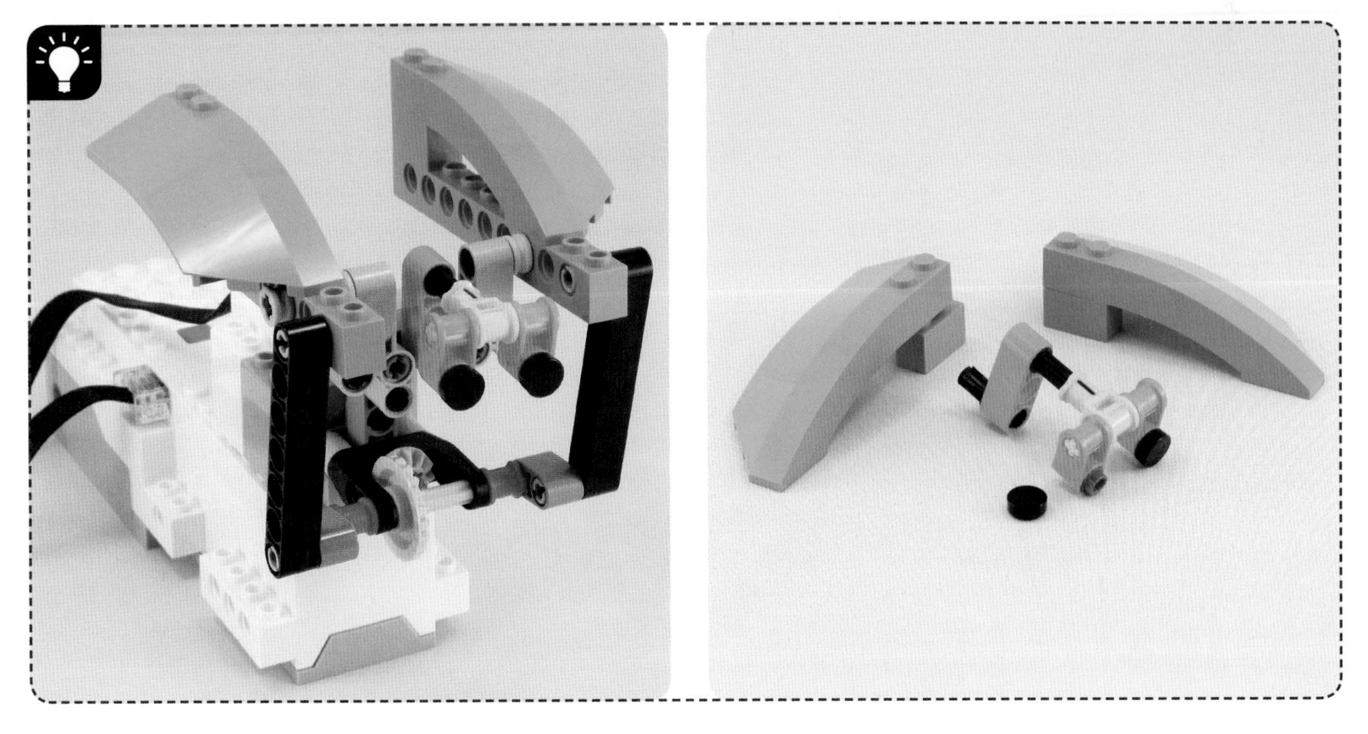

52

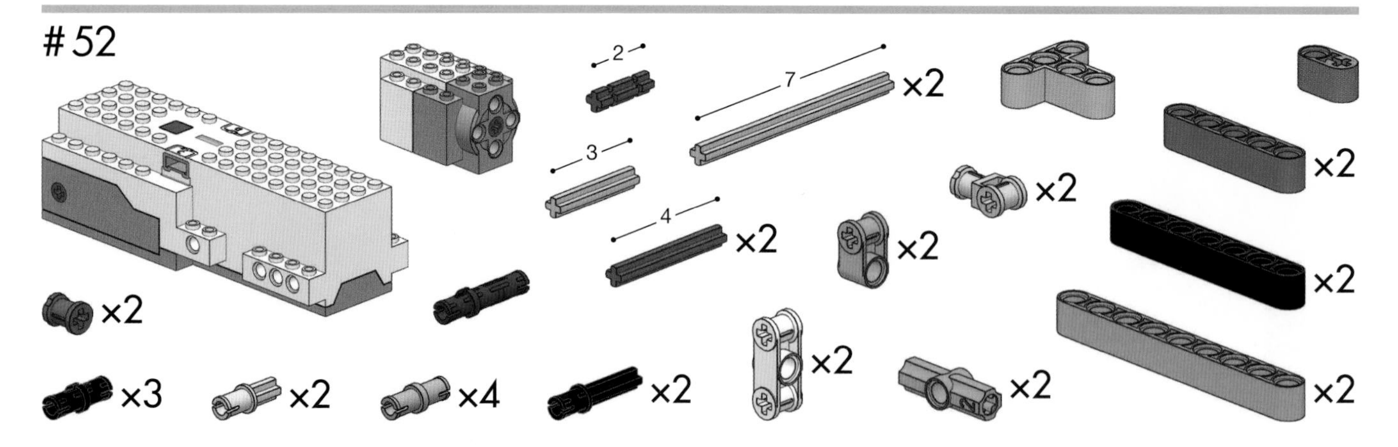

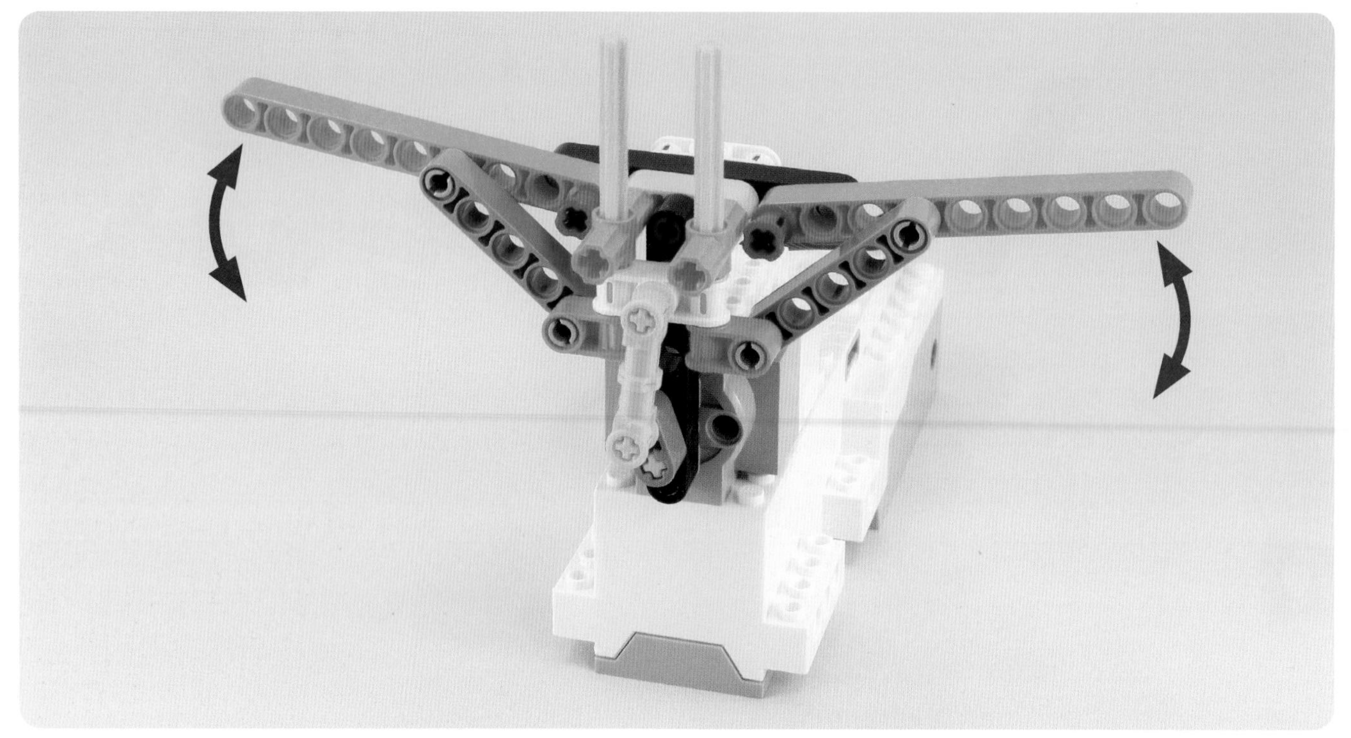

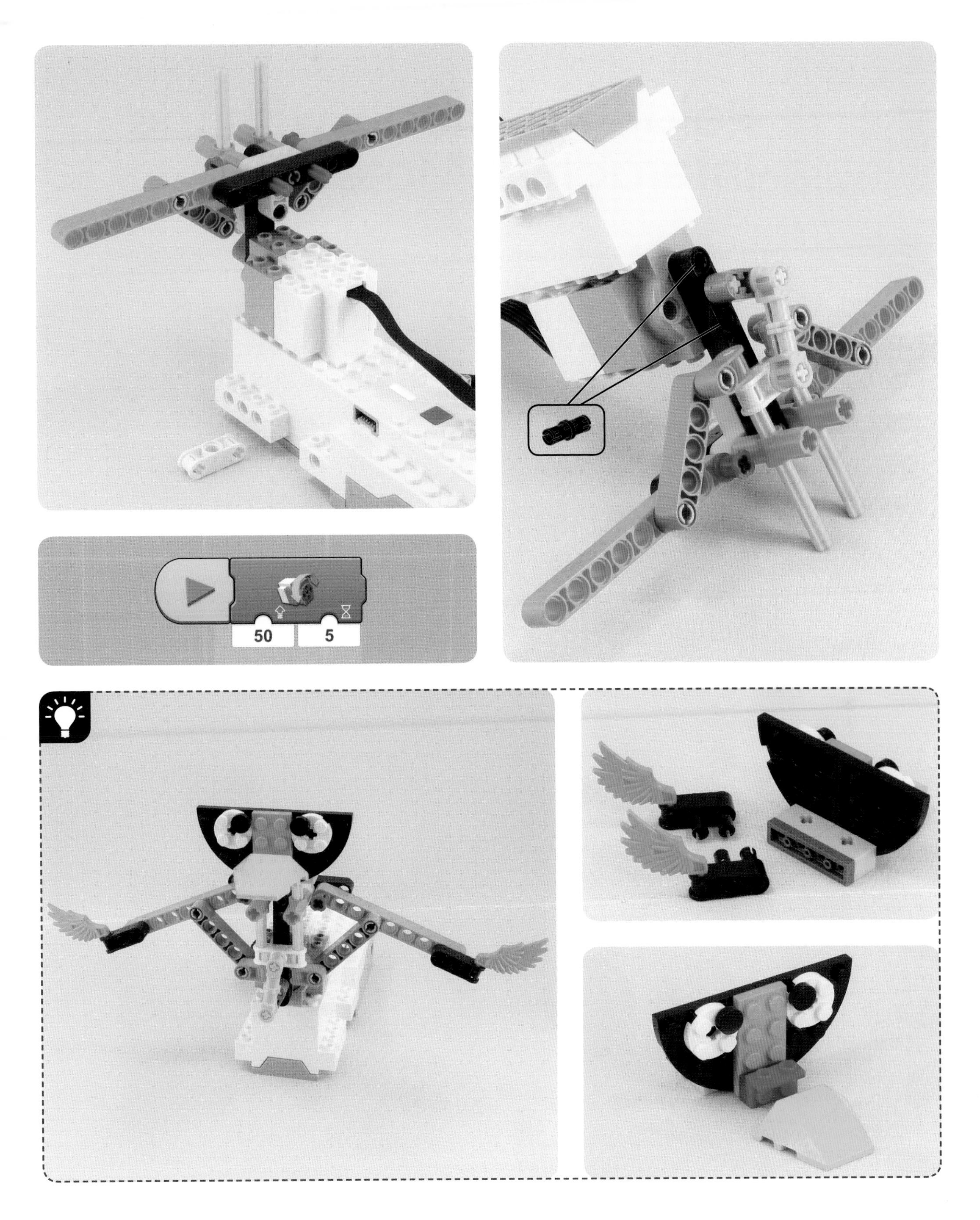
50
5

Rotating wheels with the Interactive Motor

53

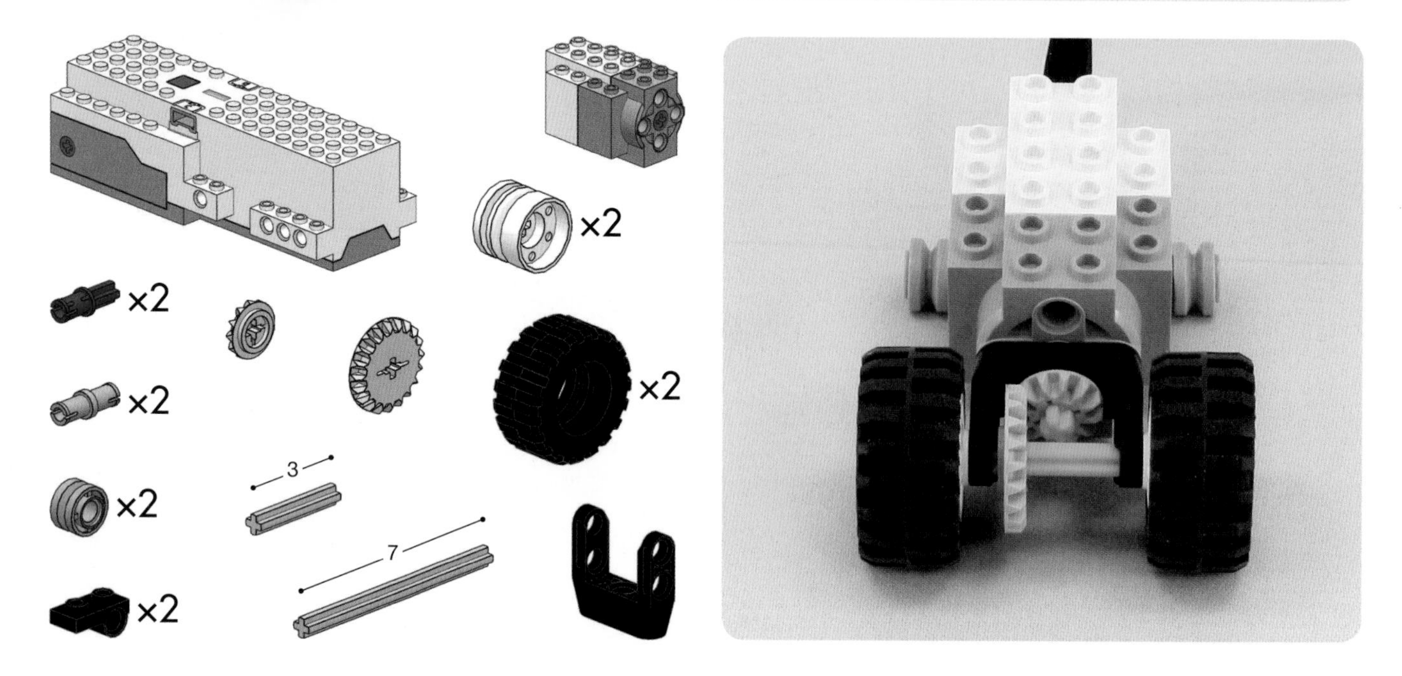

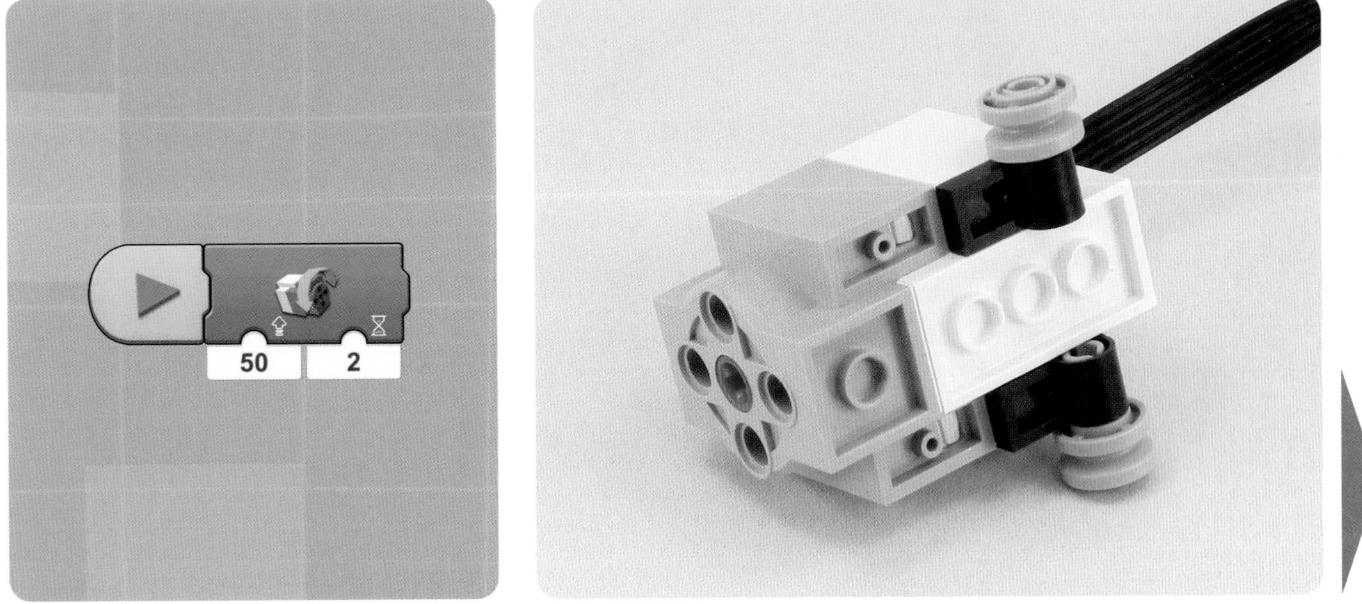
50
2

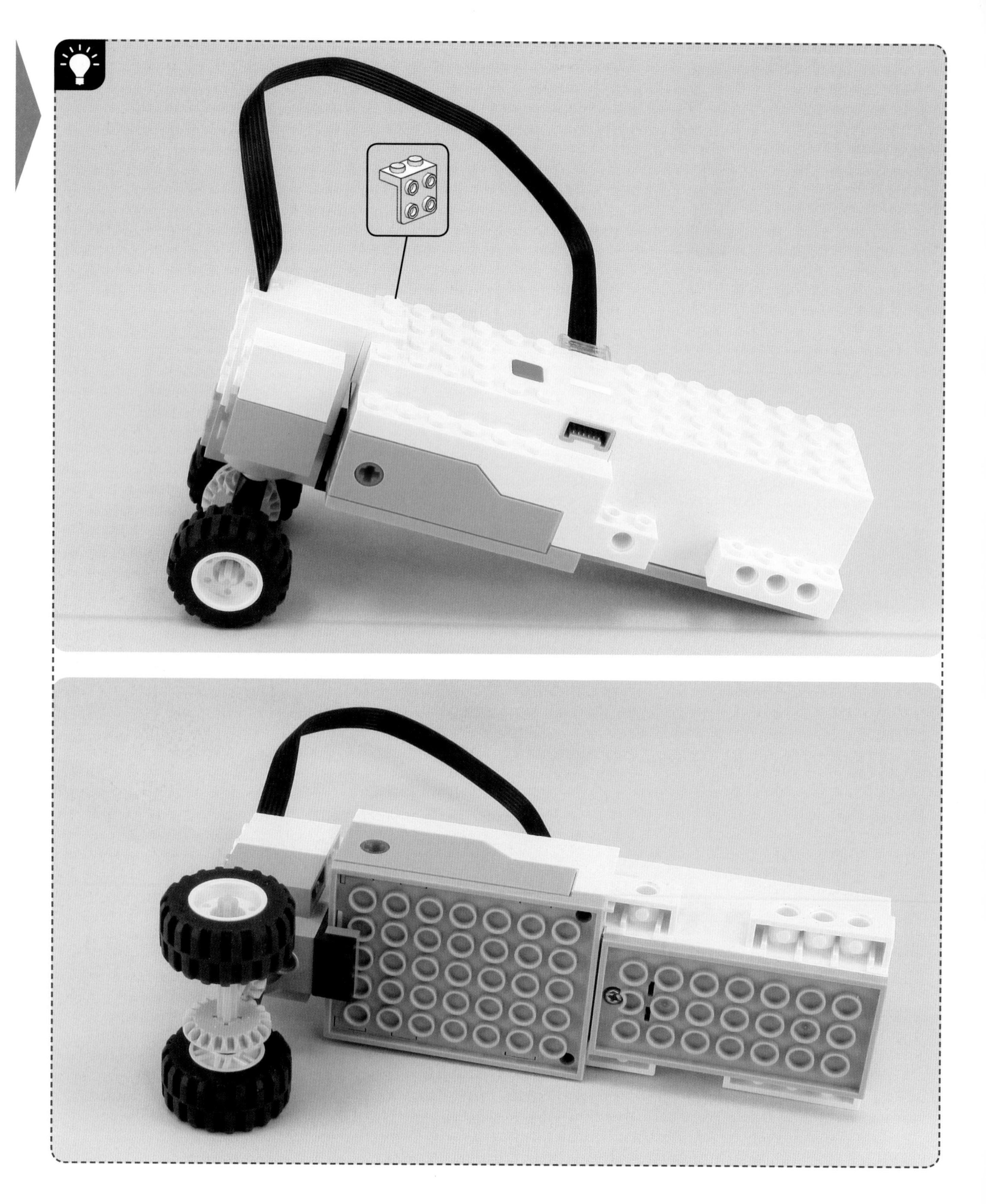

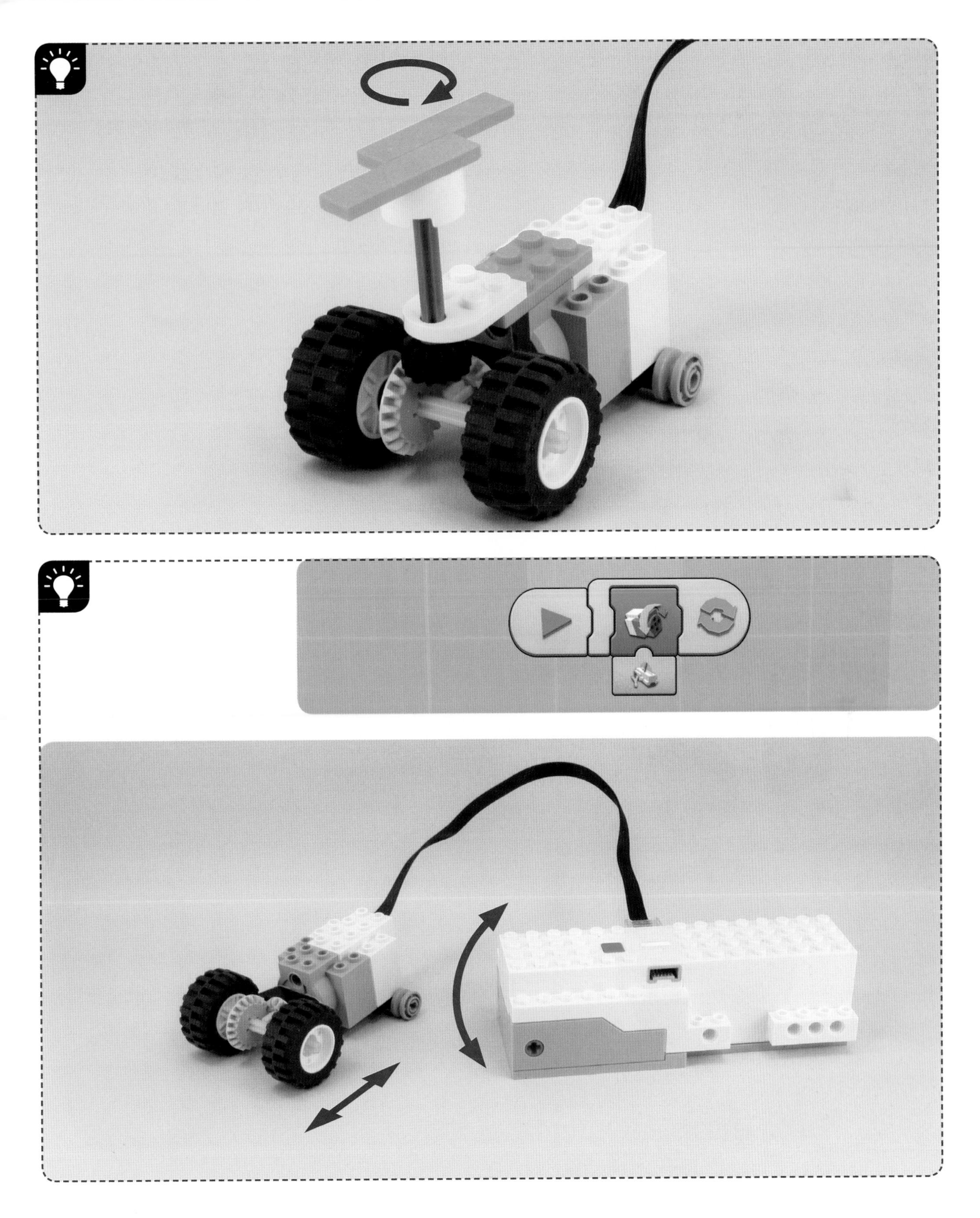

#54

×2

×6

×4

×2

3 ×2

4

5.5

7

10

×2

×2

×2

×2

×2

×2

×2

×2

×2

80 5

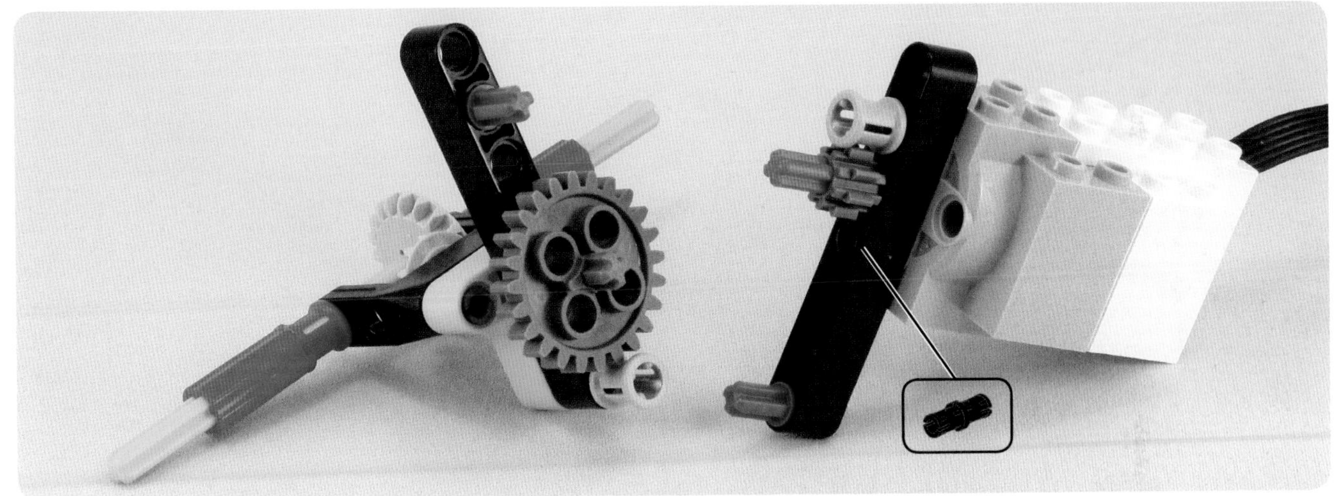

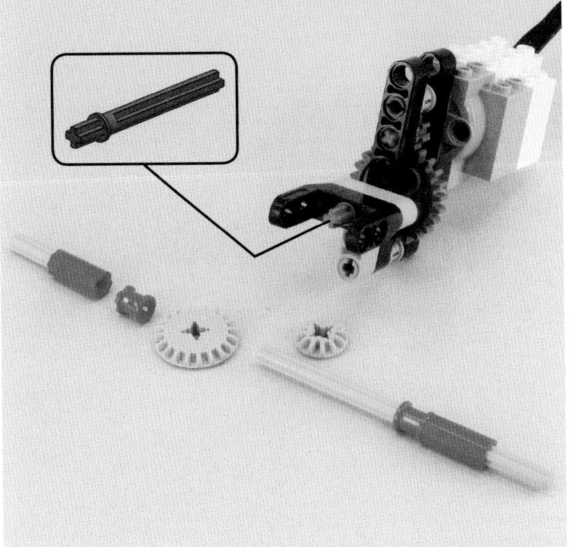

A
B
80
3
50
360
50
360

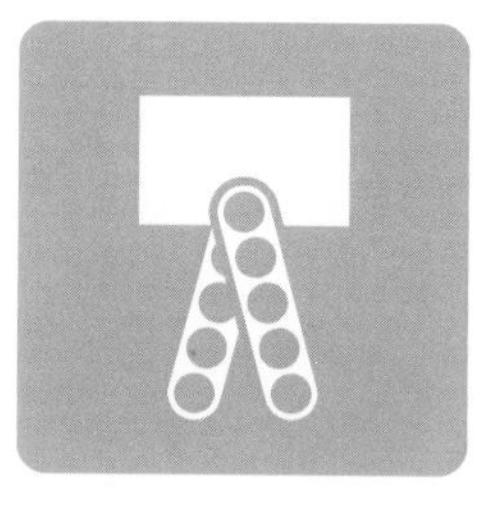

Walking with the Interactive Motor

#55

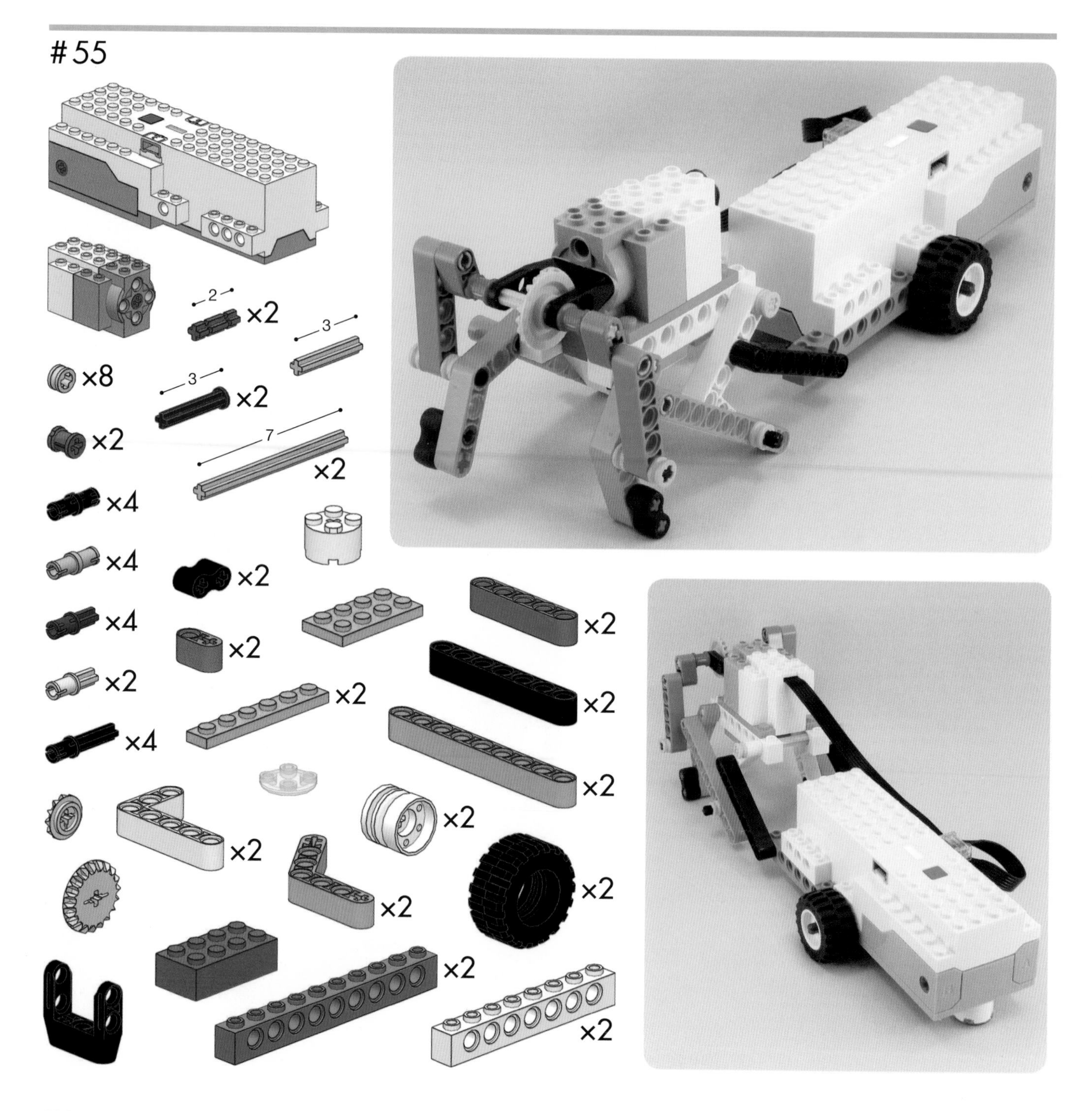

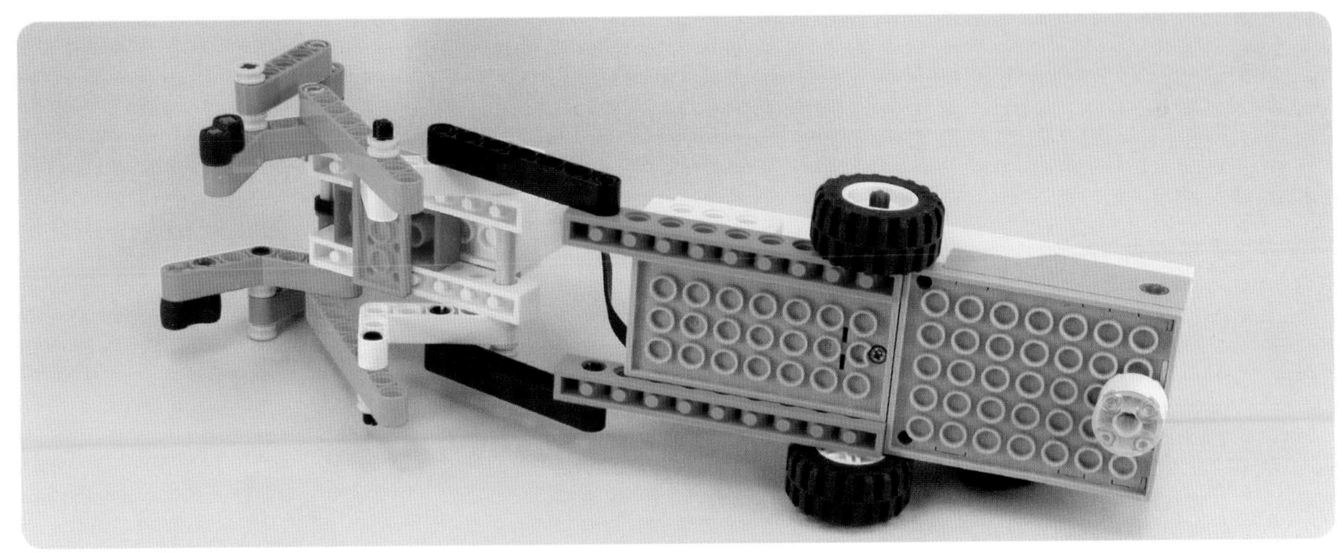

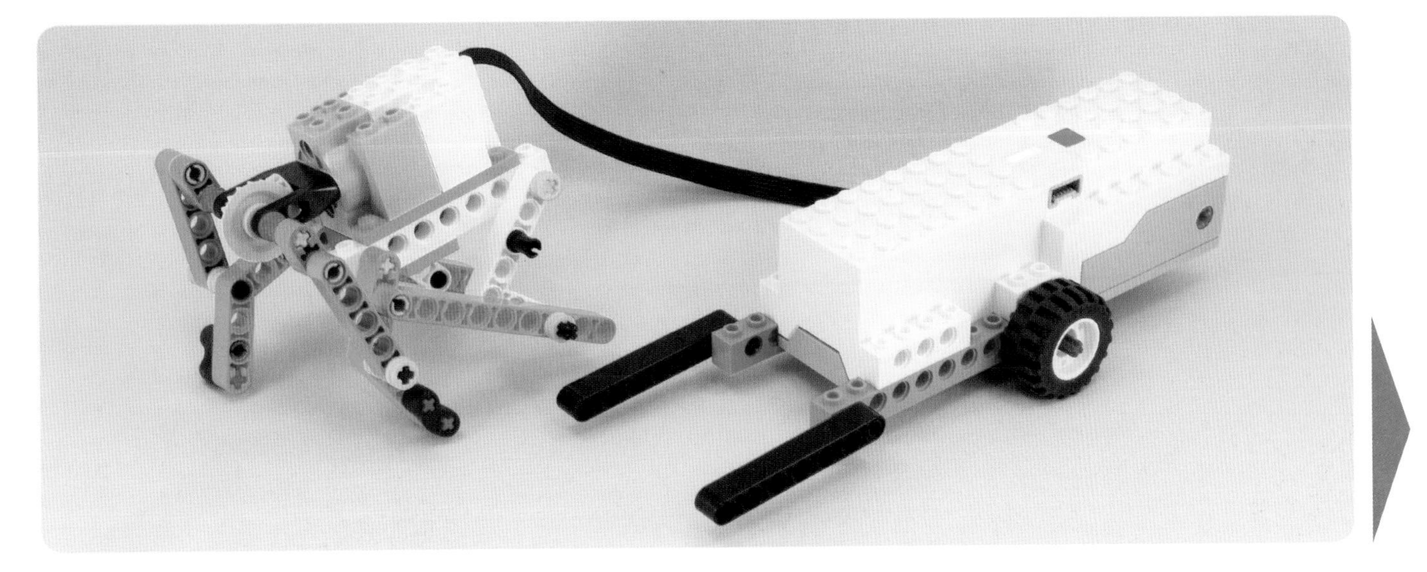

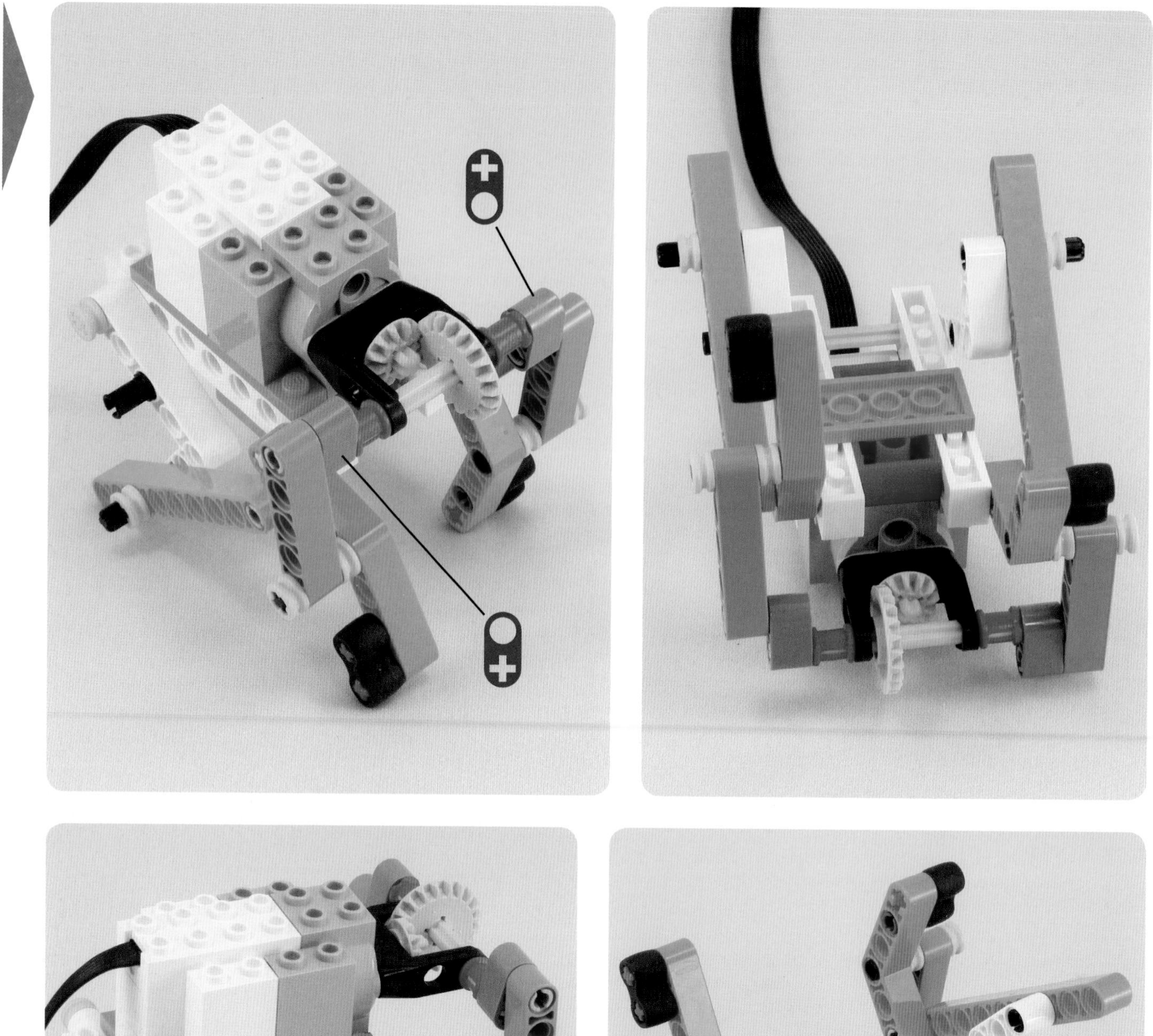

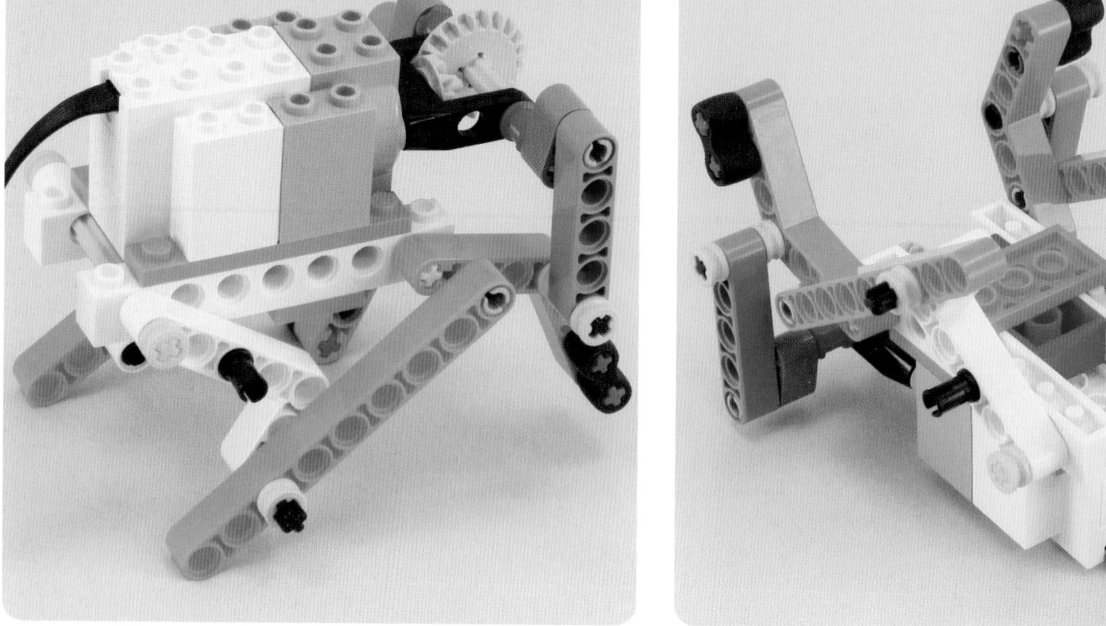

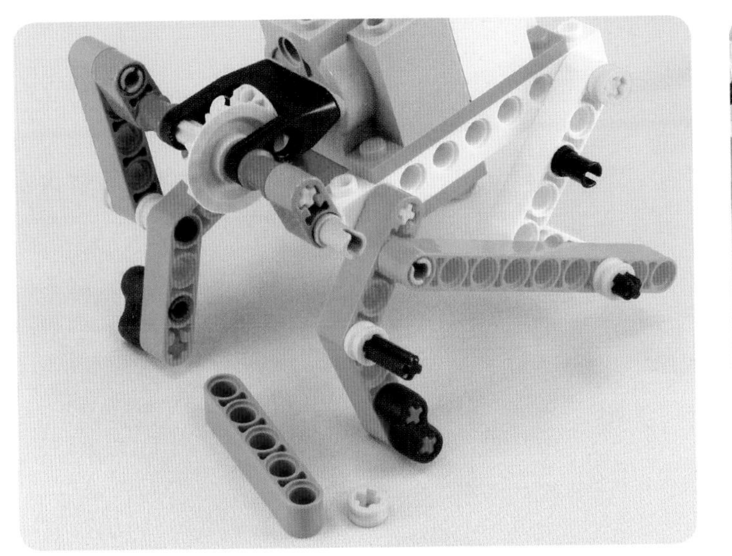

50
5

#56

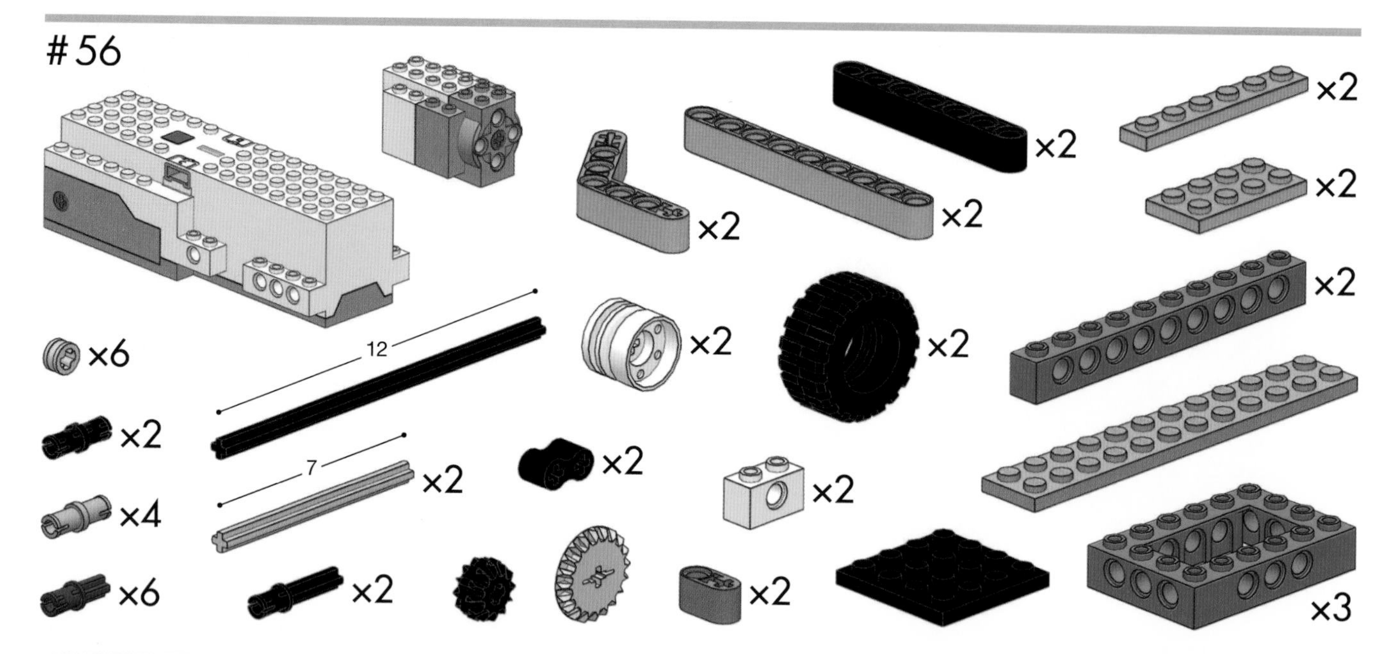

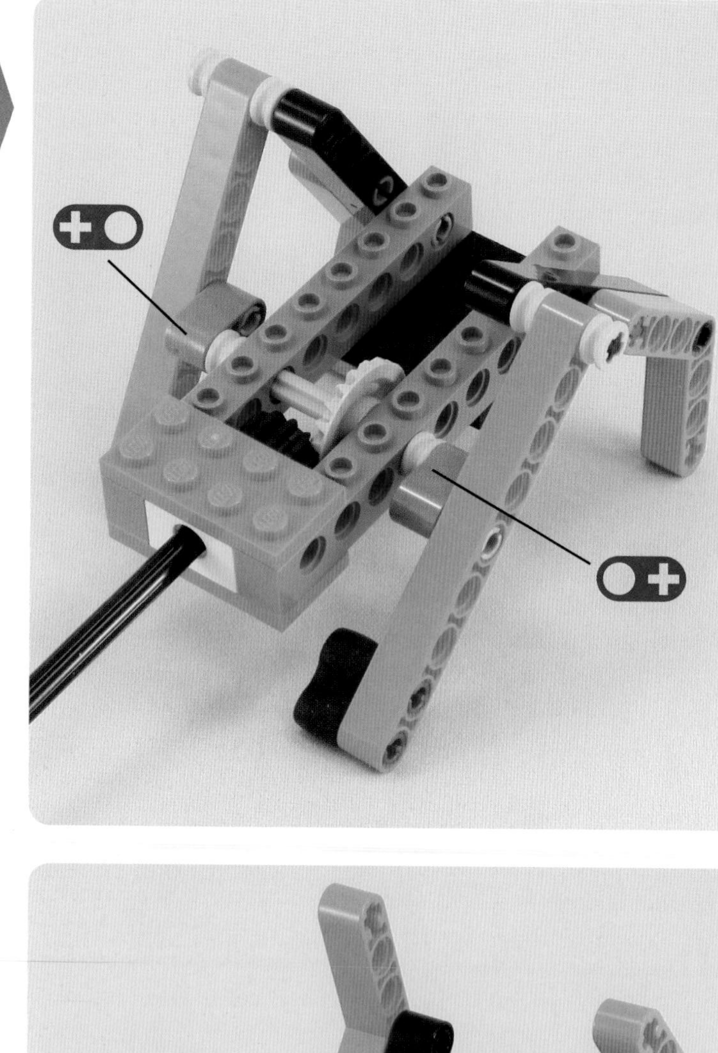

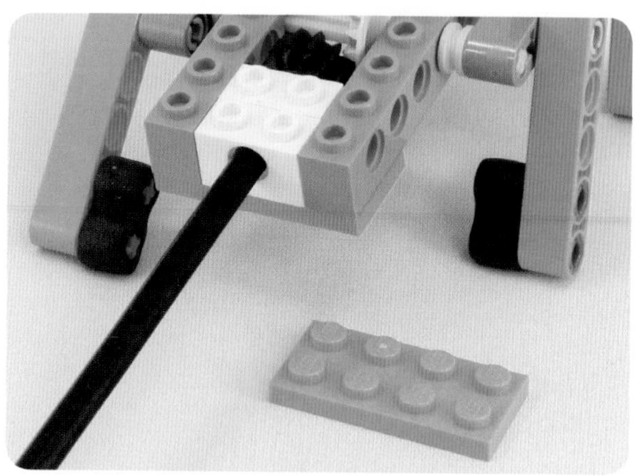

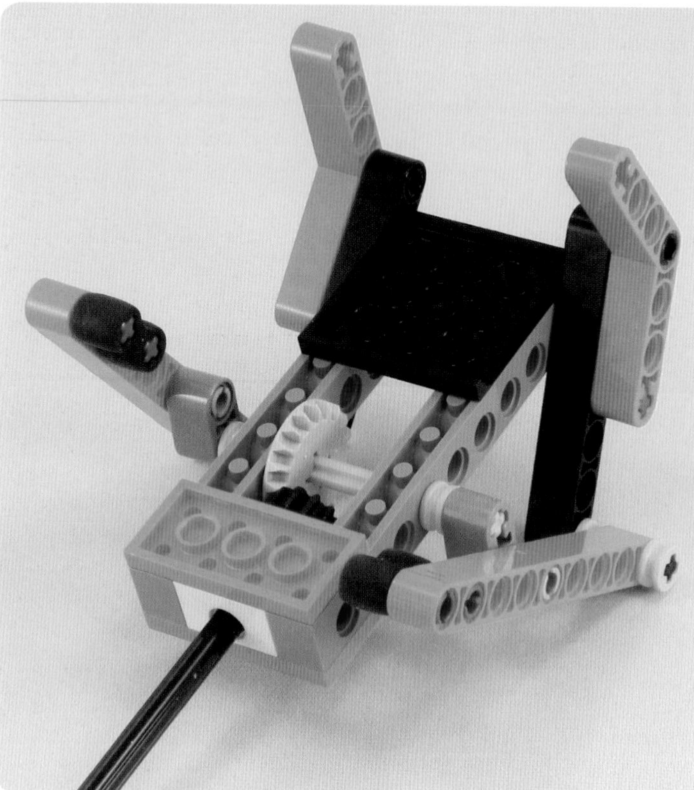

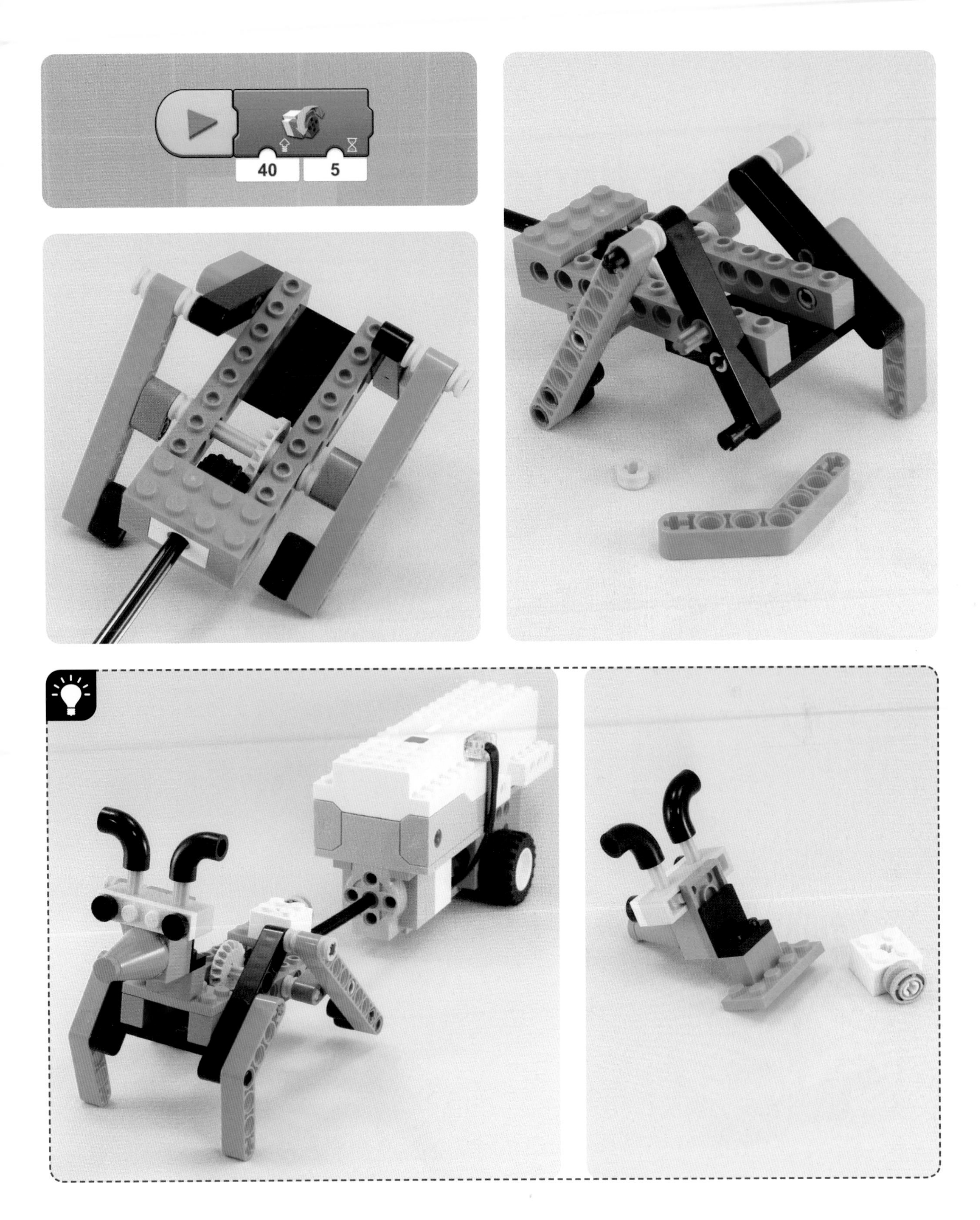
40
5

Shooting things

57

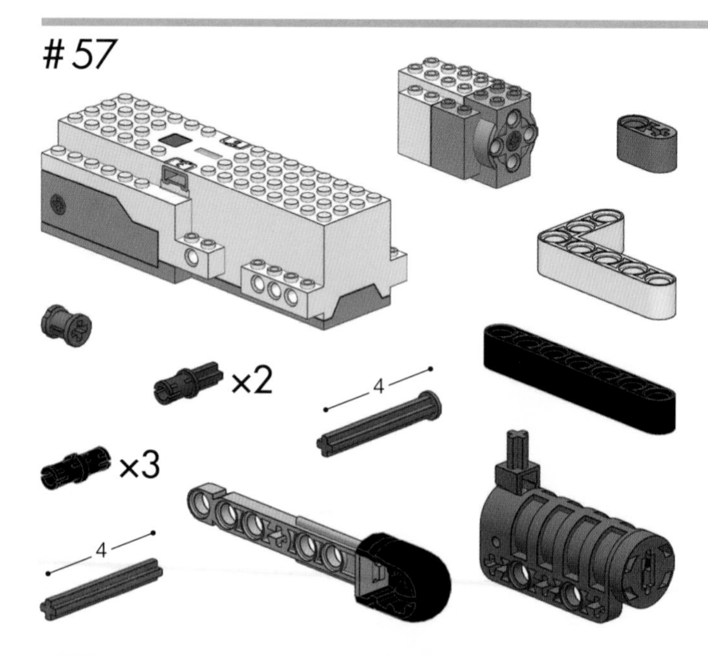

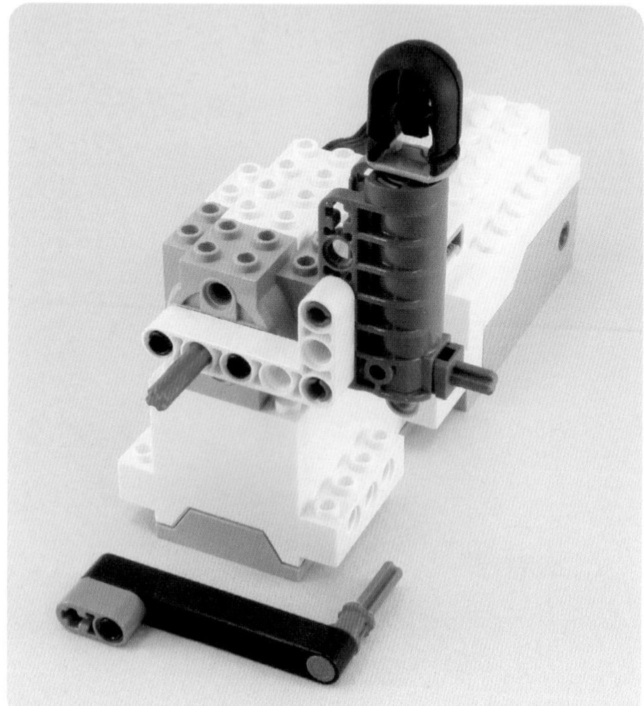

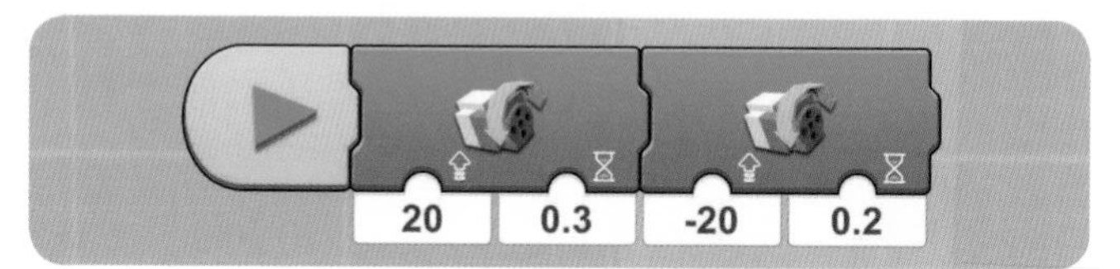

58

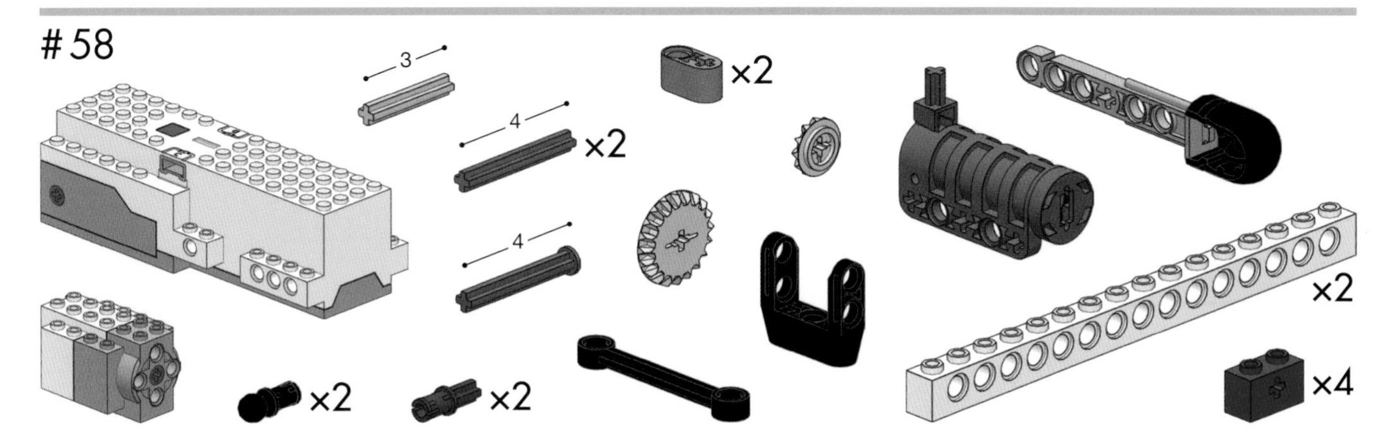

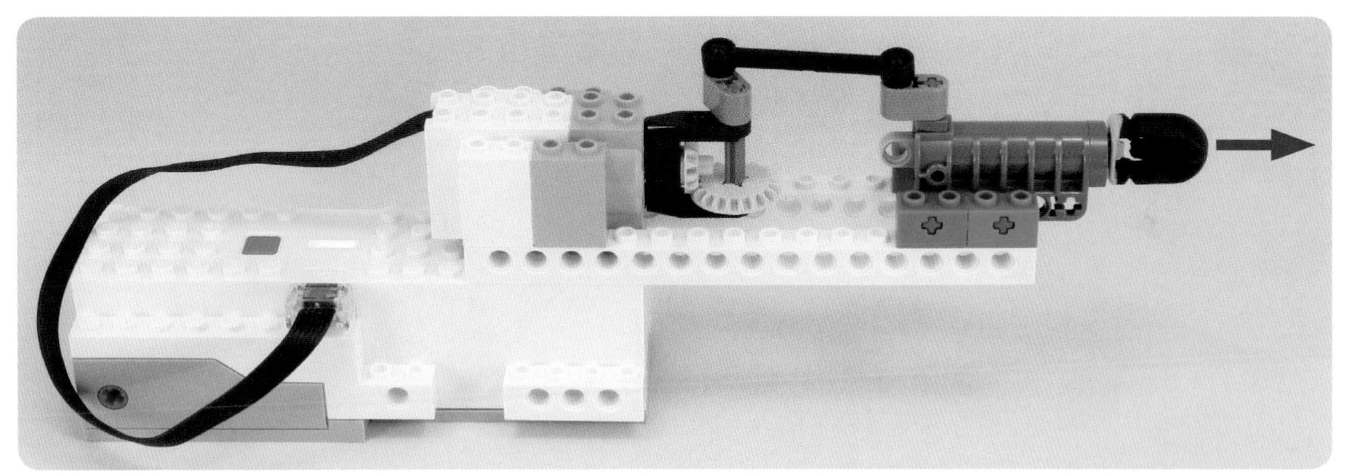

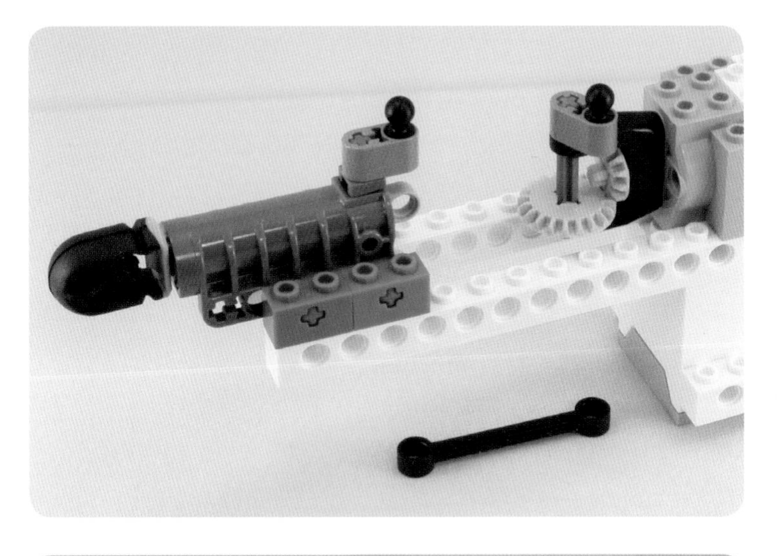

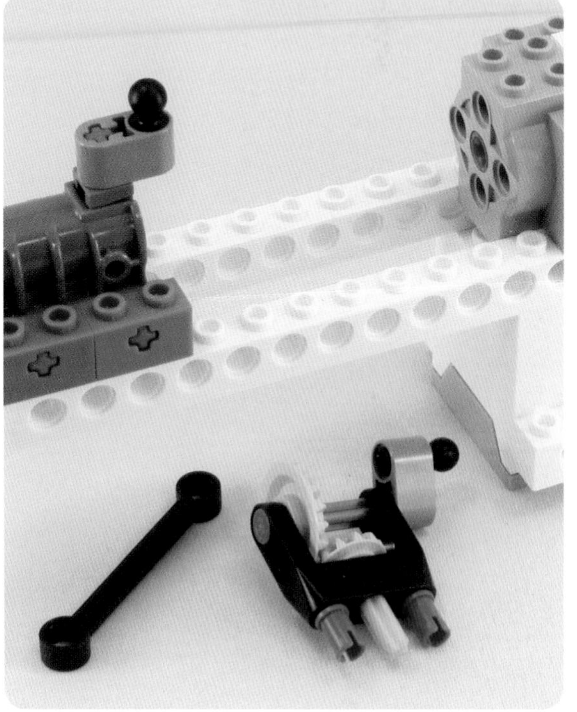

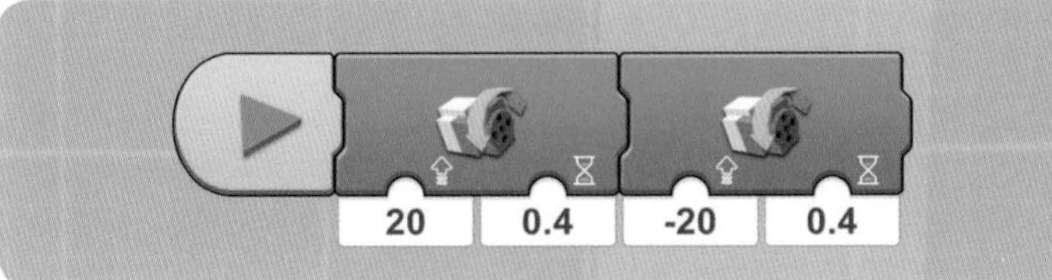

#59

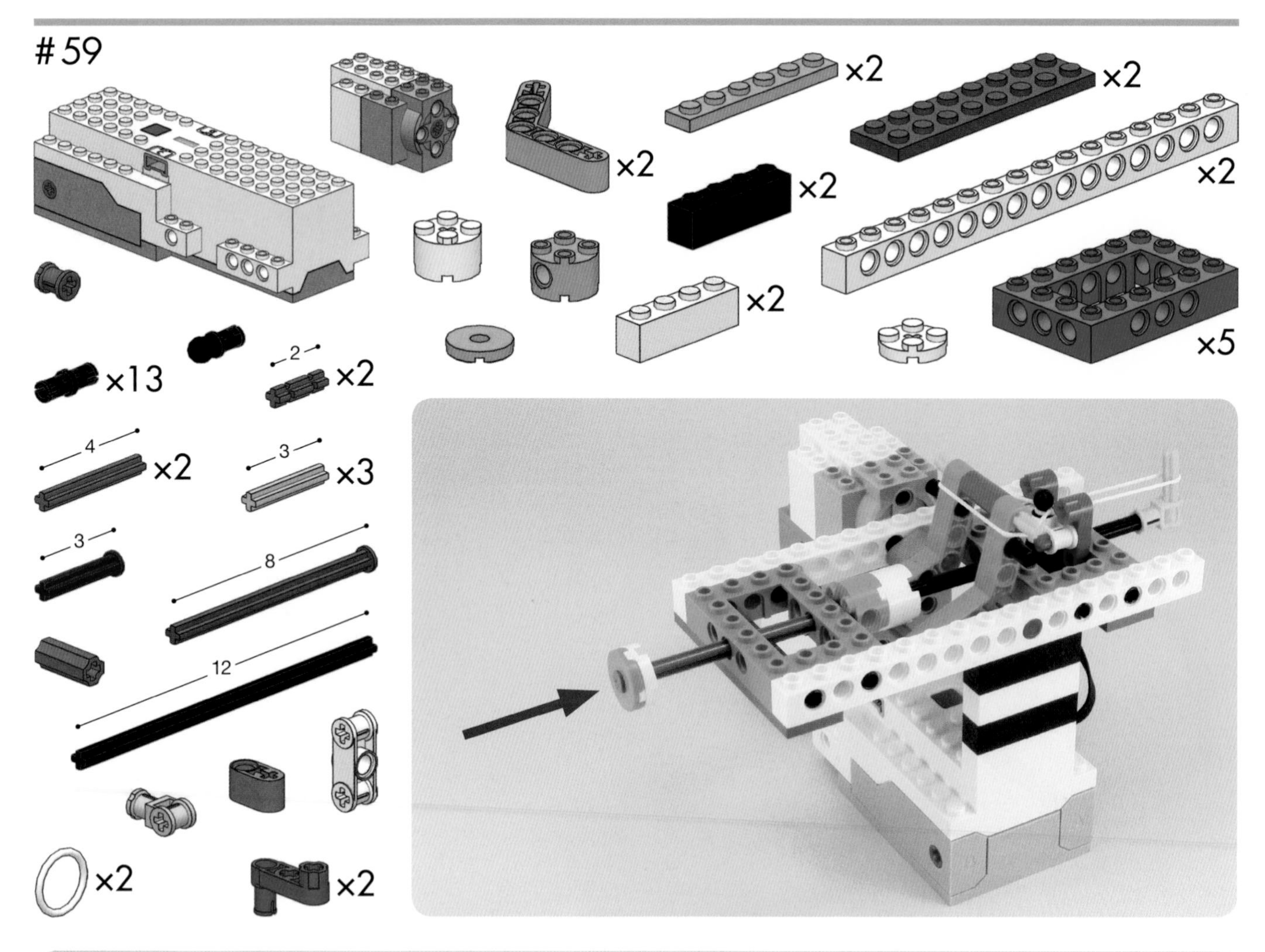

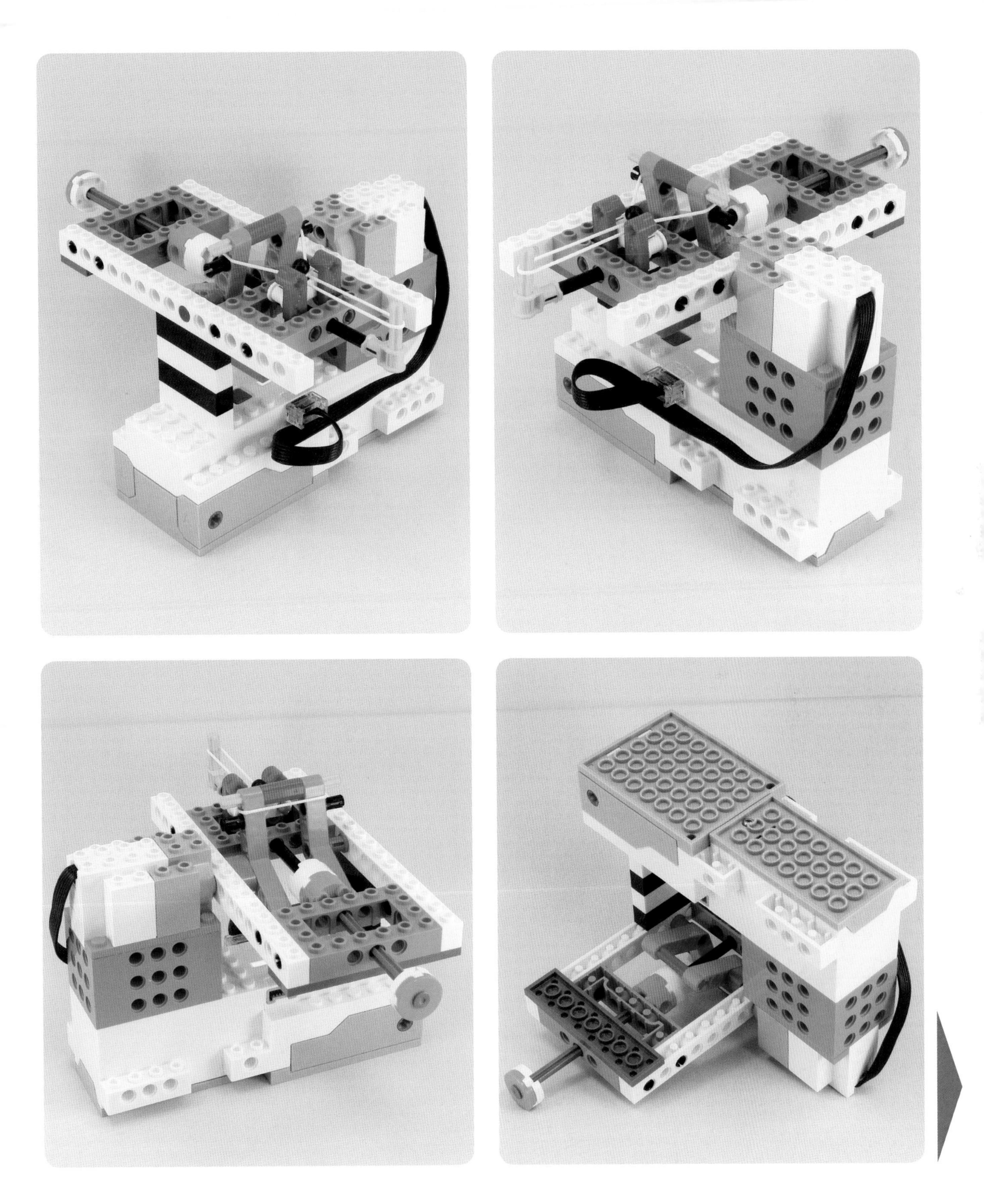

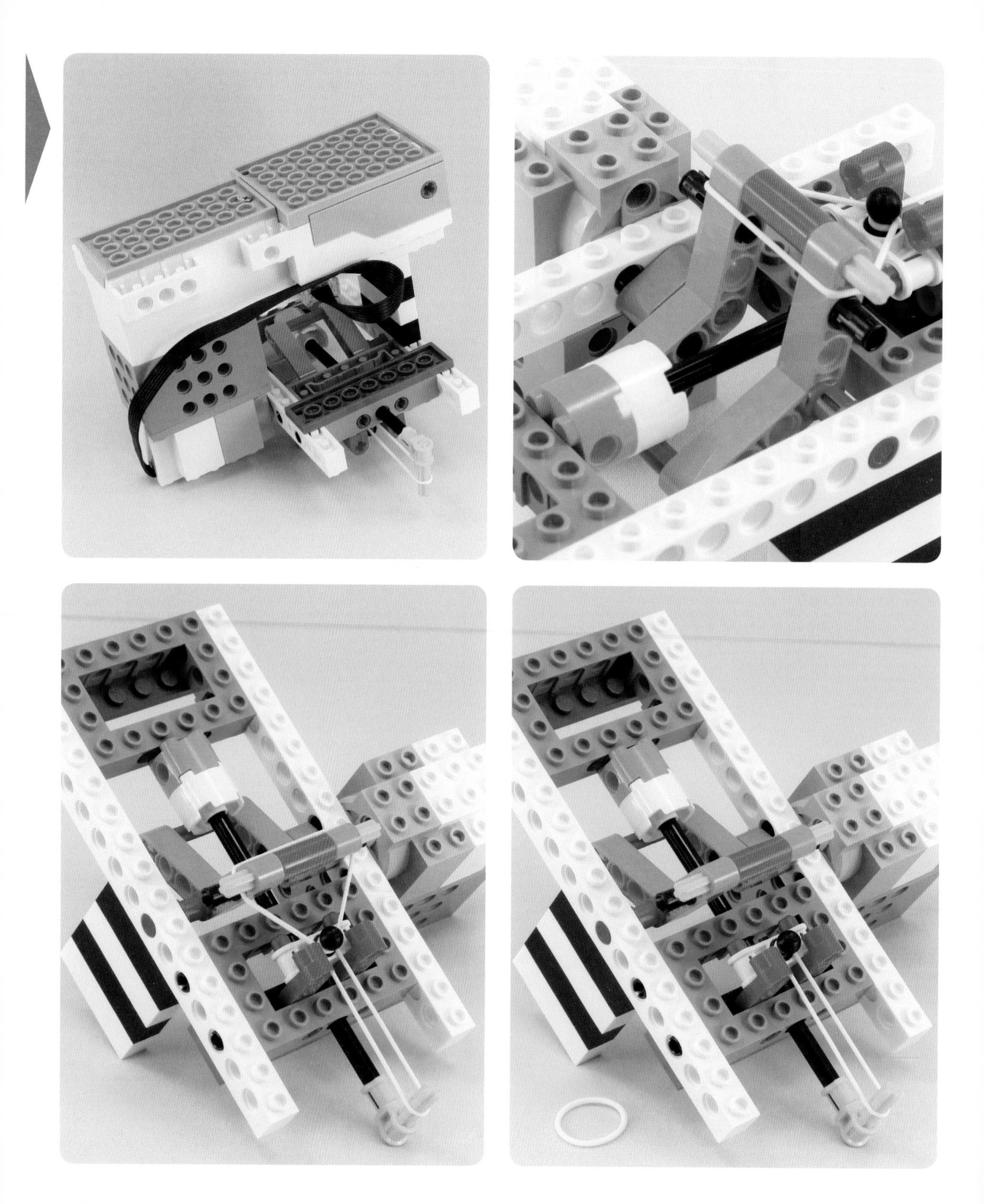

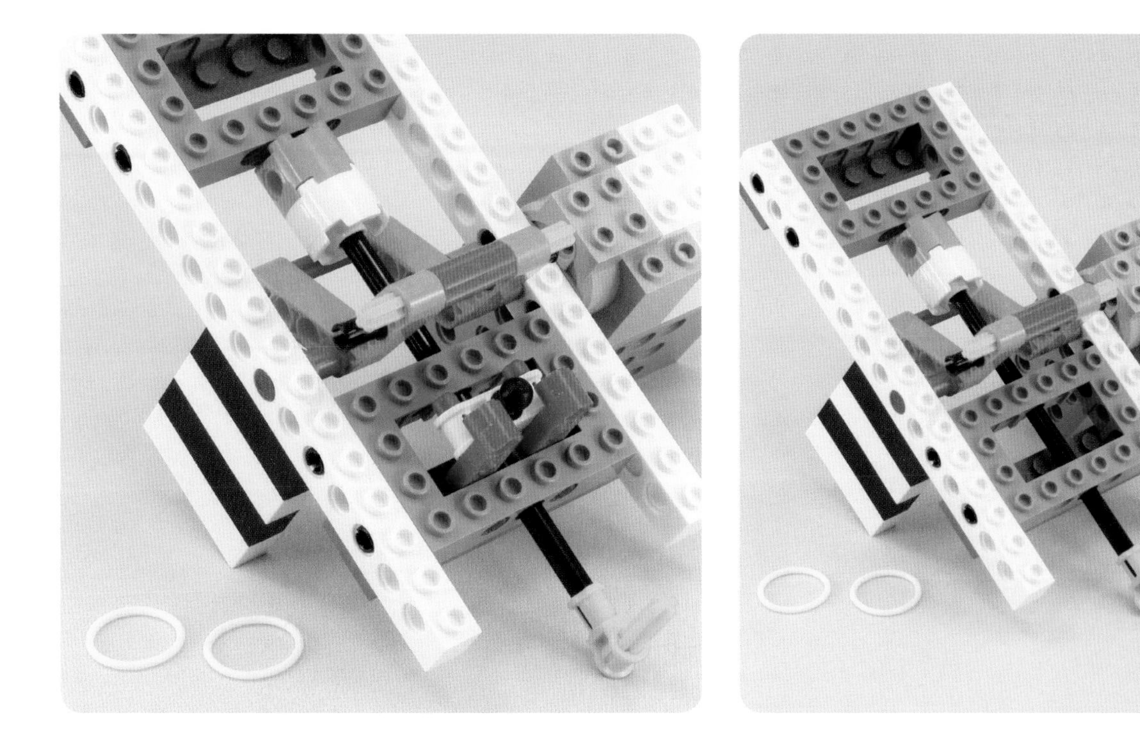

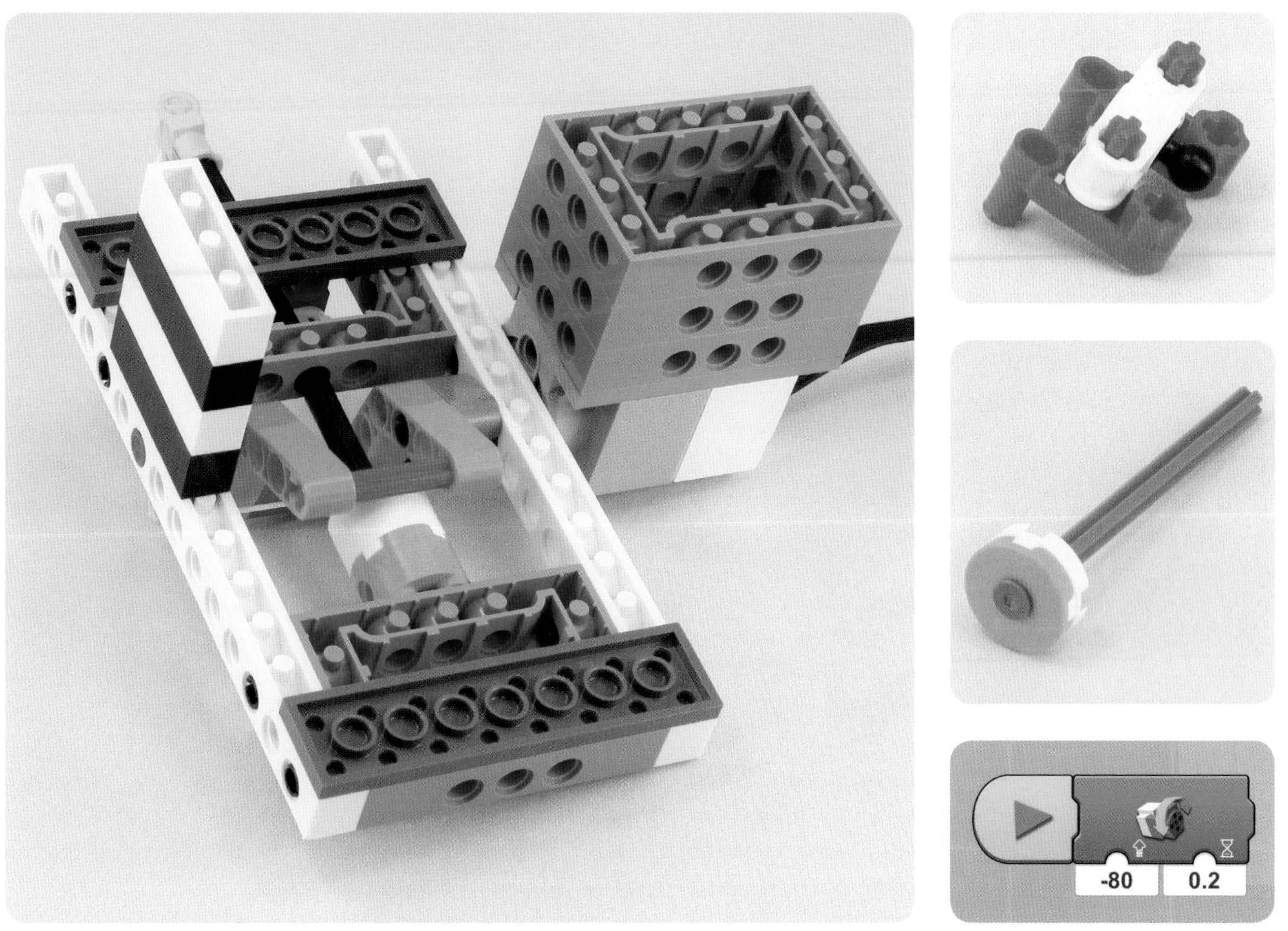
-80
0.2

#60

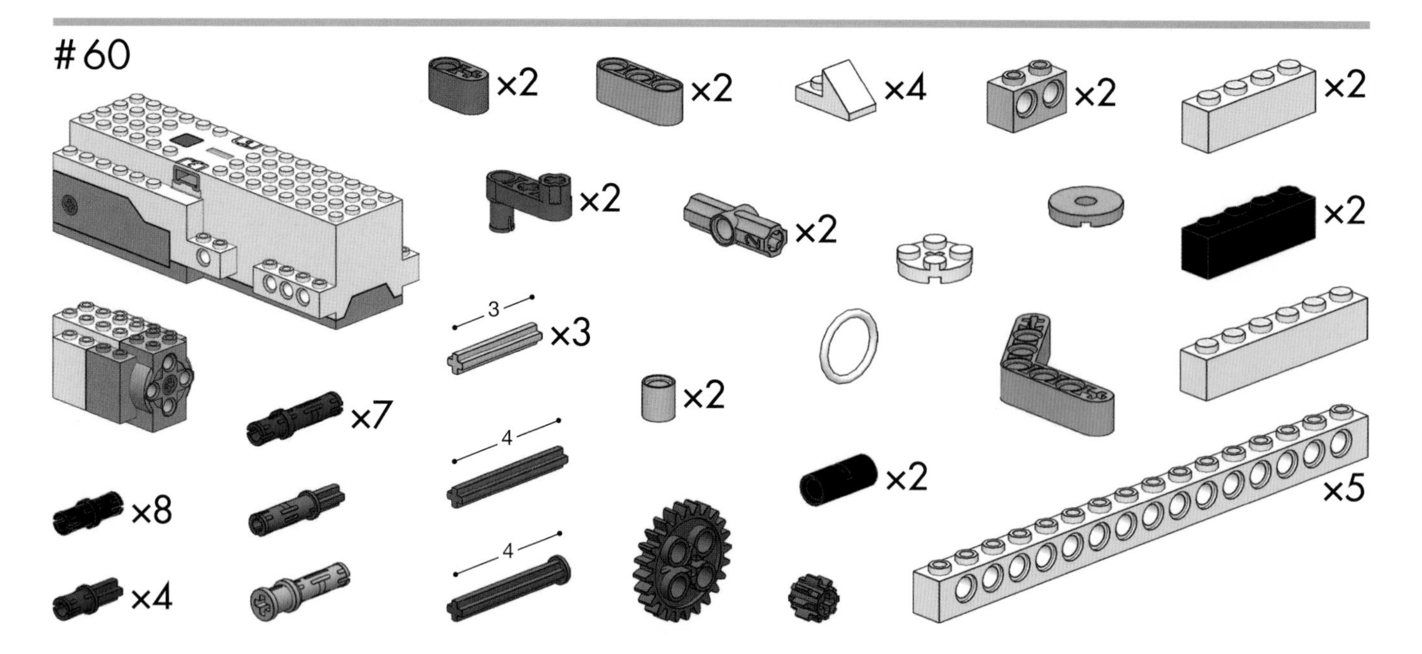

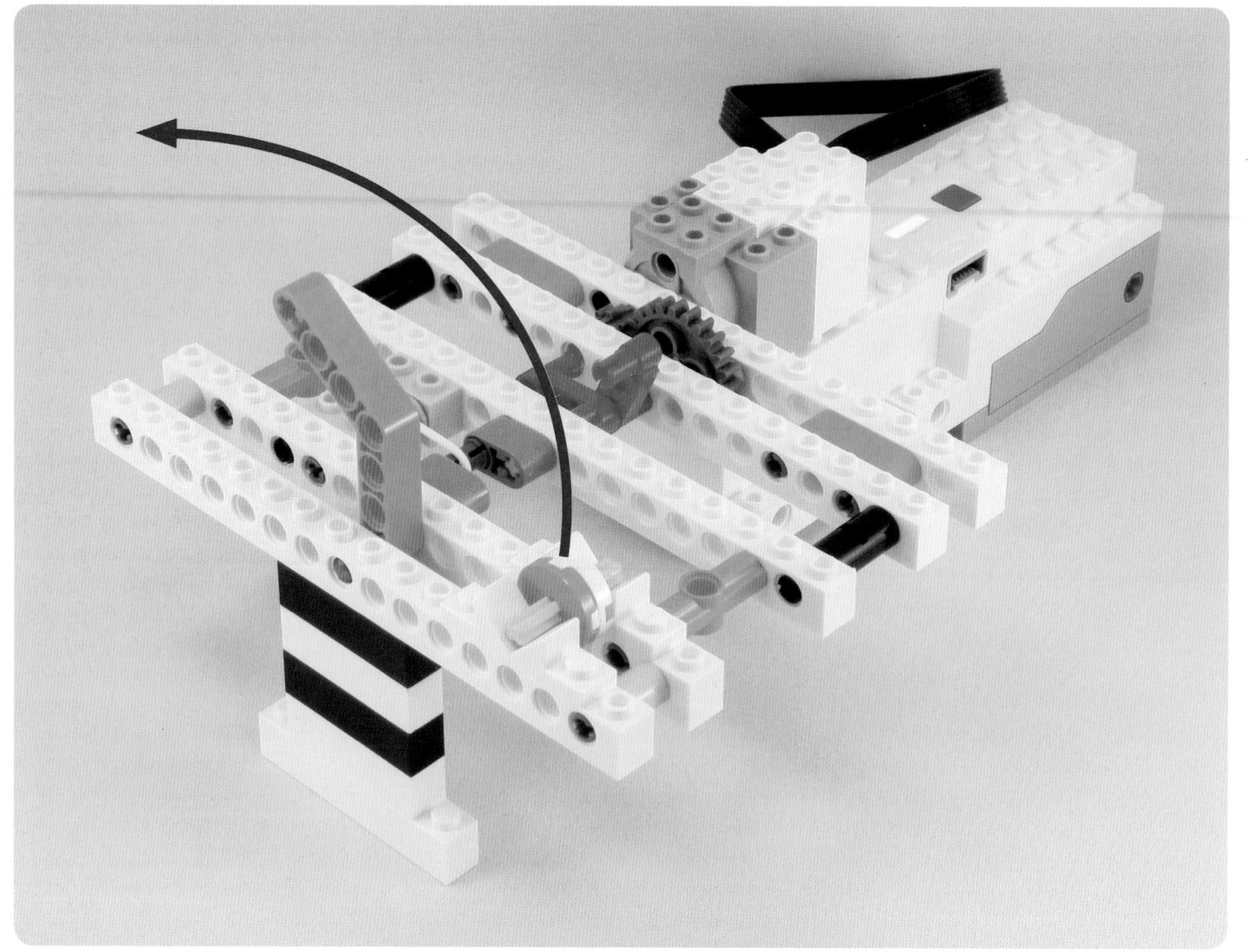

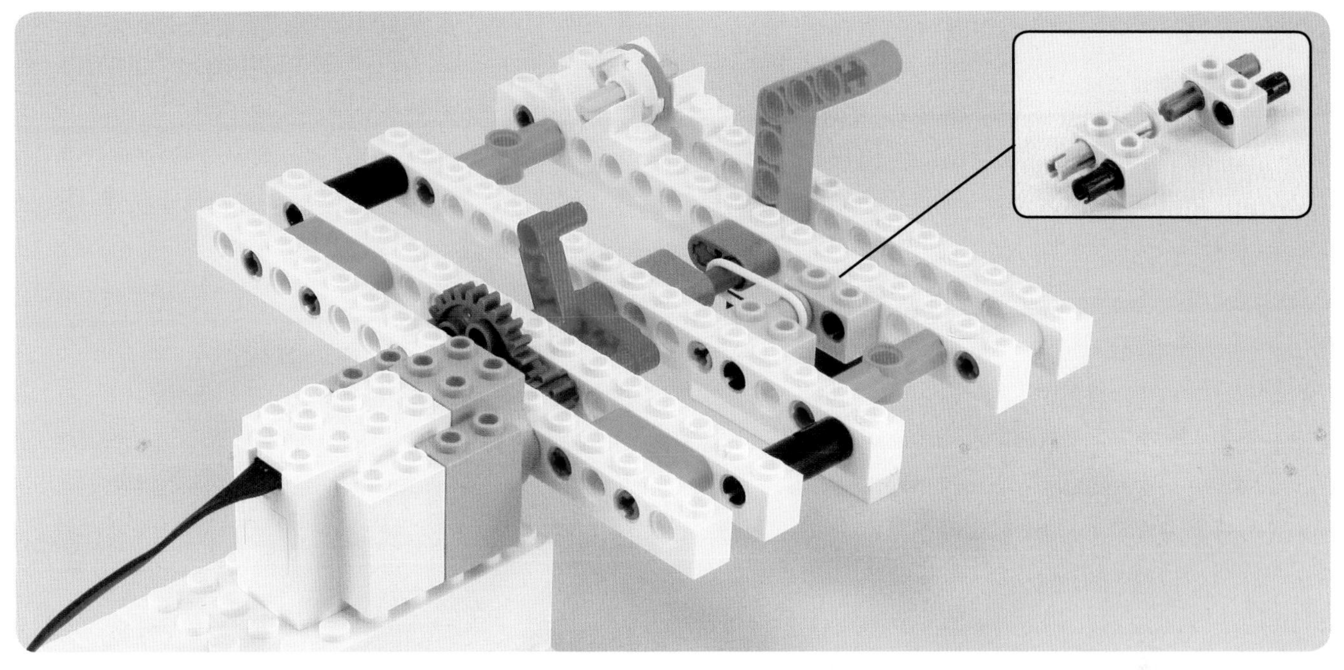

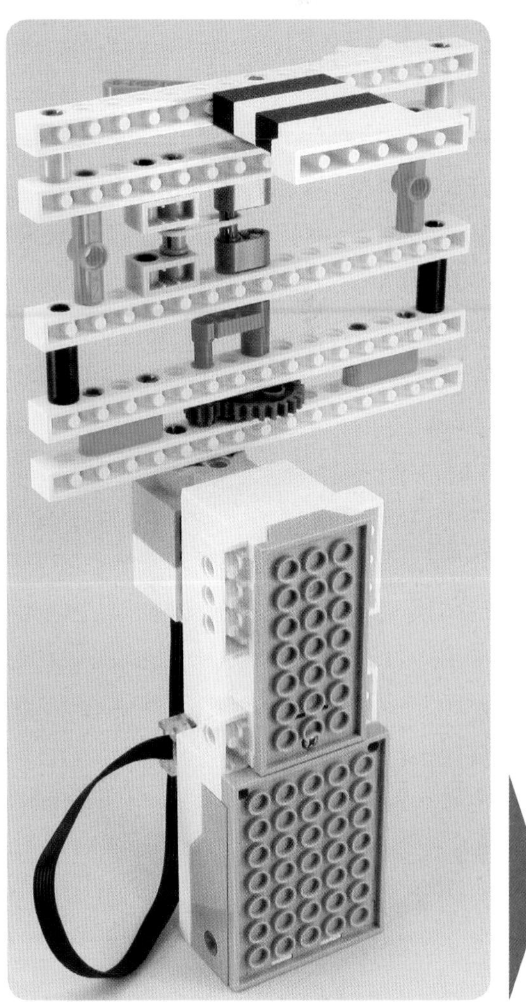

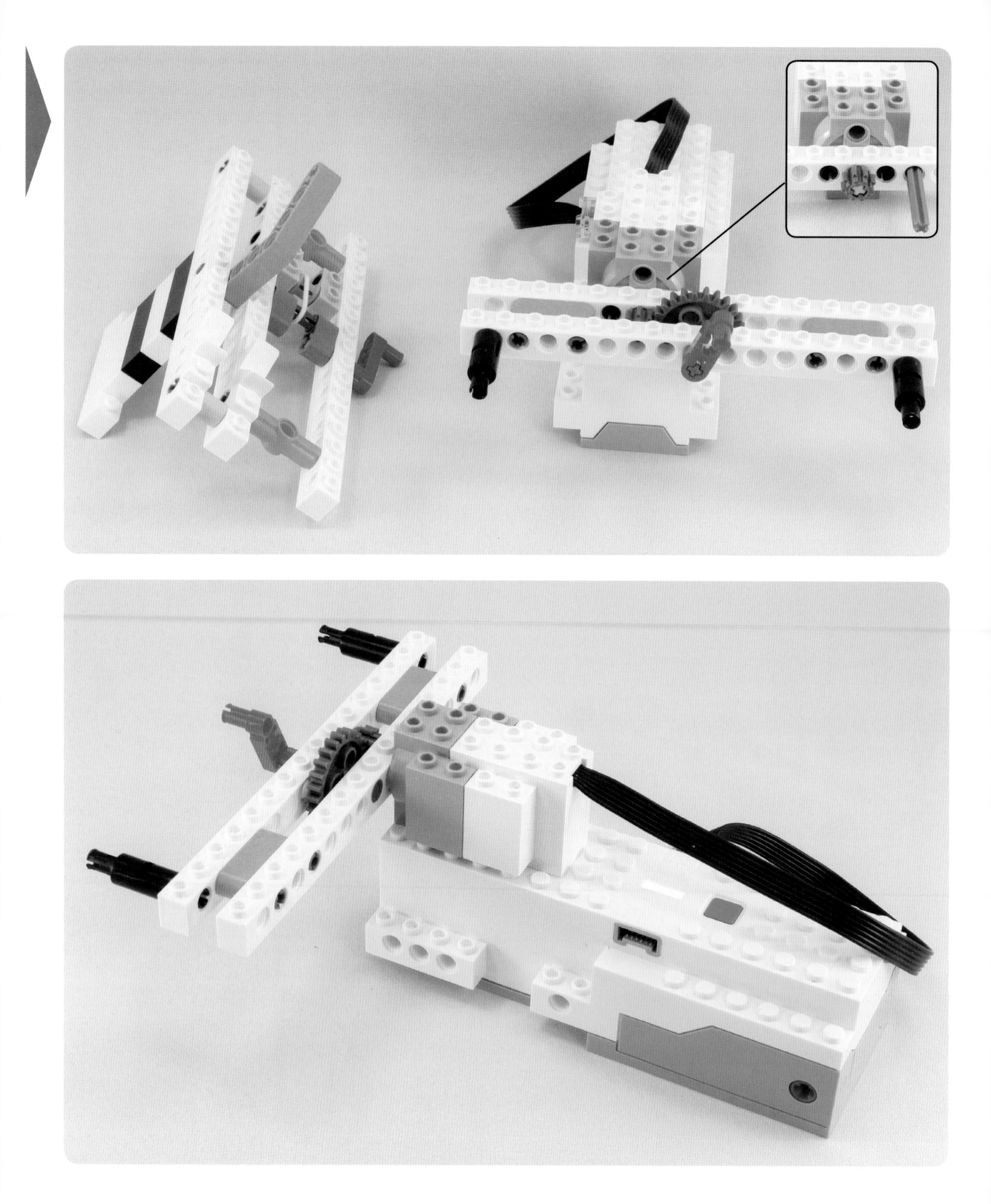

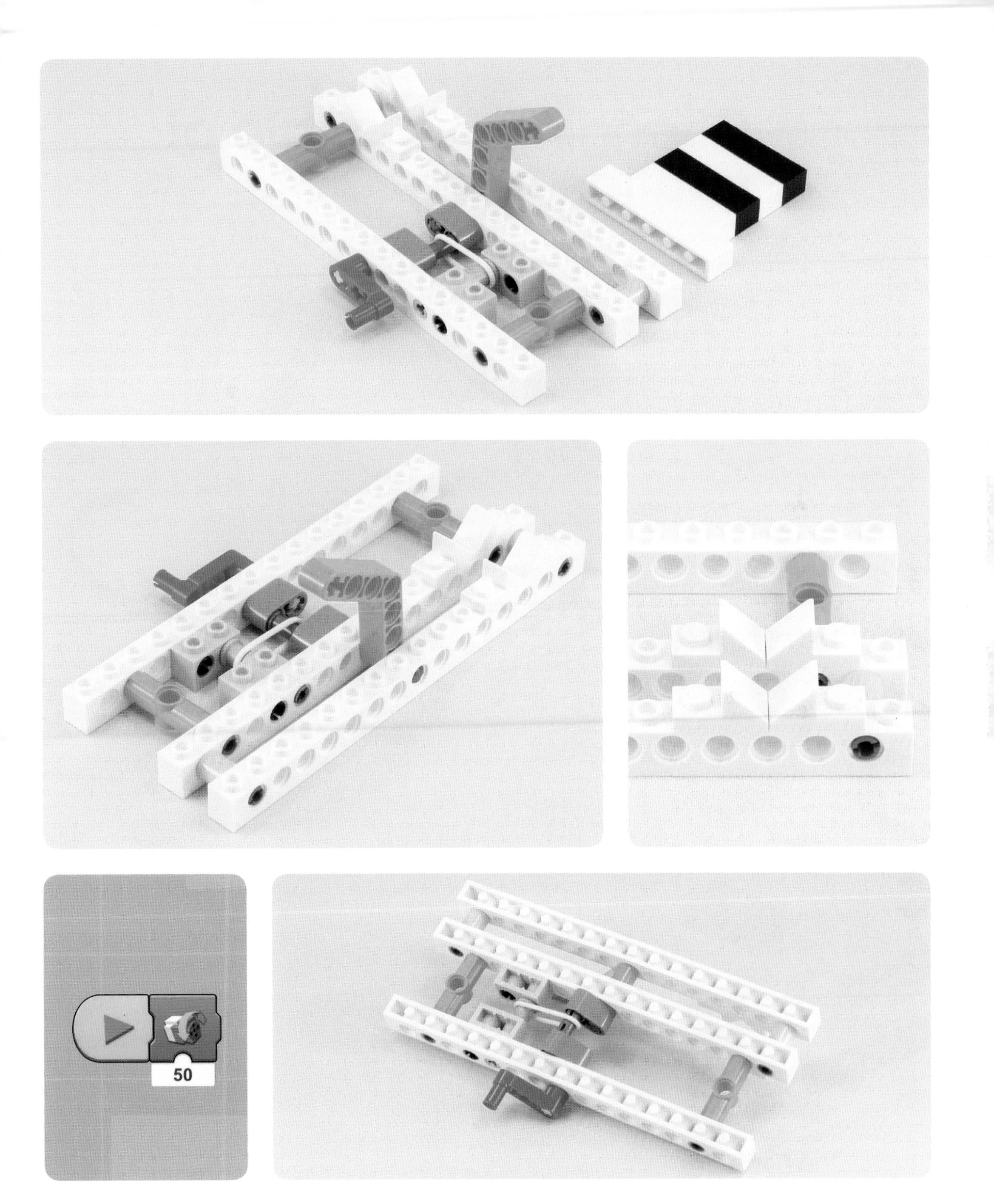
50

Changing the angle of rotation freely

61

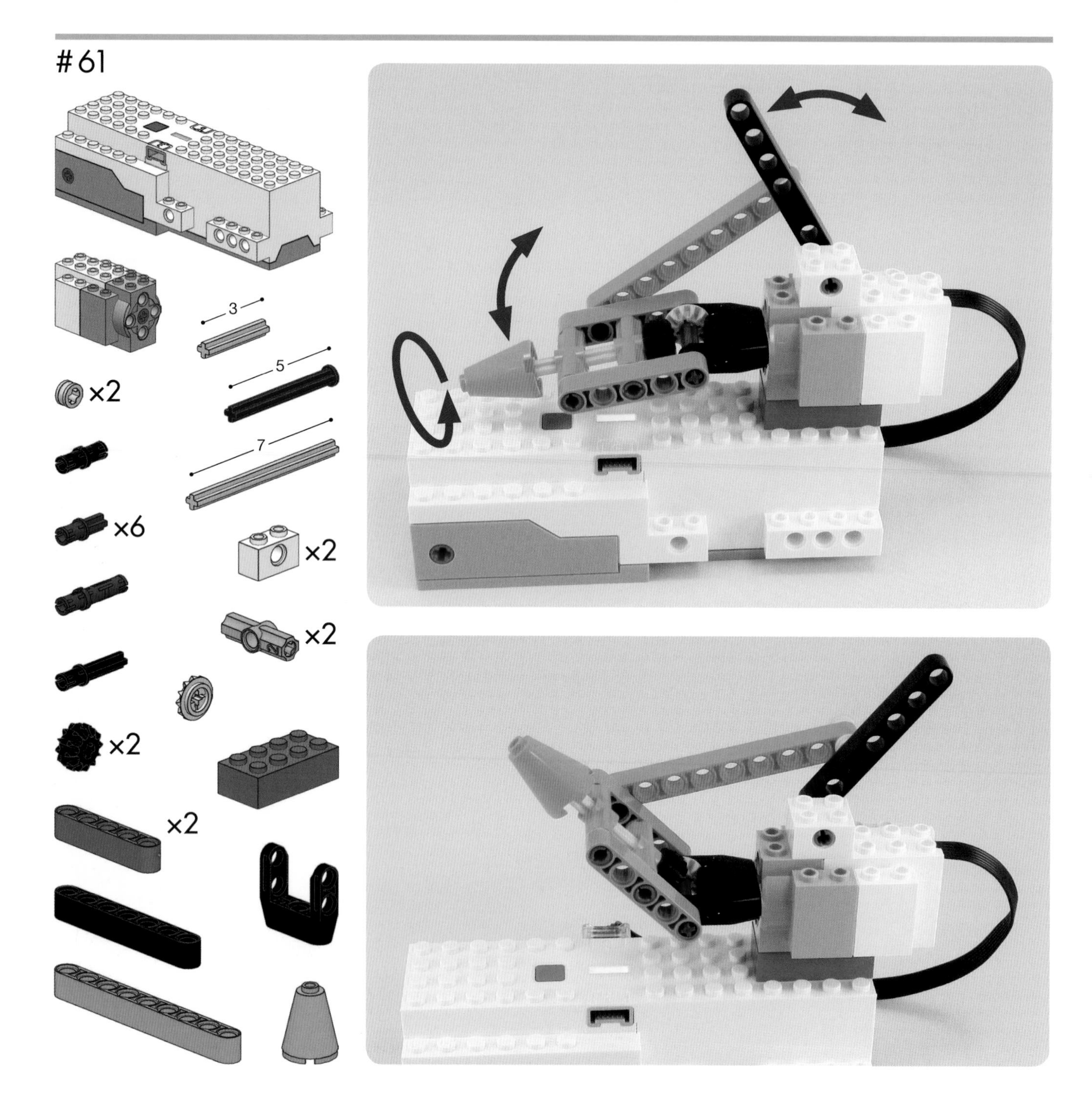

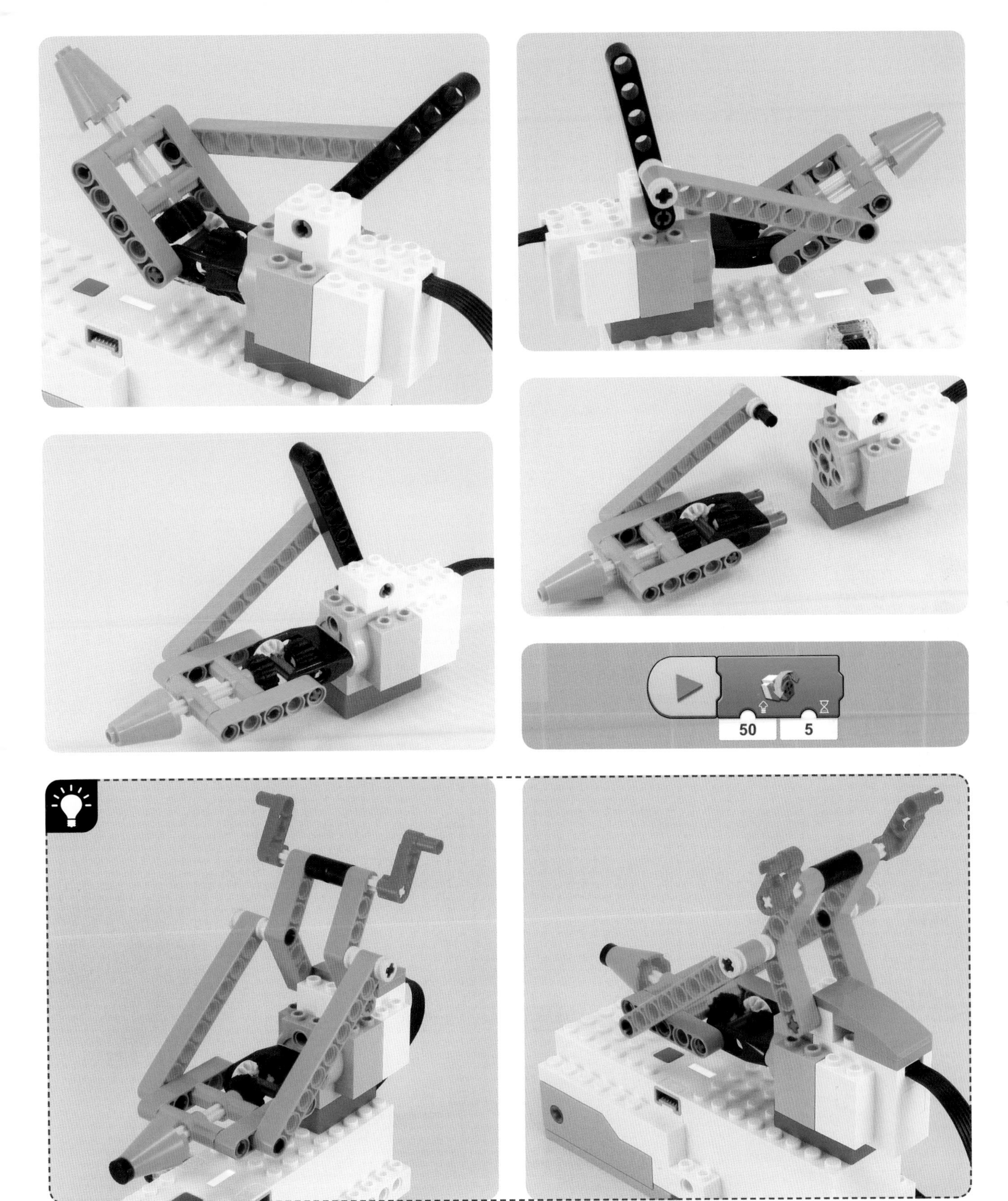
50
5

#62

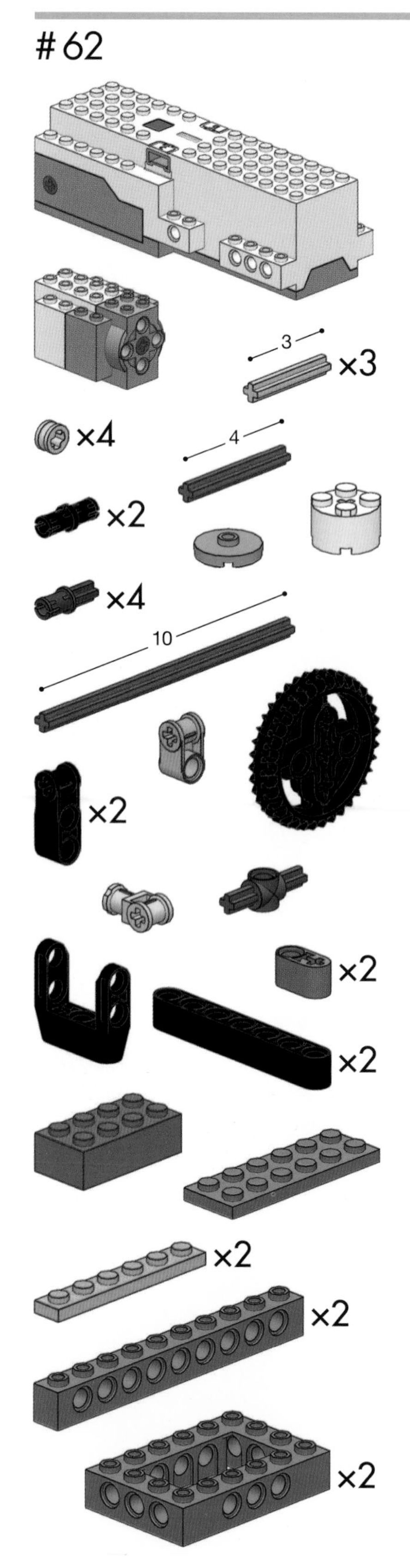

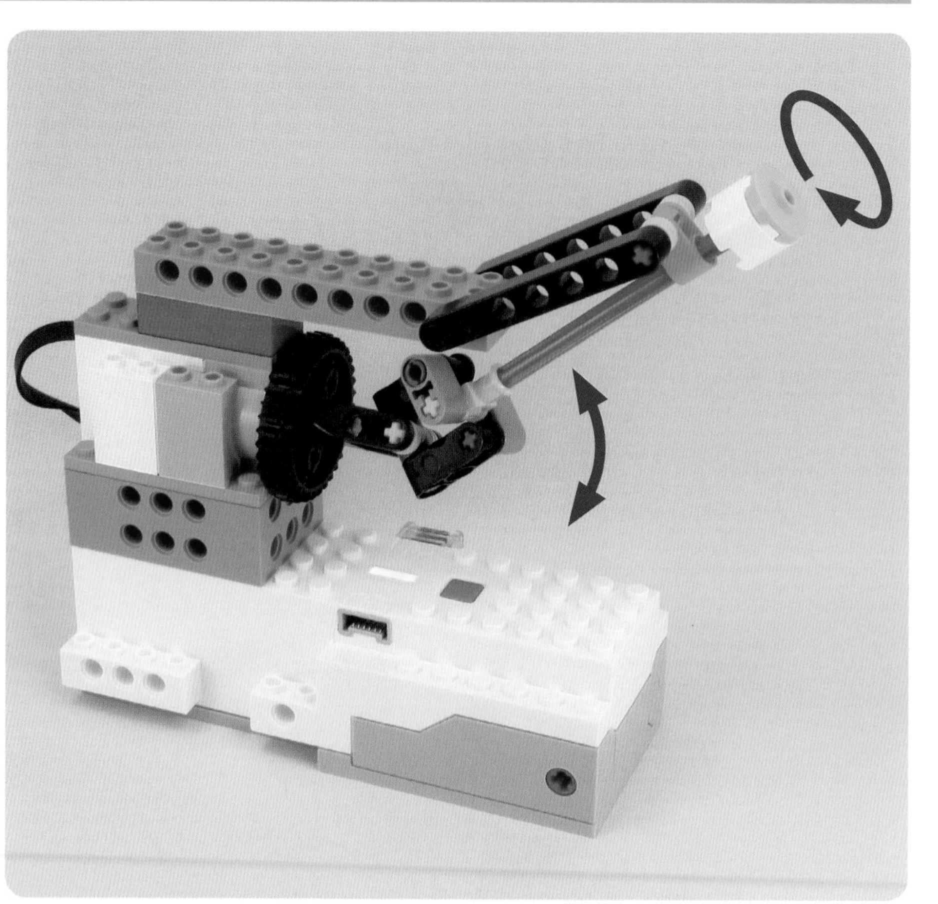

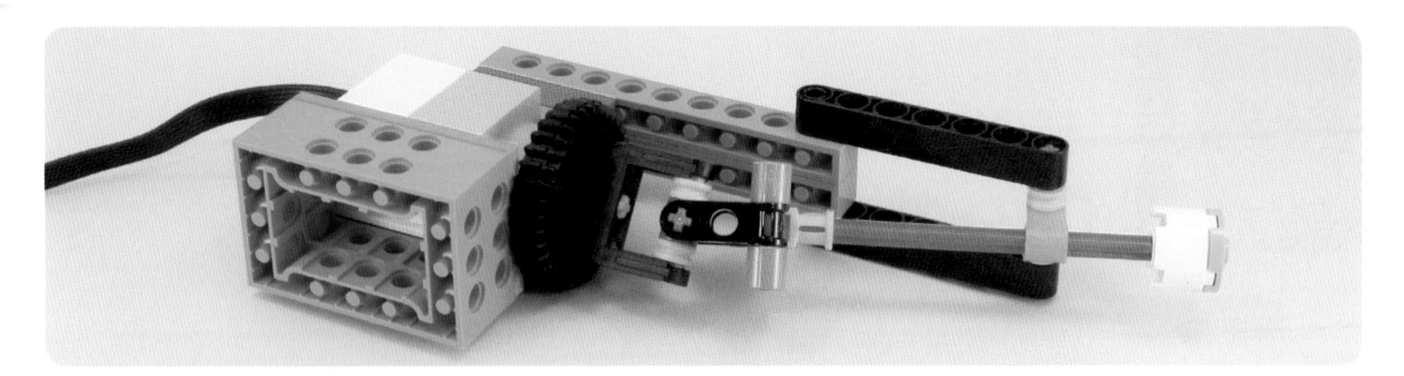

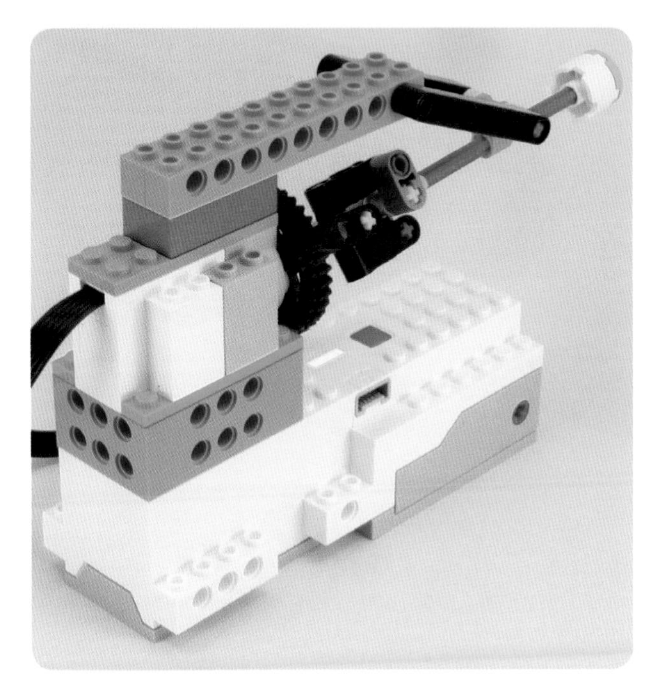

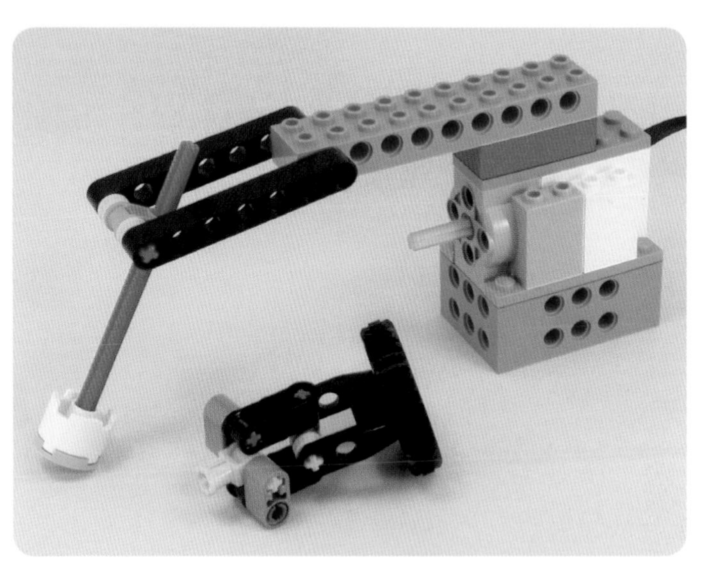

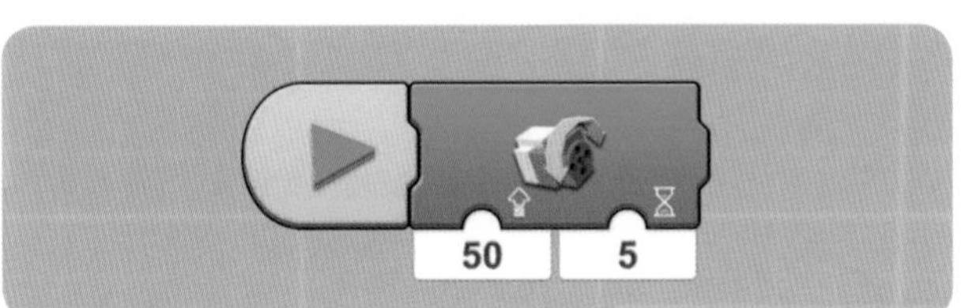
50
5

Creating wind

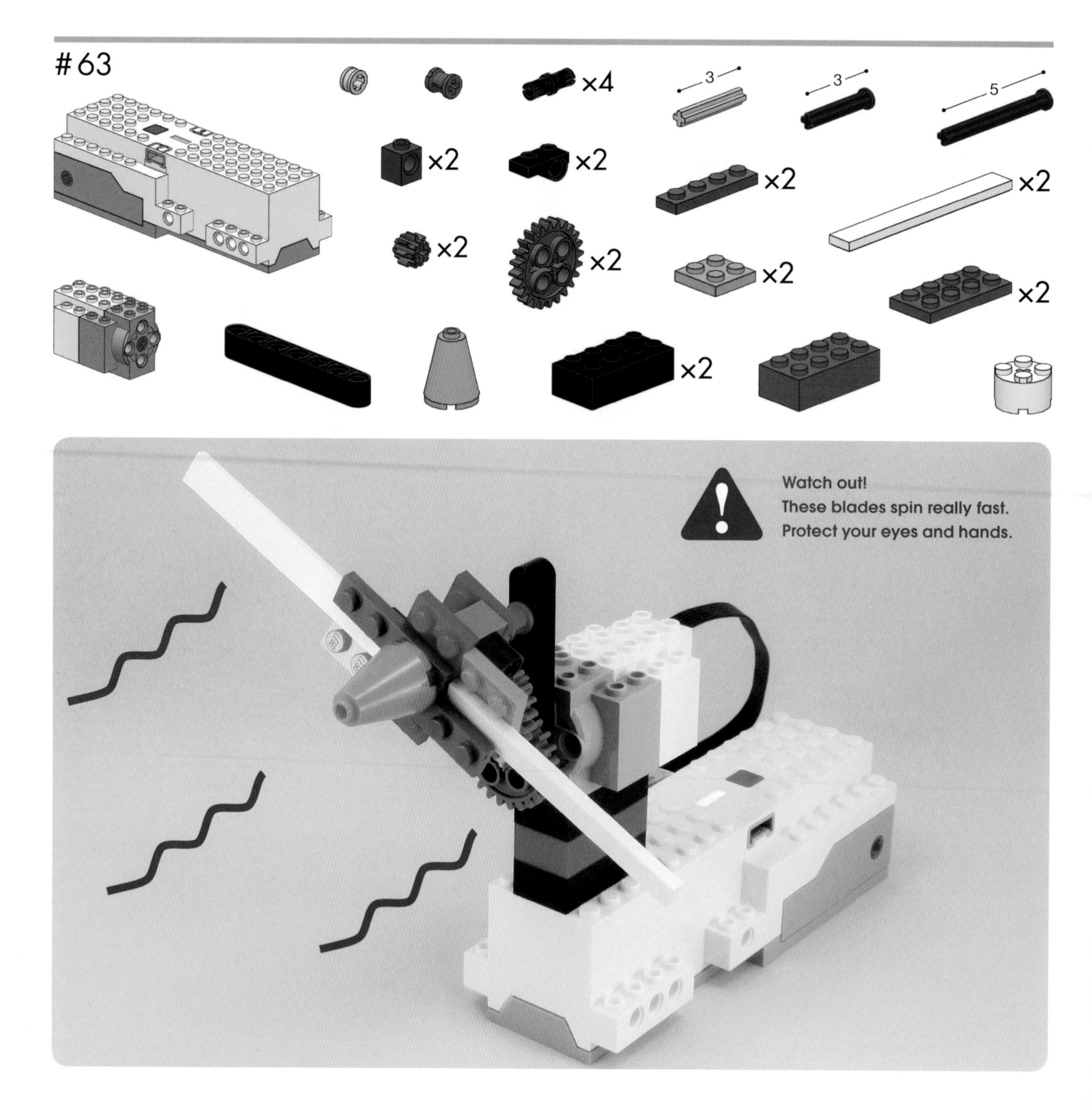

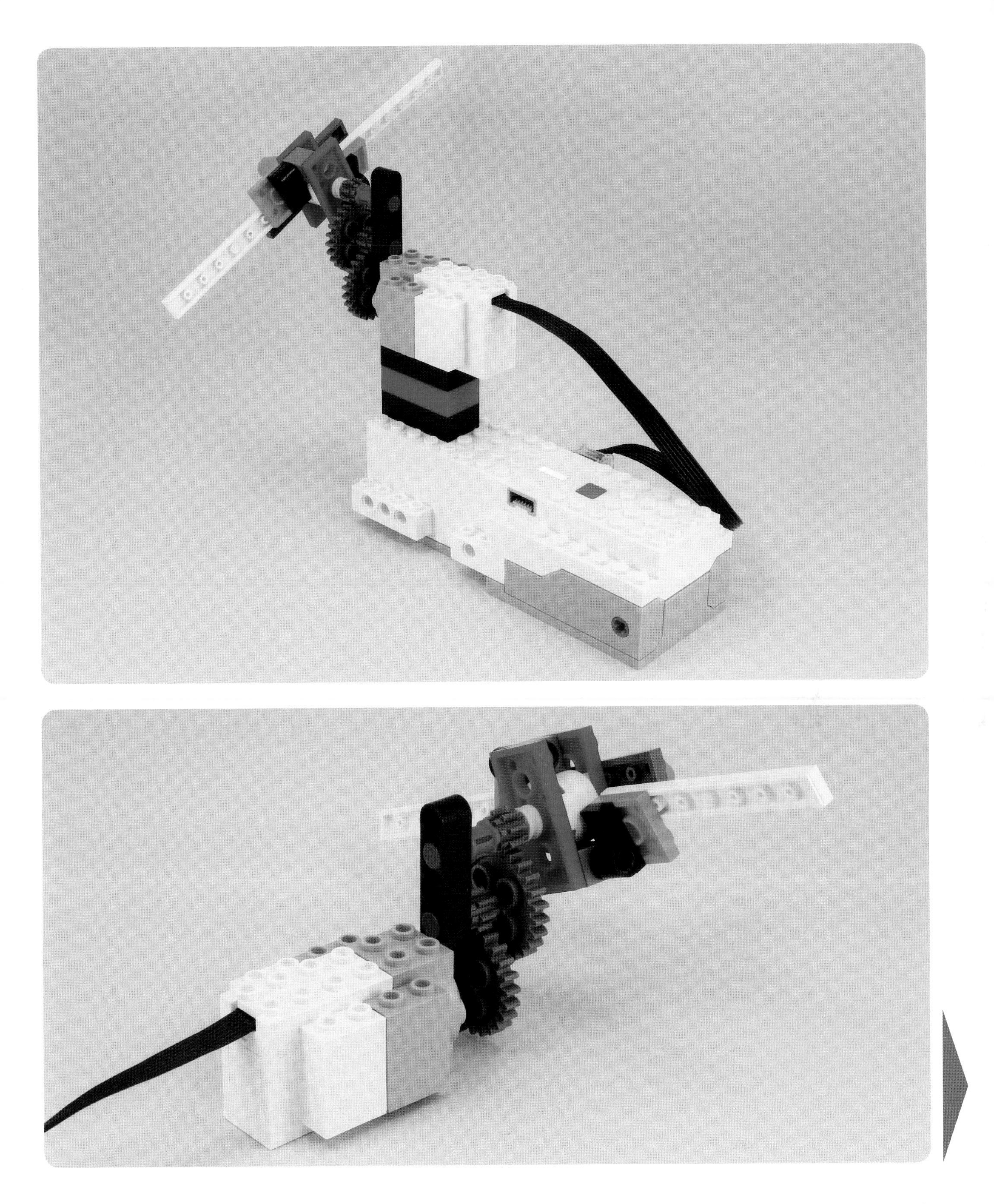

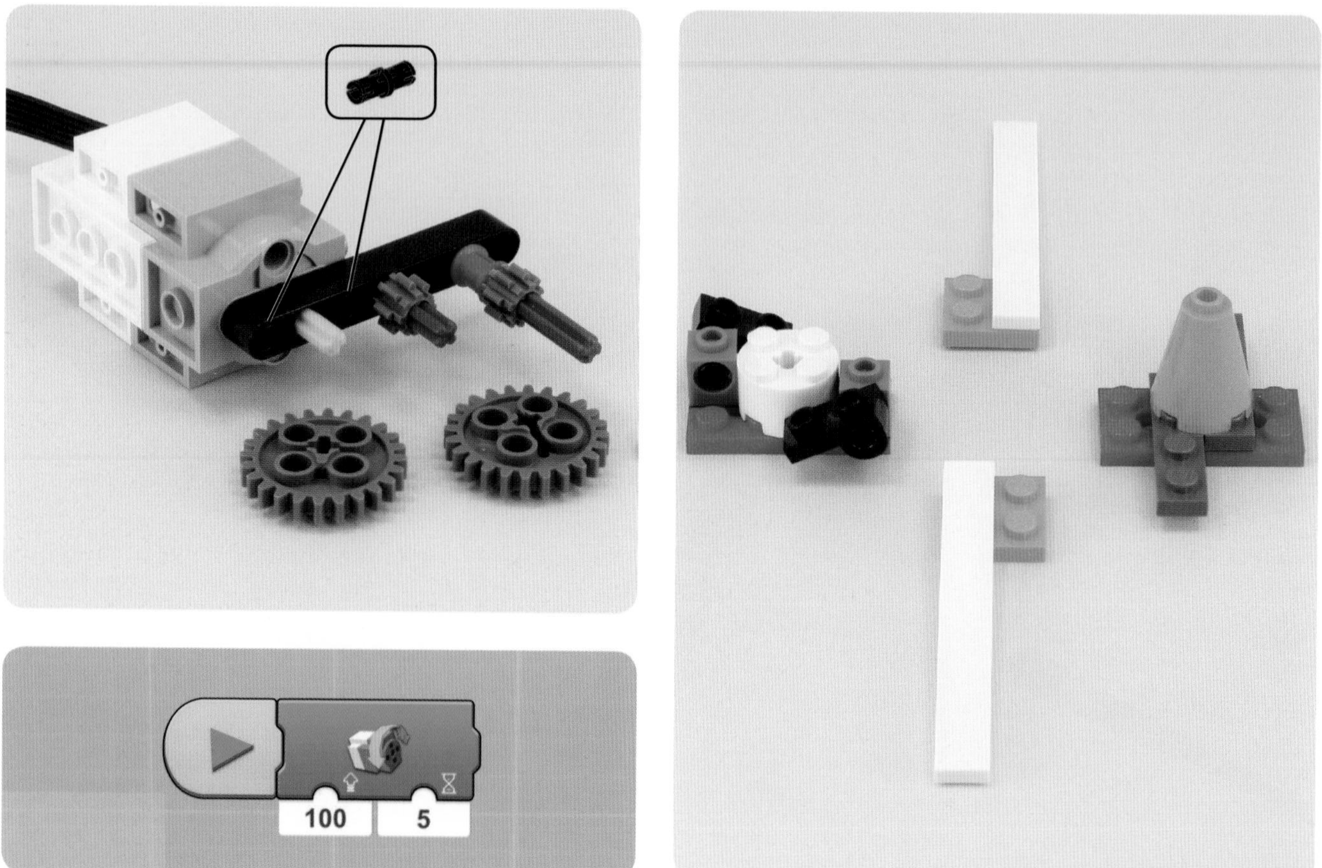
100
5

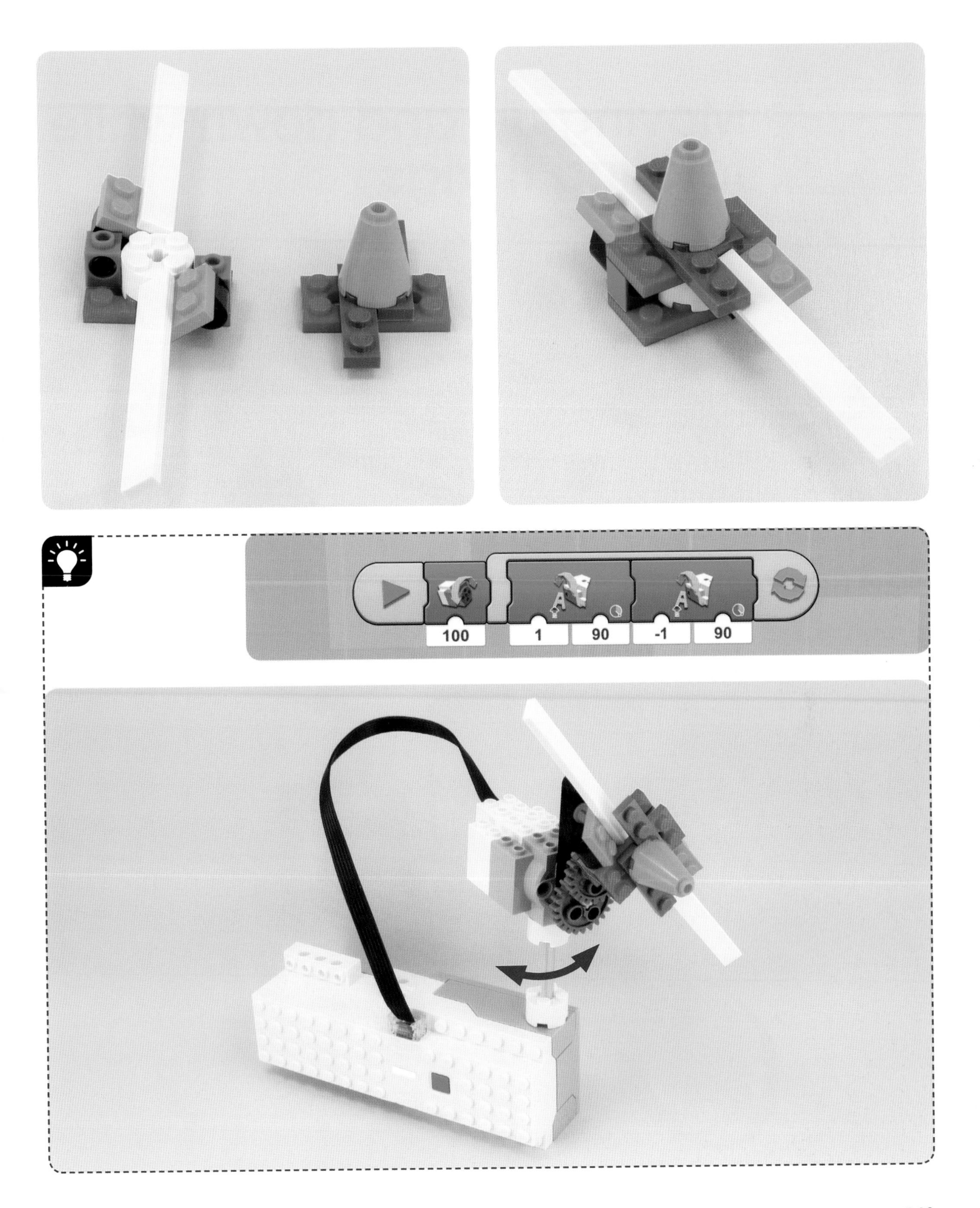
100
1
90
-1
90

Moving up and down while rotating things

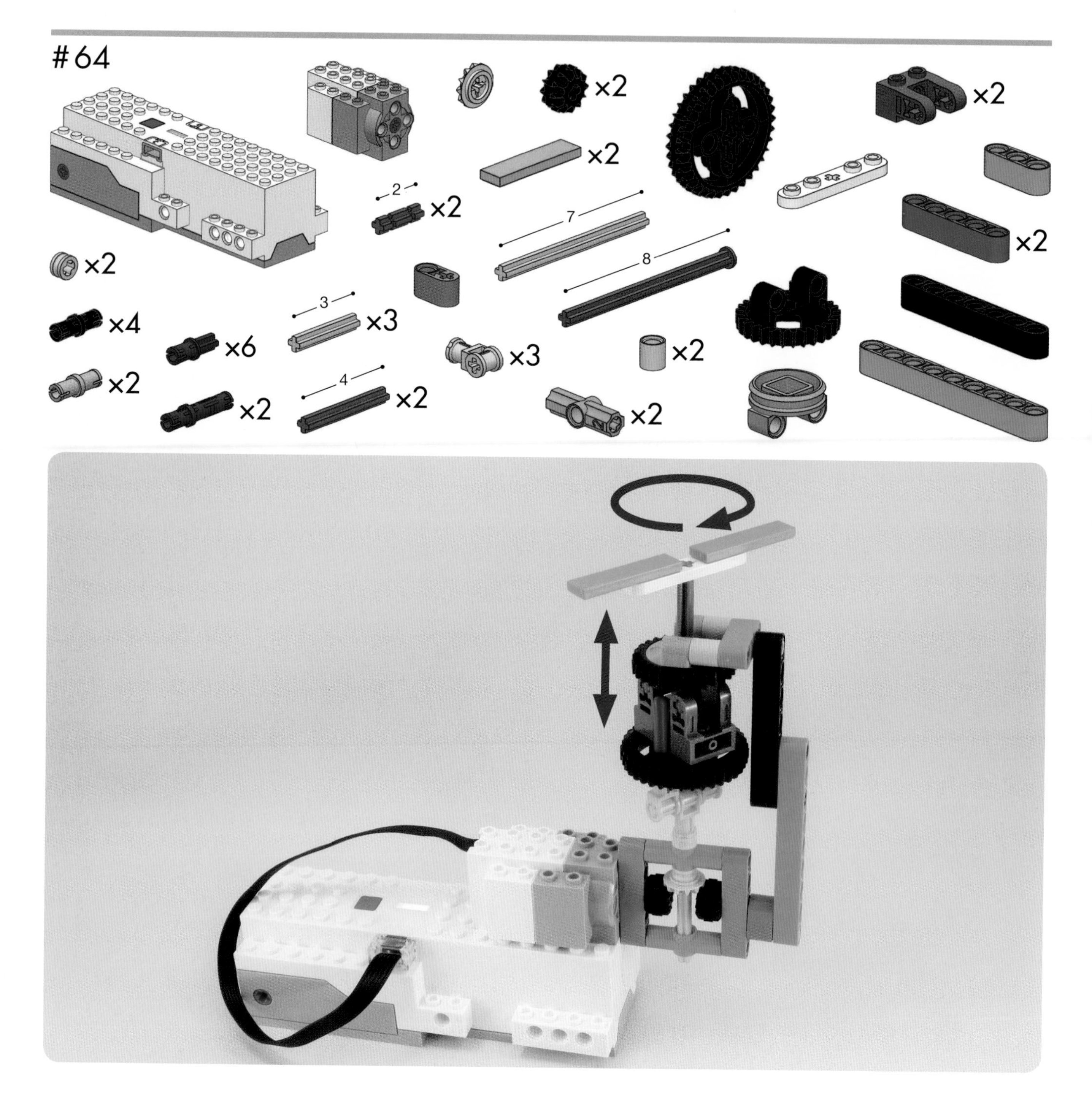

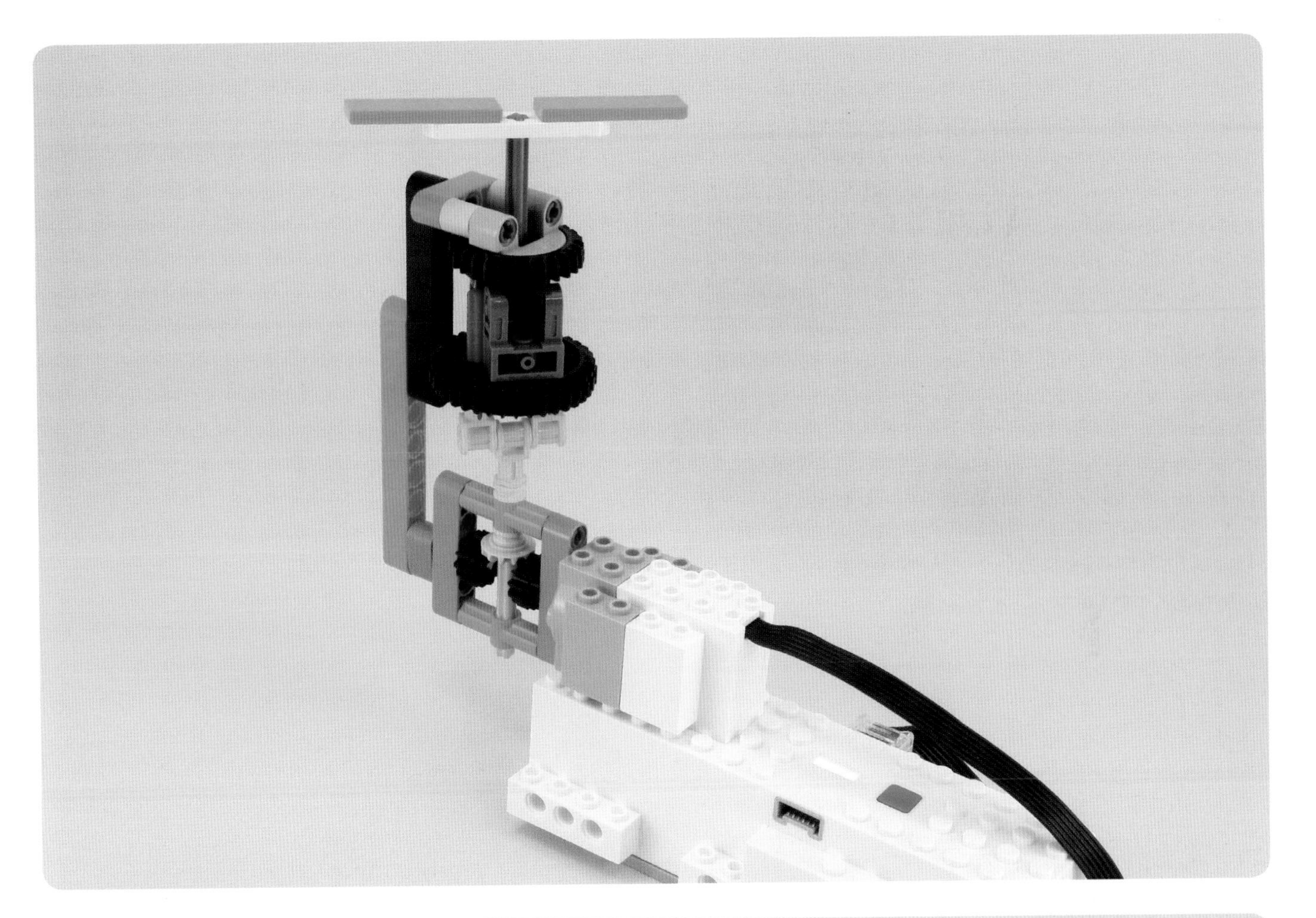

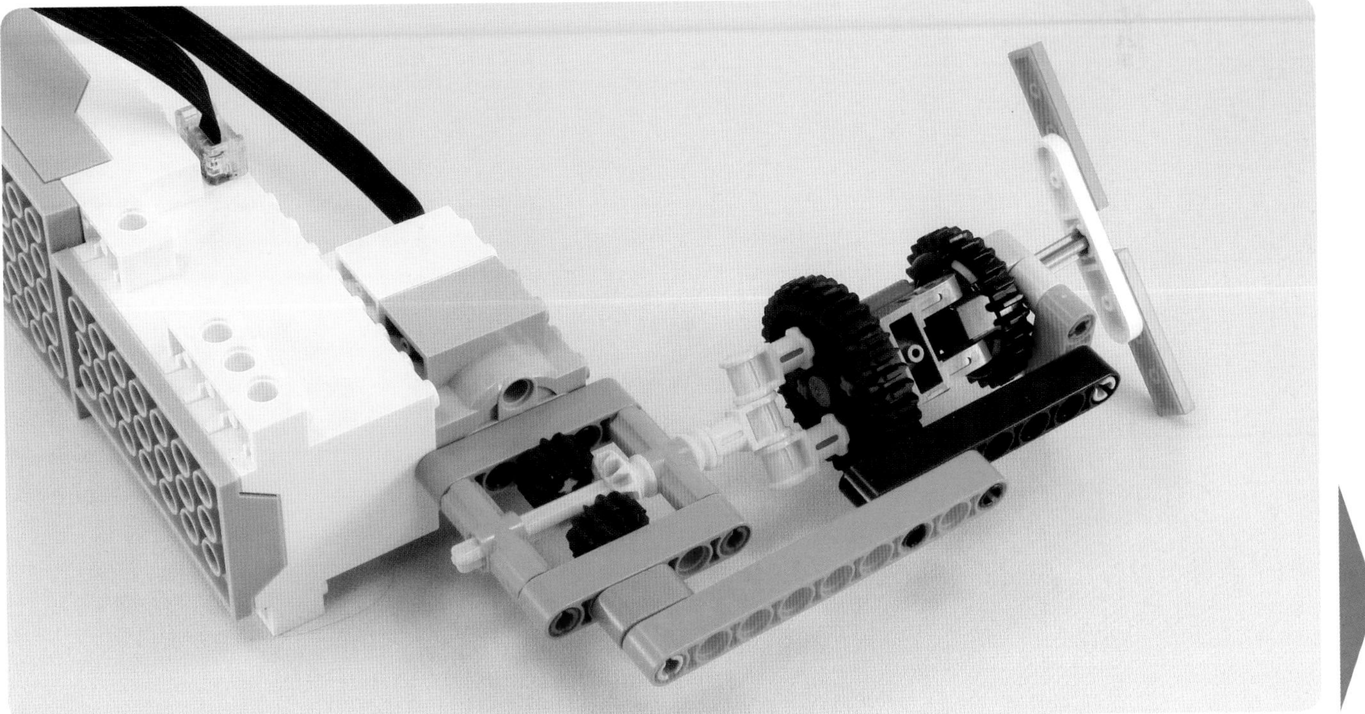

50
5

Stepper mechanism

#65

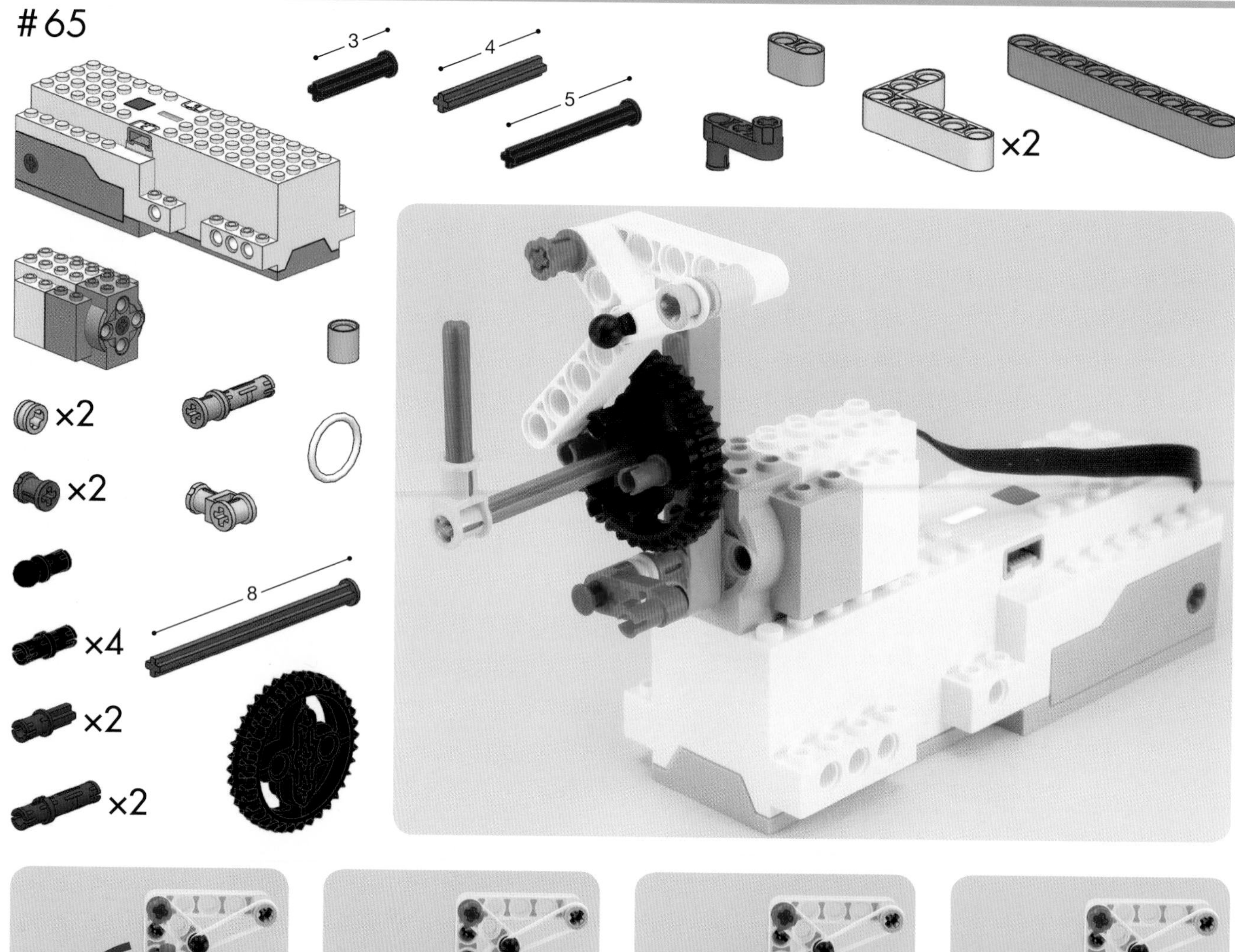

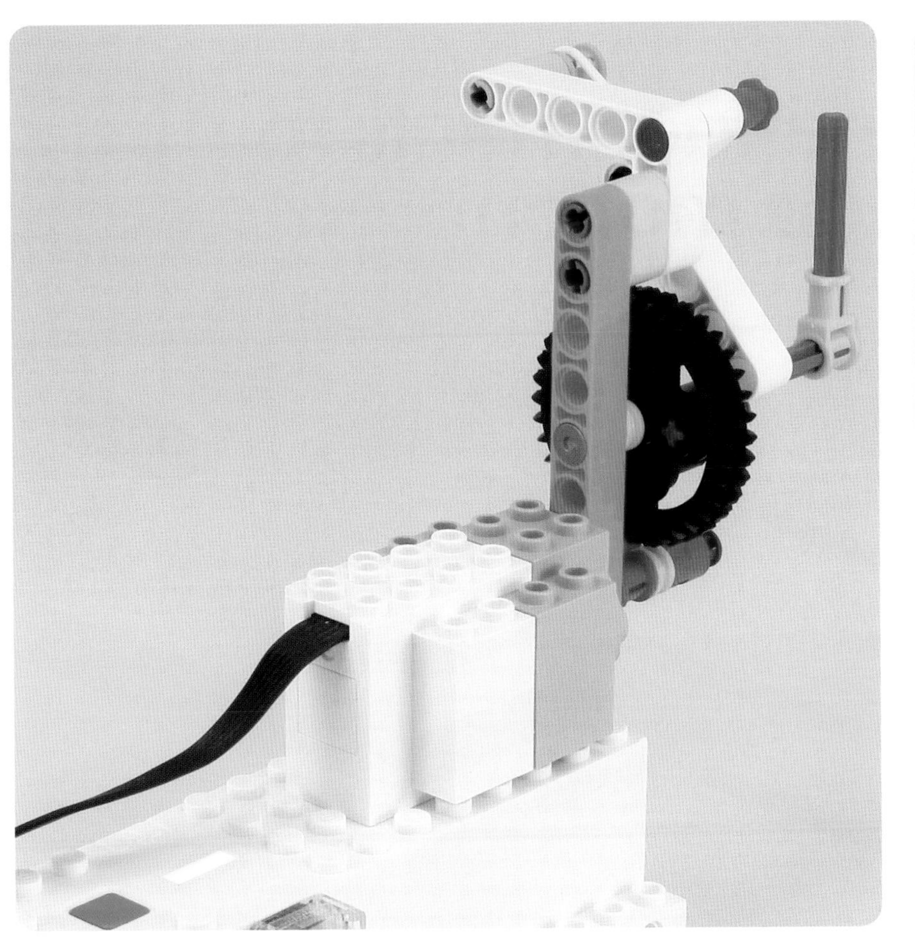

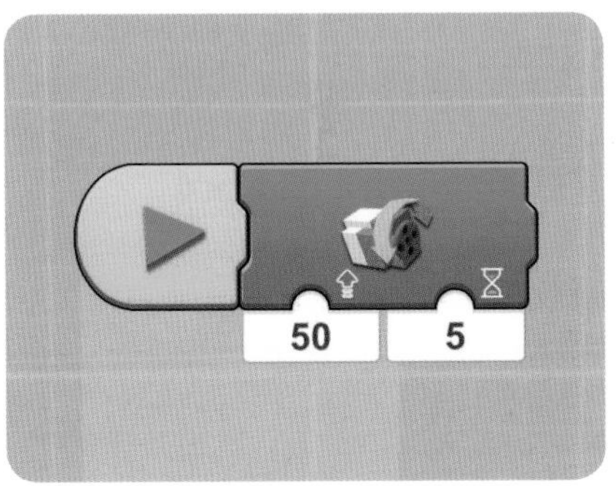
50
5

Using attachments to change motion

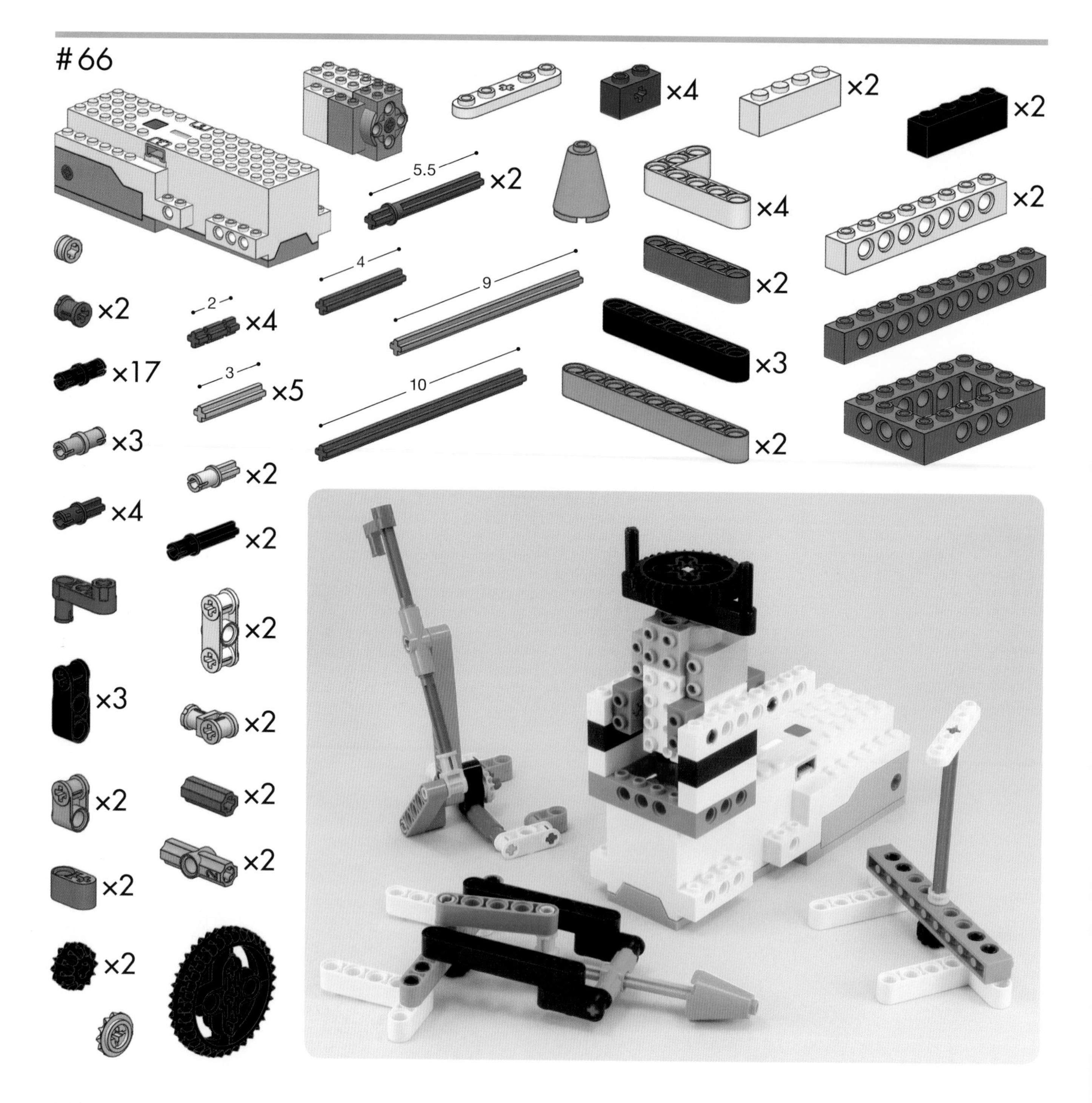

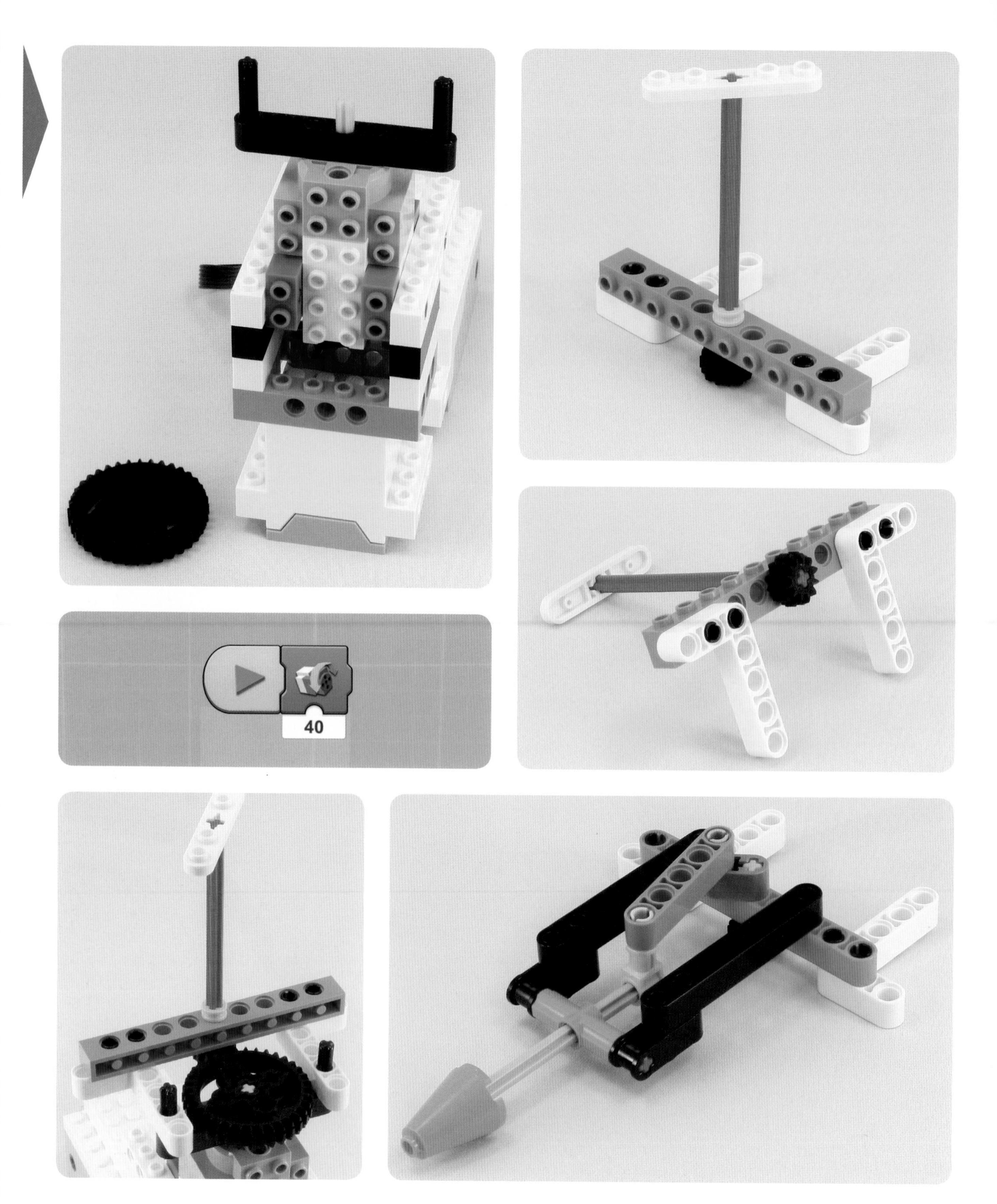
40

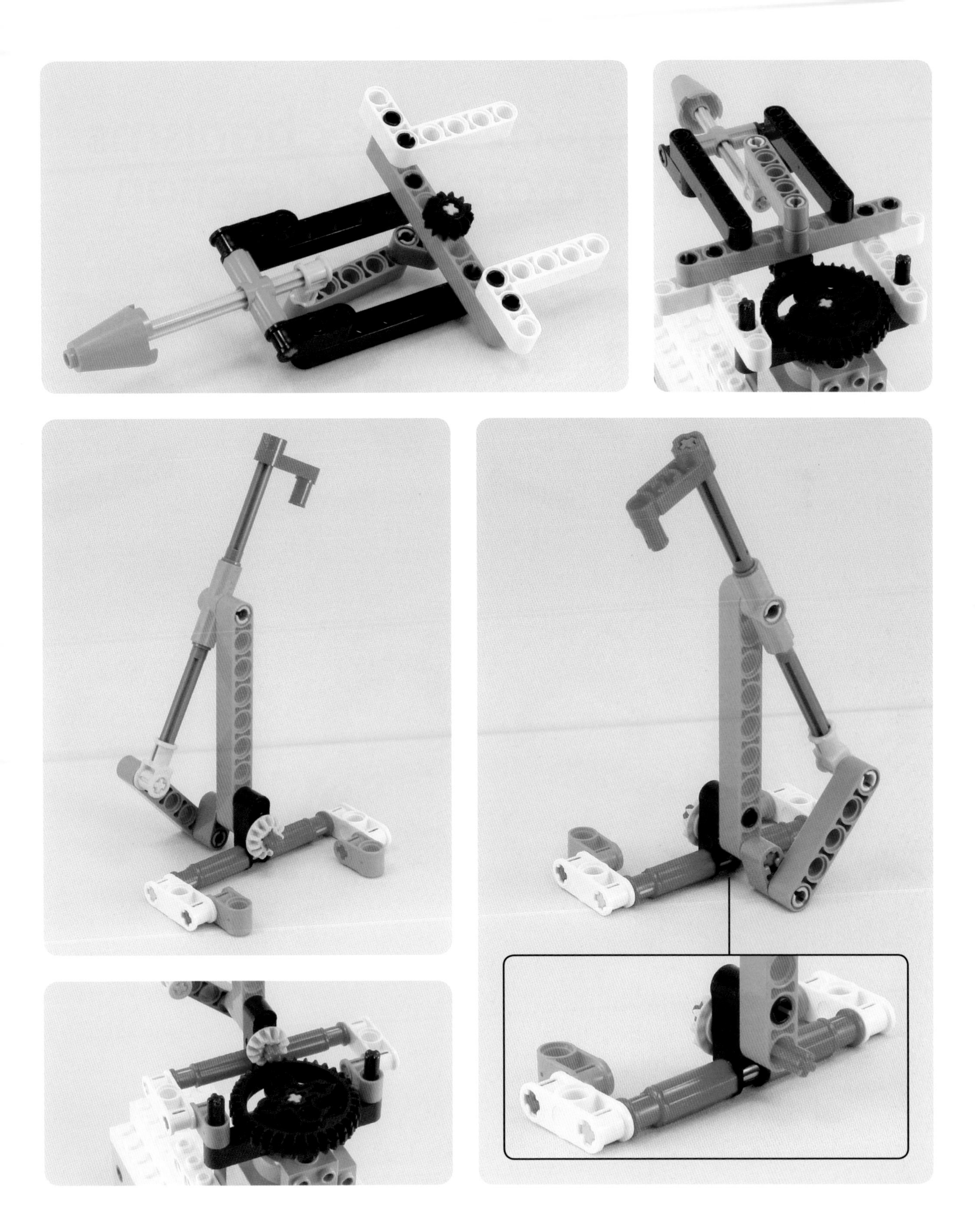

Changeover mechanisms using rotational direction

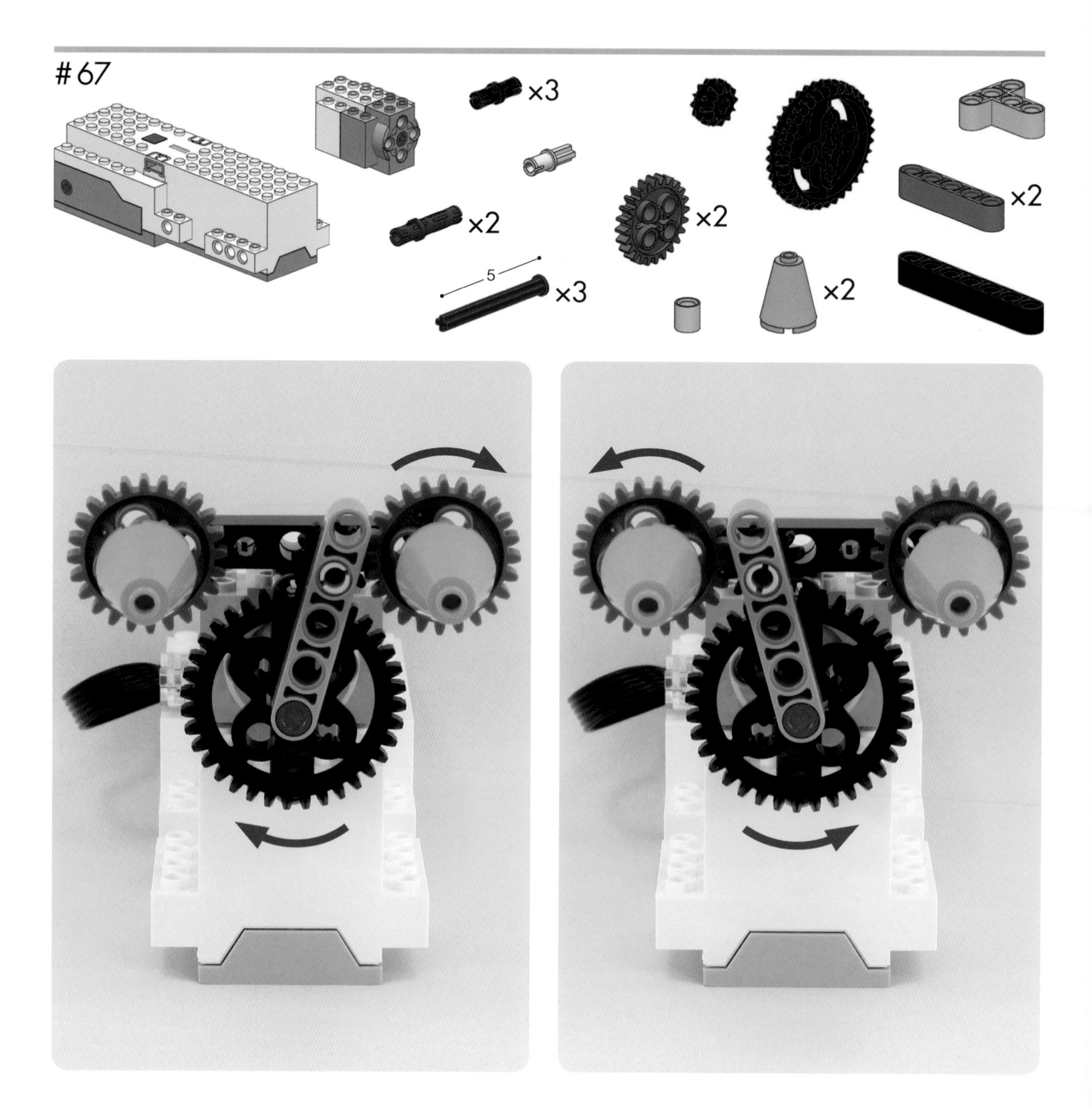

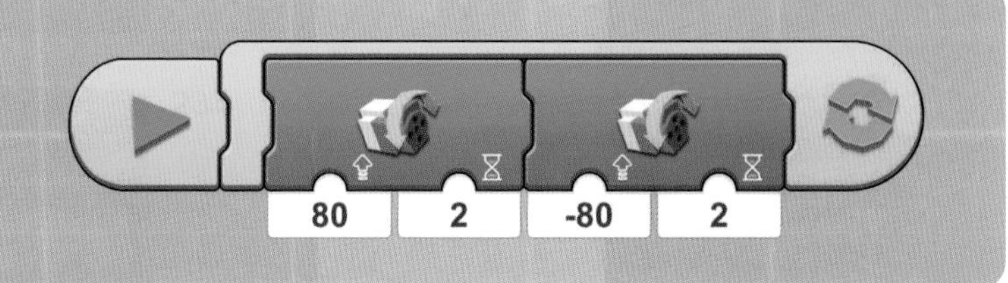
80
2
-80
2

Using the Color and Distance Sensor

#68

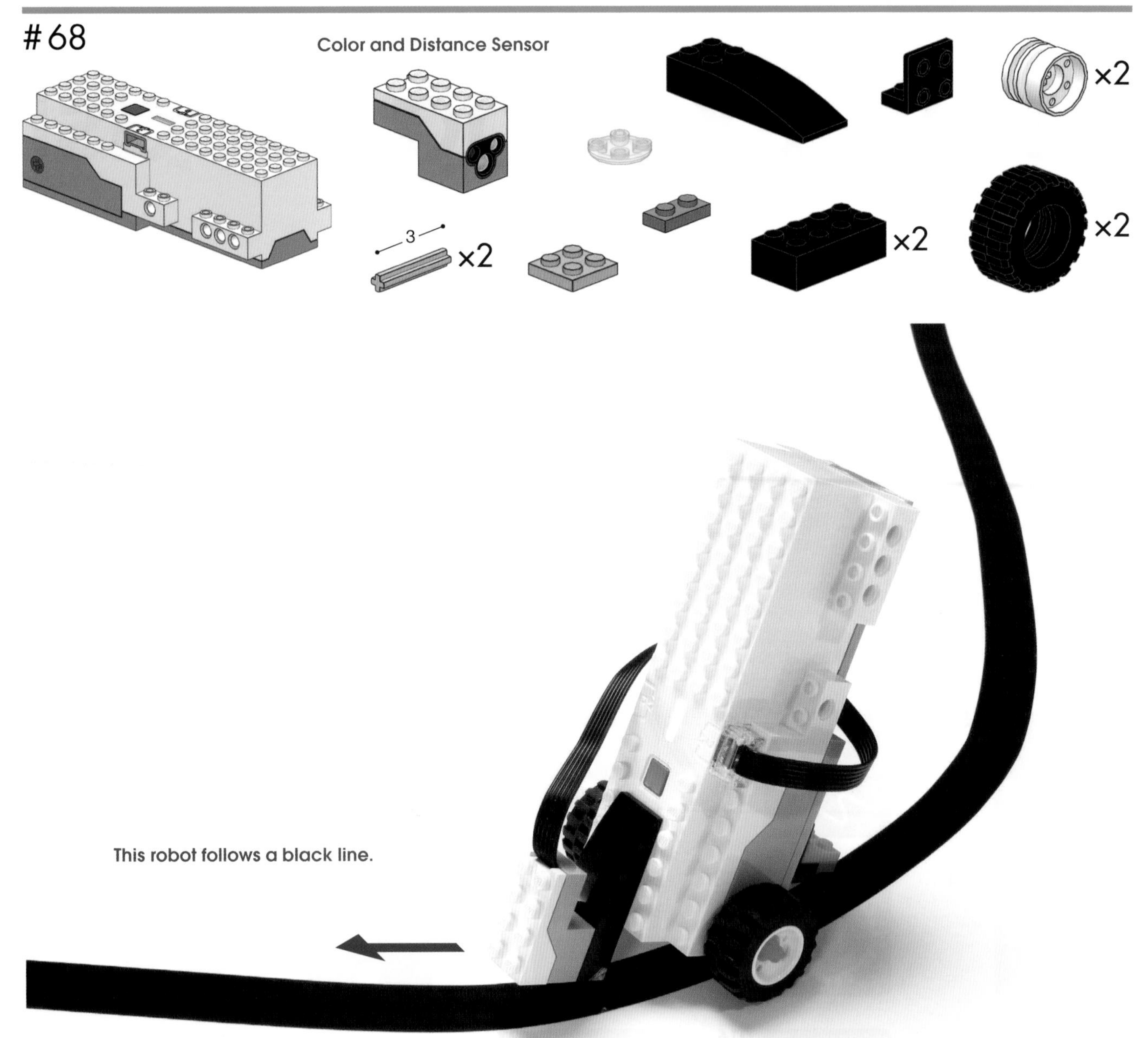

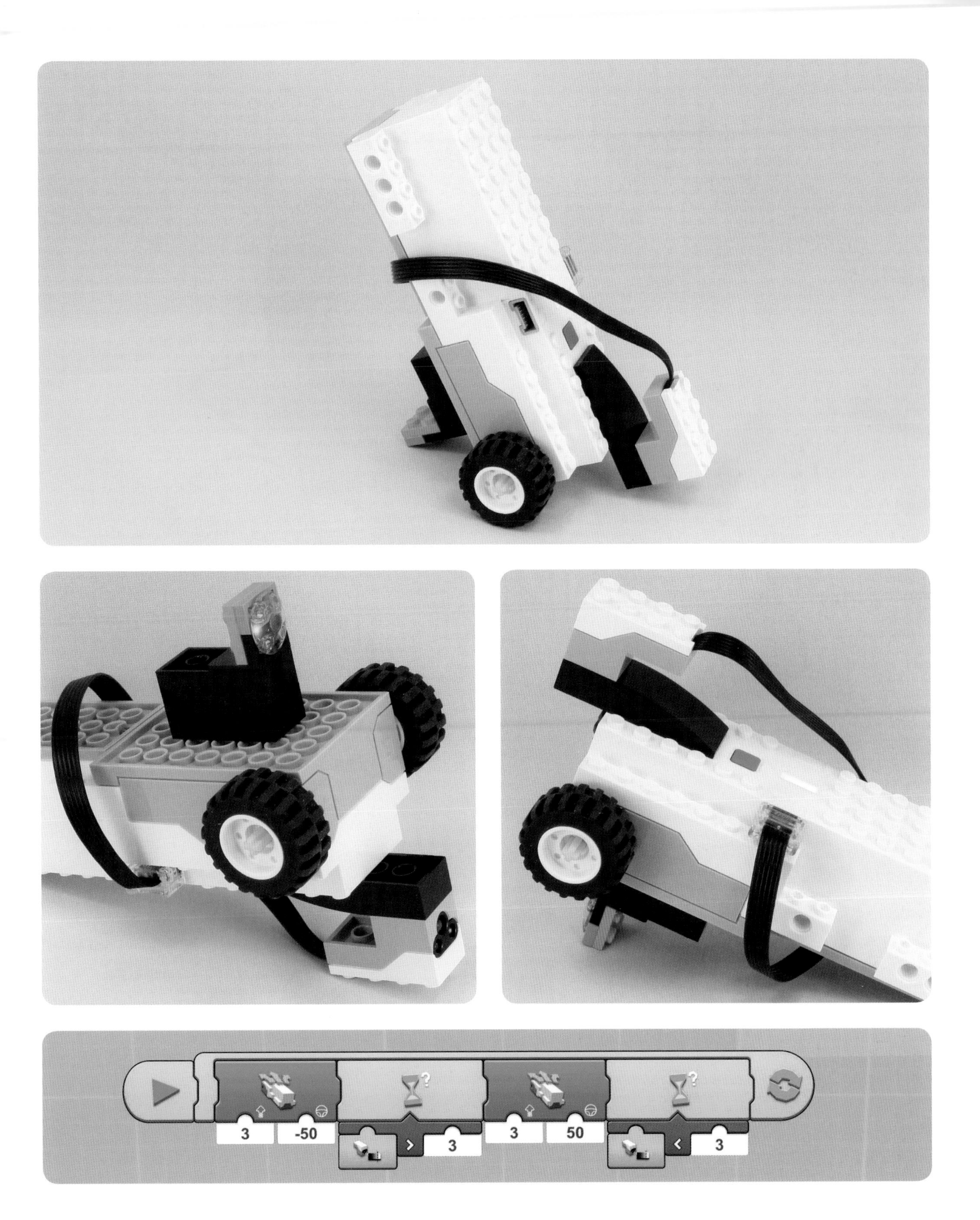
3
-50
>
3
3
50
<
3

#69

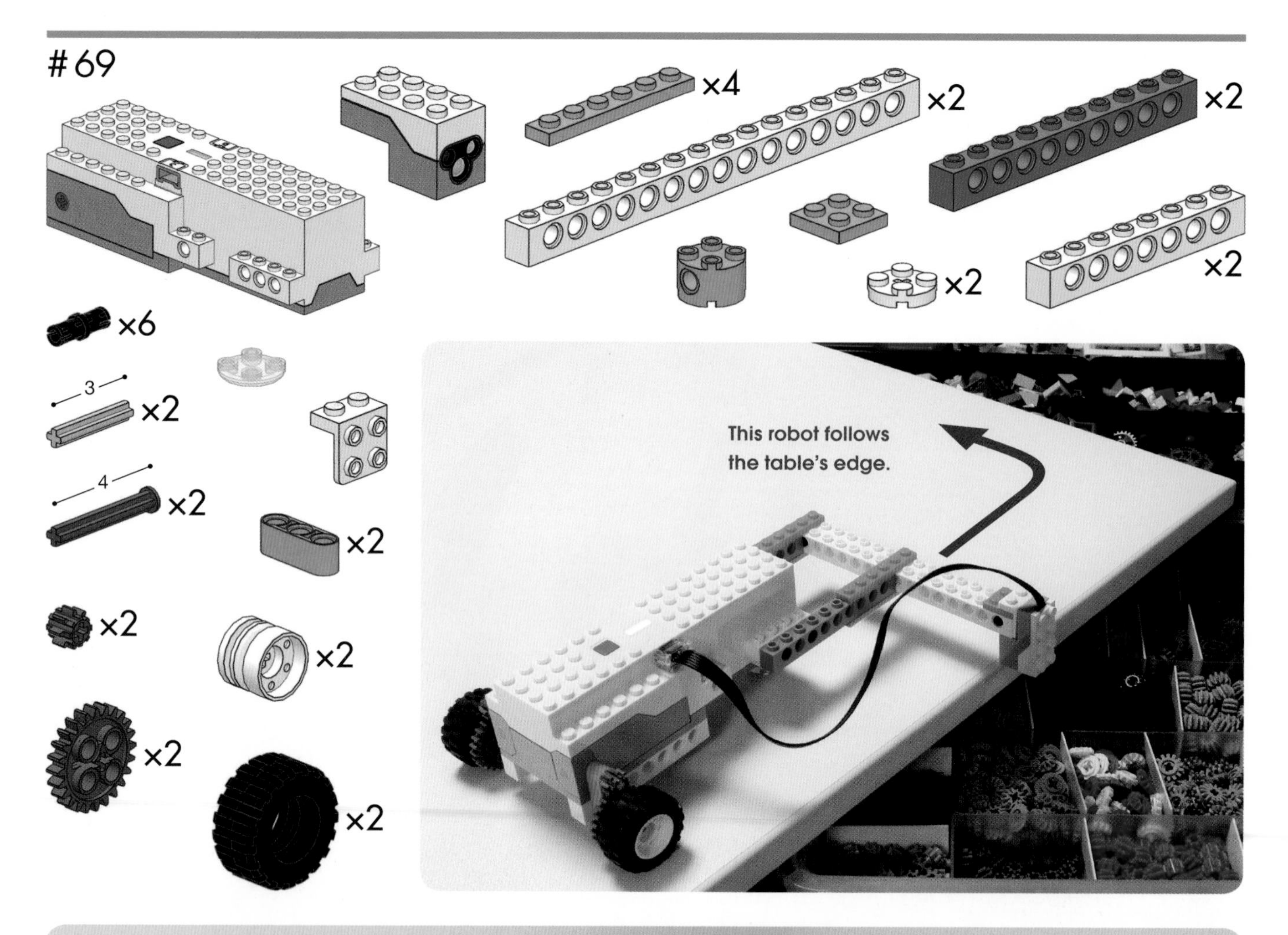

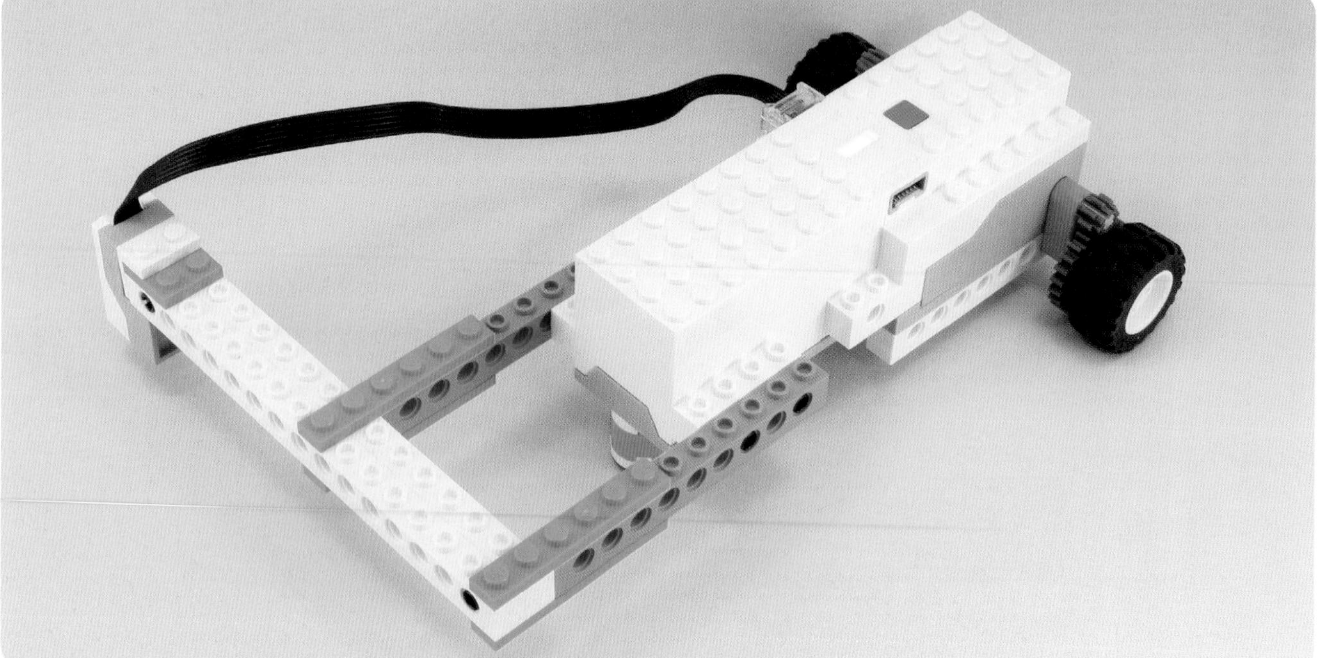

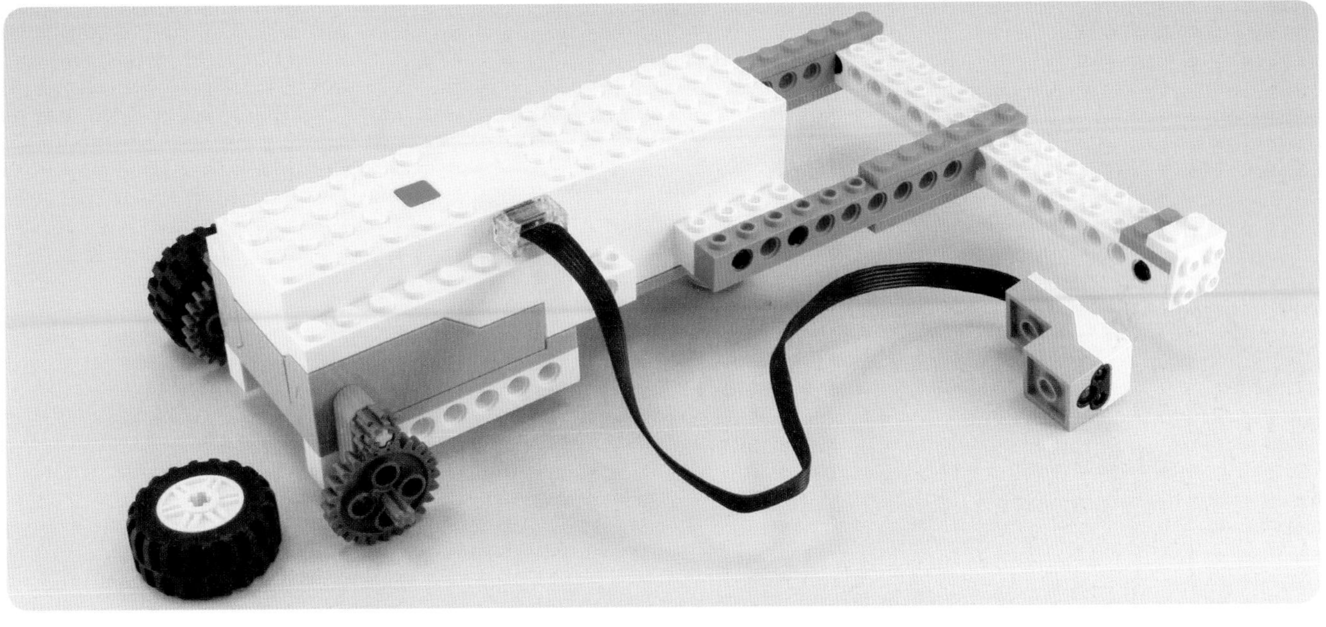

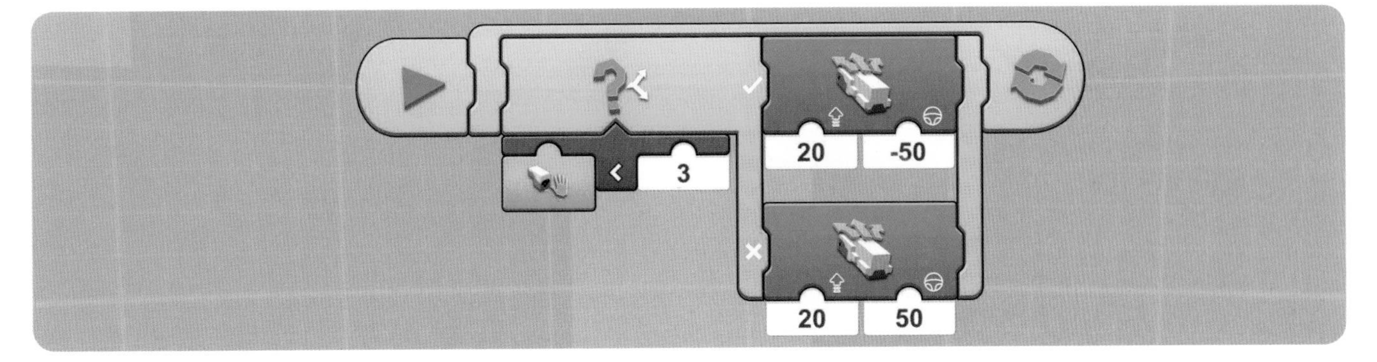
20
-50
3
20
50

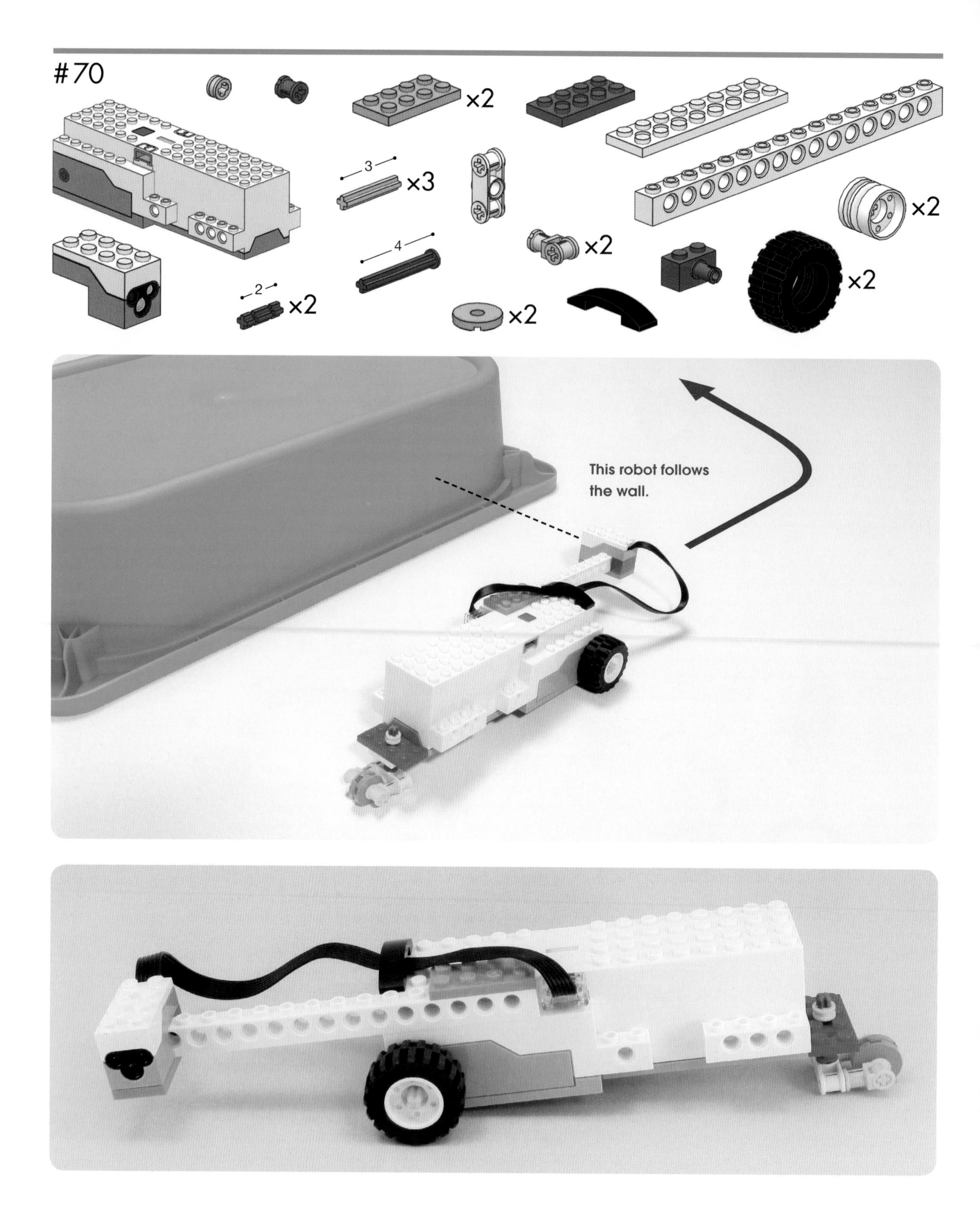
#70
3
×3
4
2
×2
×2
×2
×2
×2
×2
×2
This robot follows
the wall.

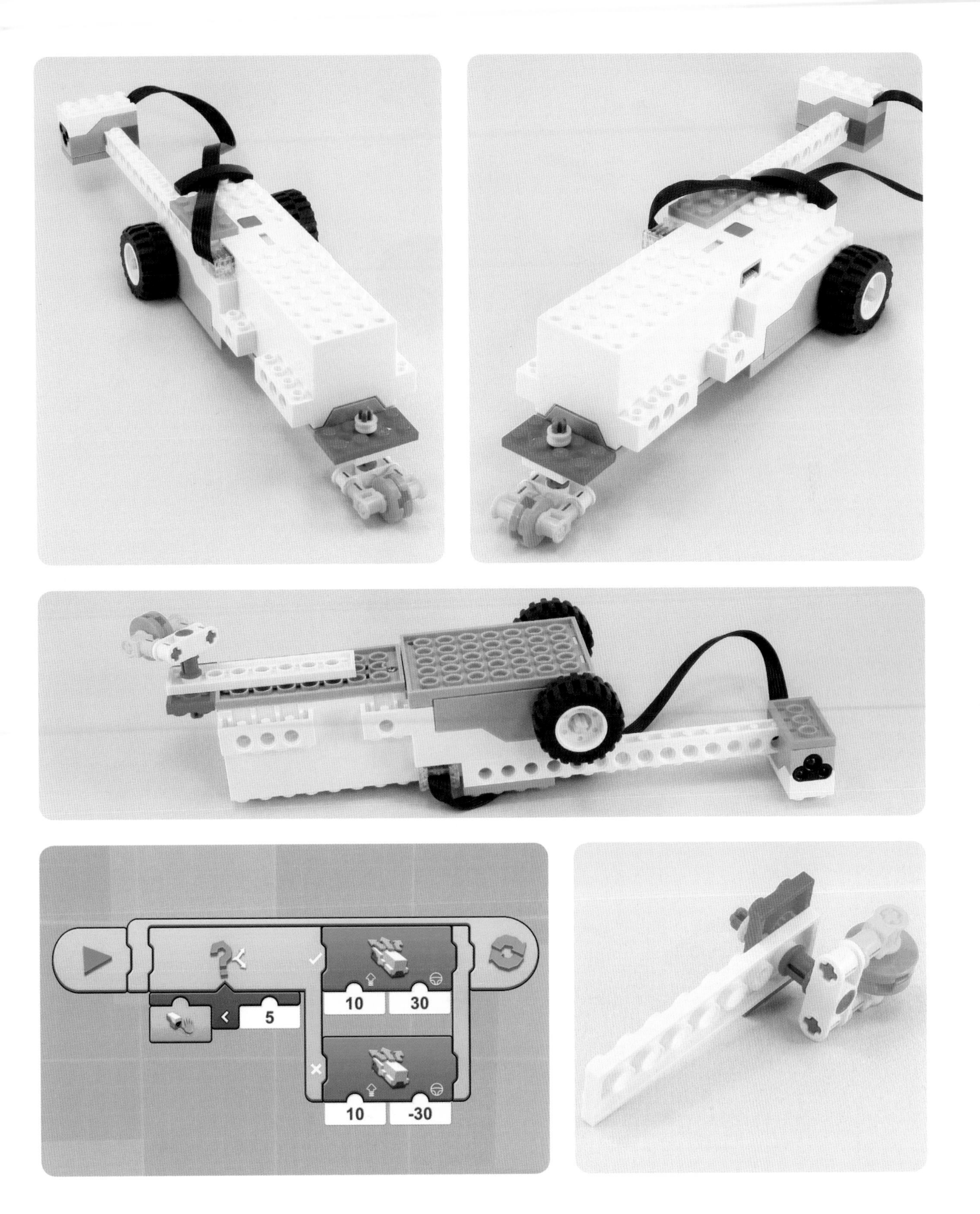
10
30
5
10
-30

#71

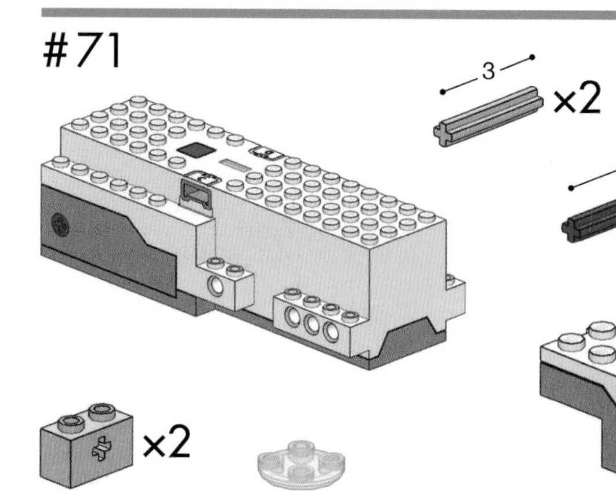

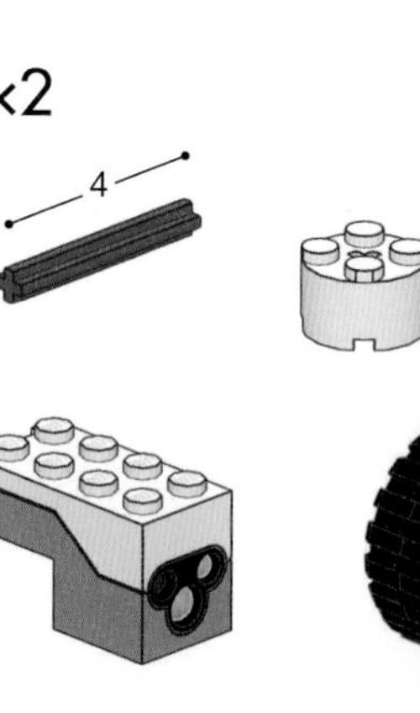

Place your hand above the Sensor to make the car go forward, turn, or stop.

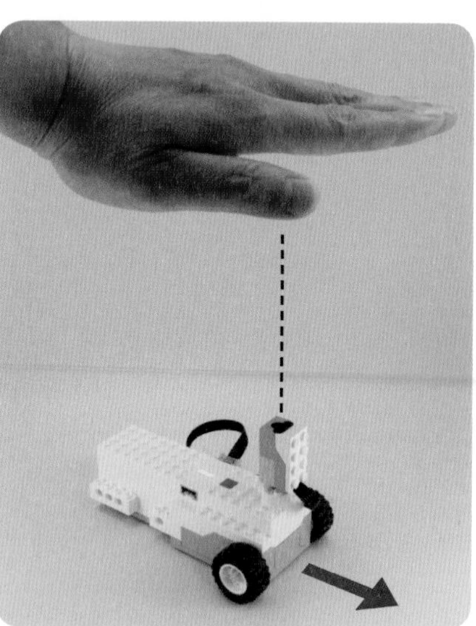

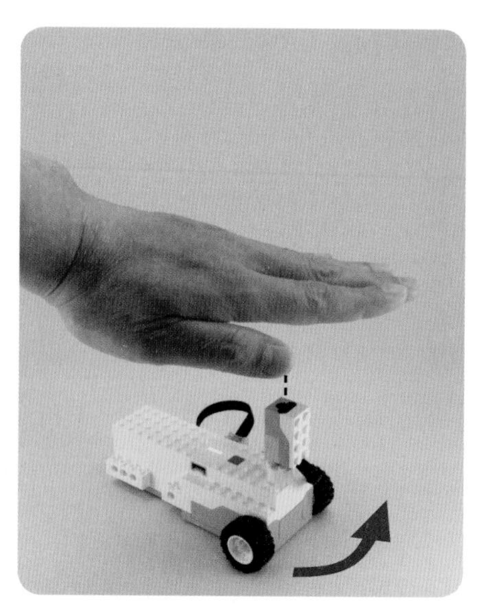

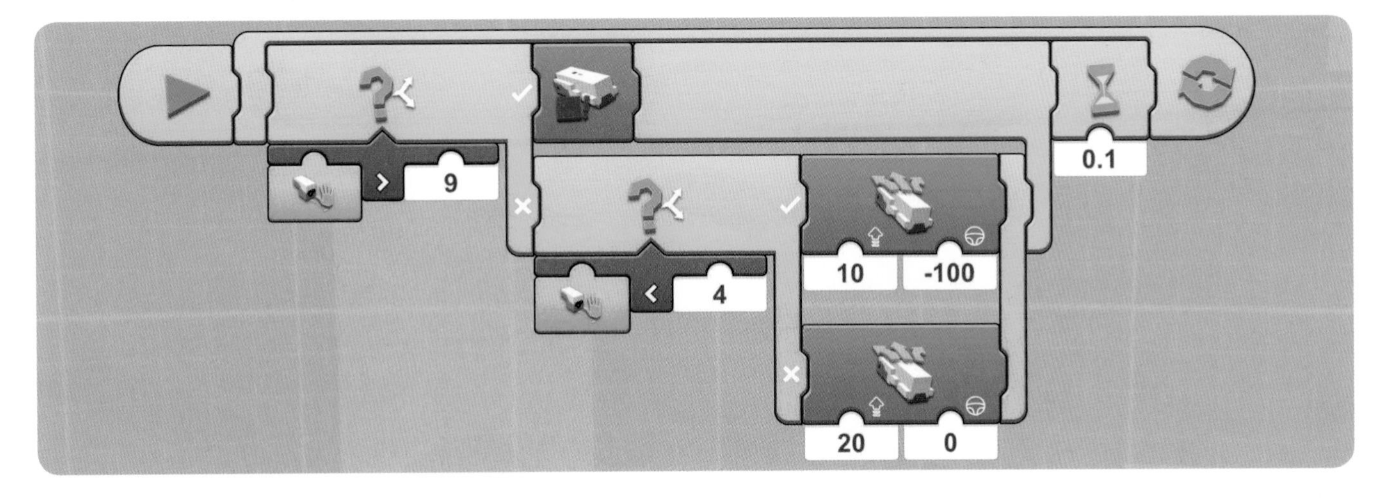
9
4
10
-100
20
0
0.1

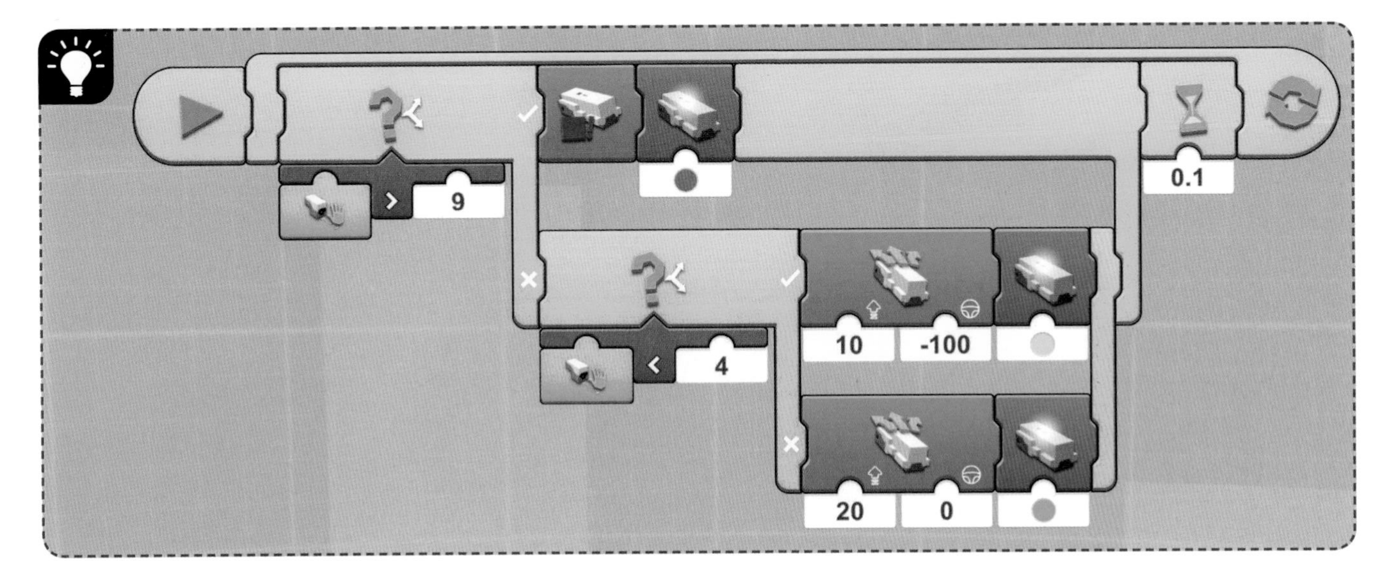
9
4
10
-100
20
0
0.1

#72

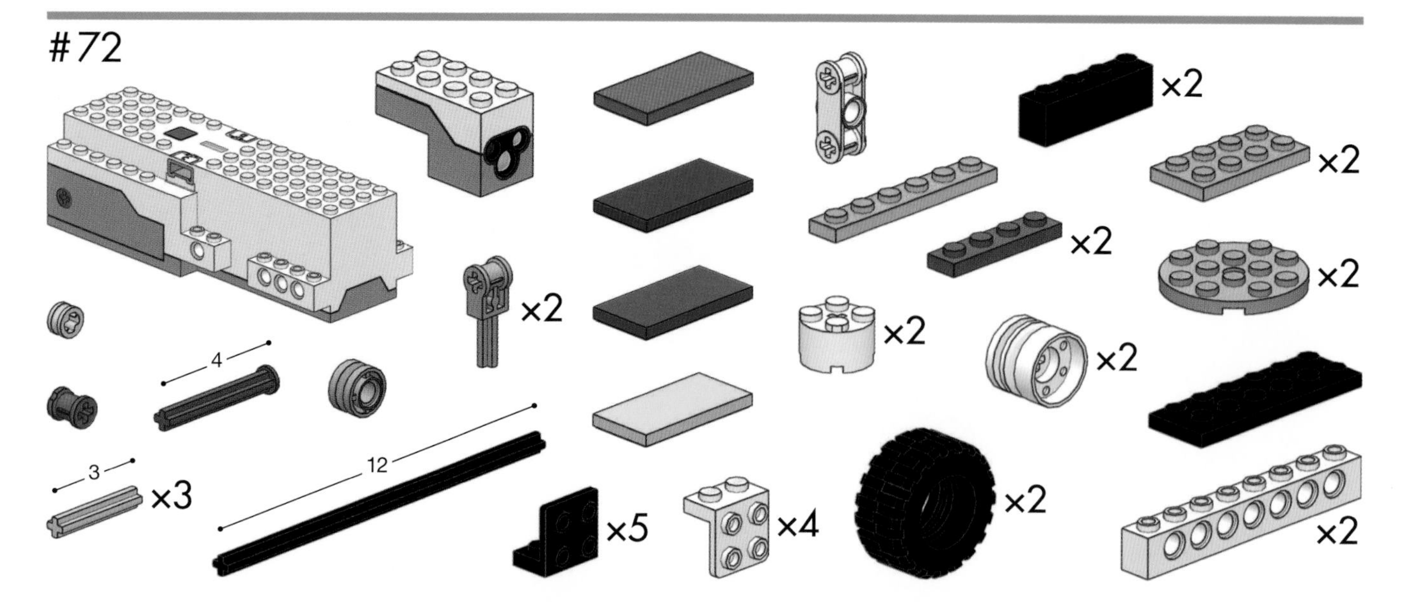

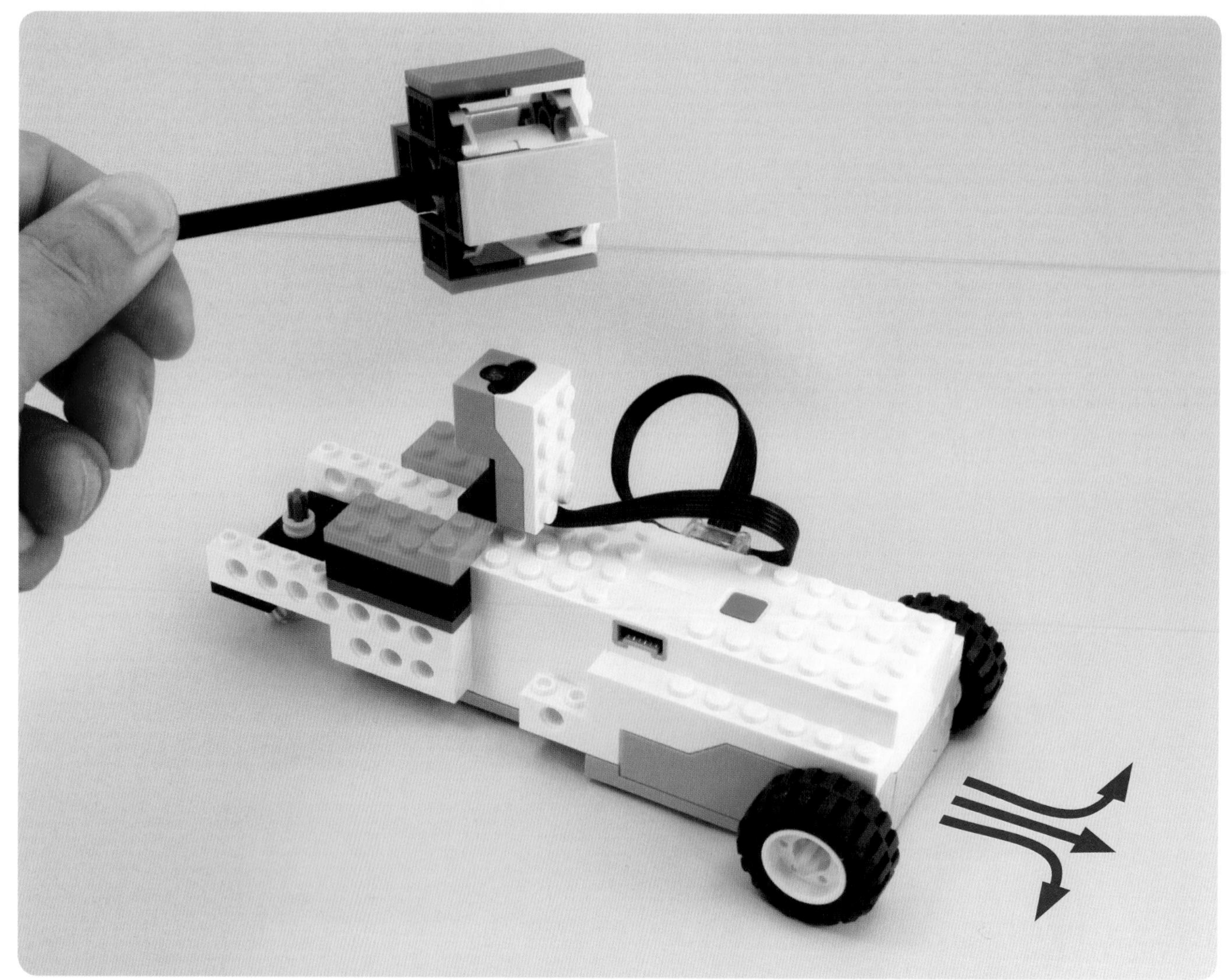

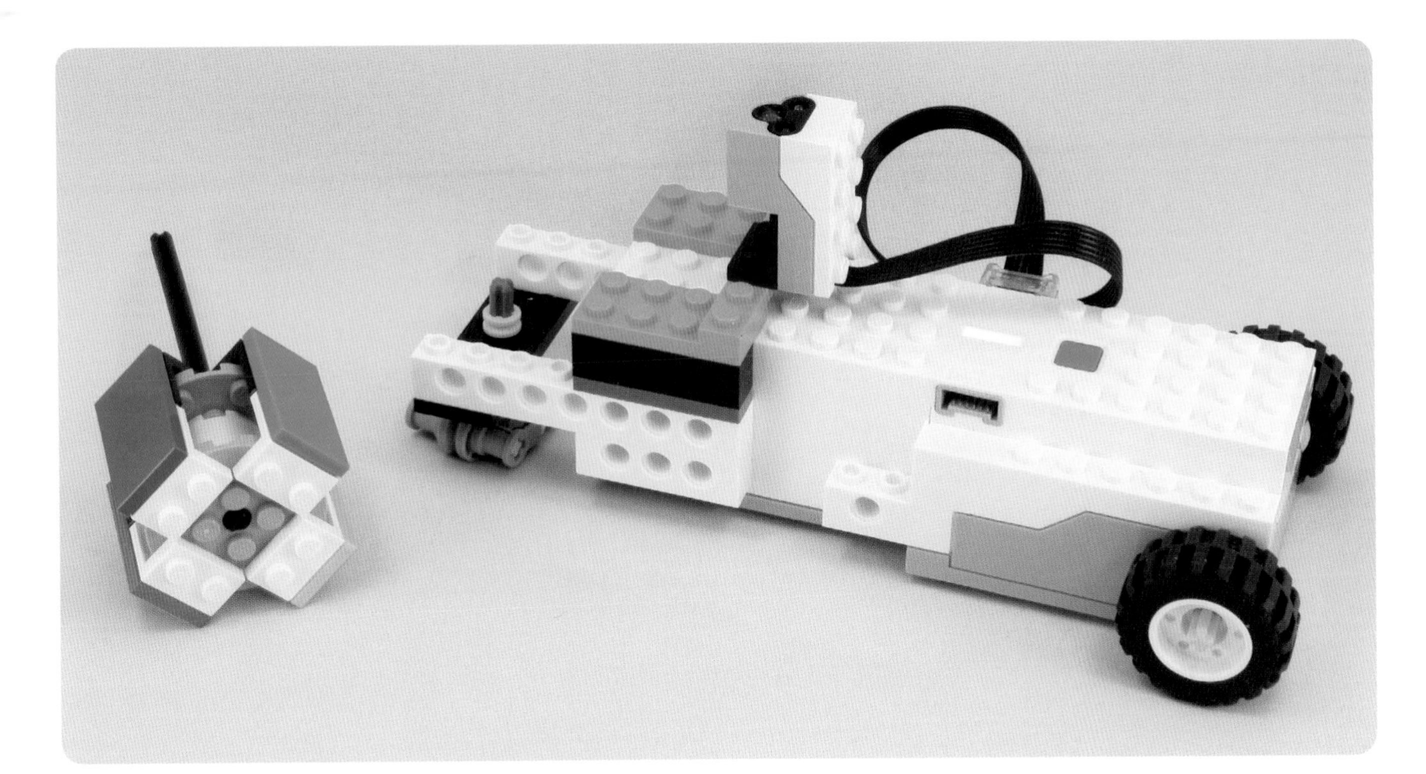

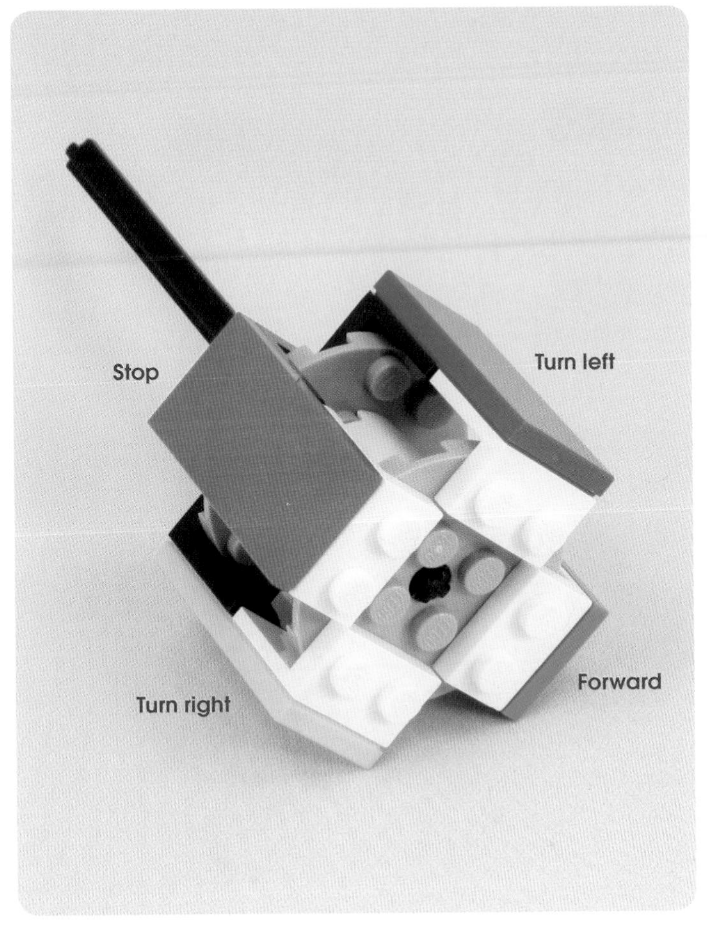
Stop
Turn left
Turn right
Forward

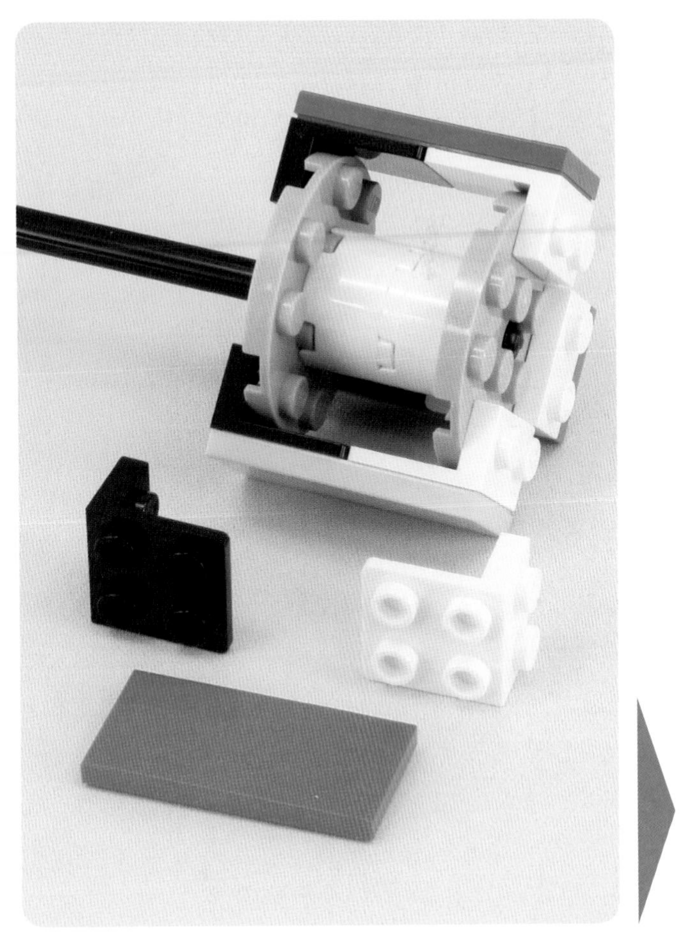

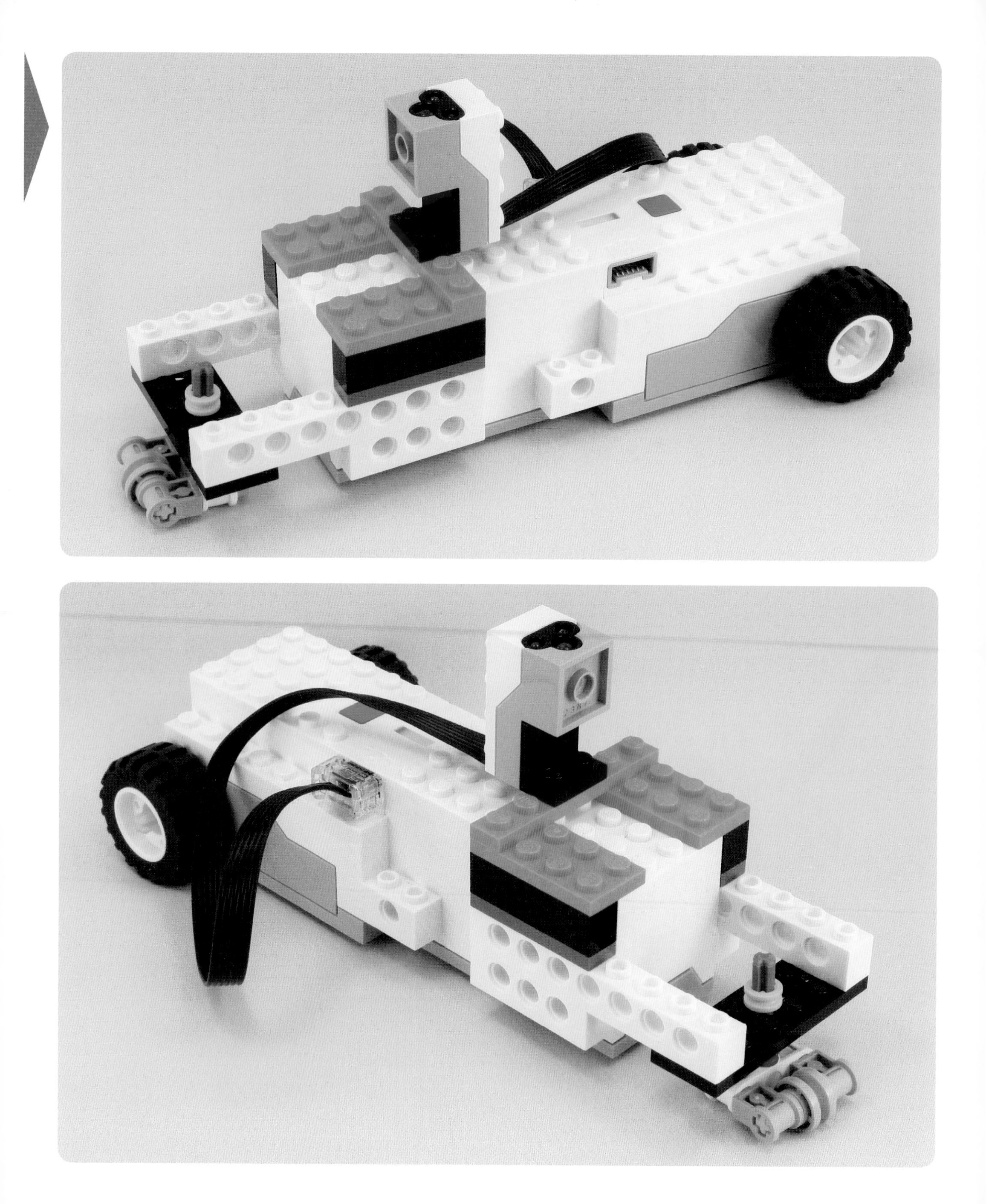

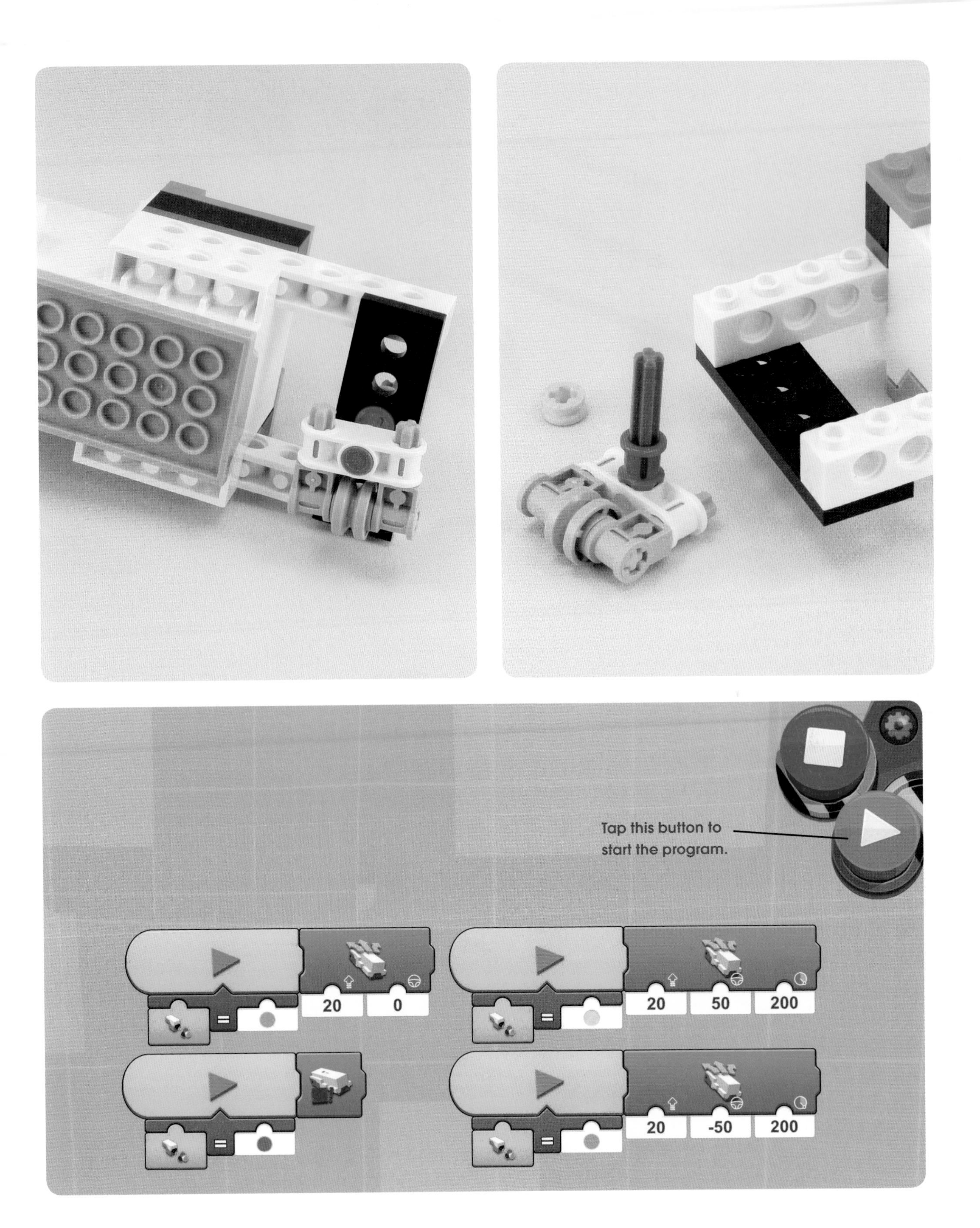
Tap this button to
start the program.
20
0
20
50
200
20
-50
200

#73

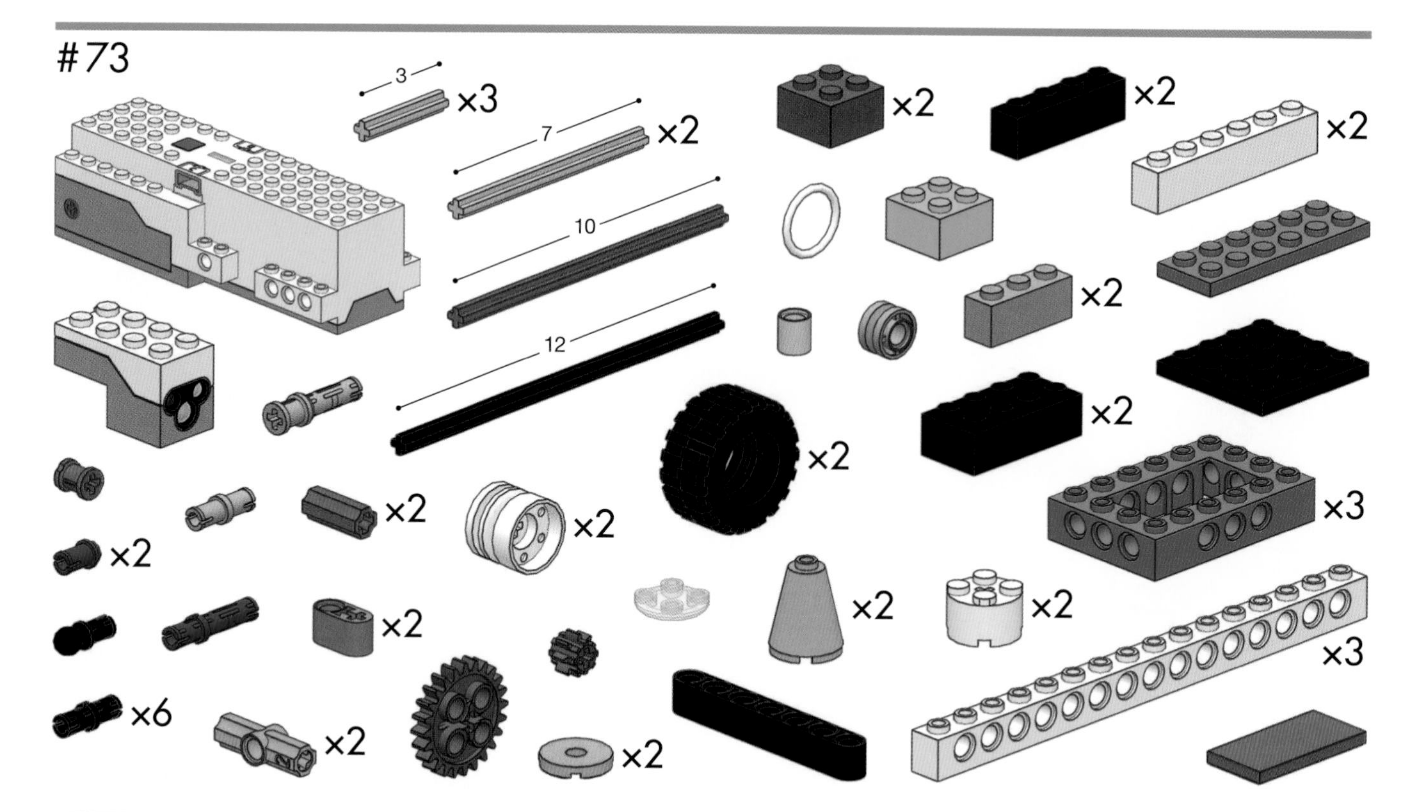
3
×3
7
×2
10
12
×2
×2
×2
×2
×2
×2
×2
×2
×2
×3
×2
×2
×2
×2
×2
×3
×6
×2
×2

Detecting obstacles
in front and behind

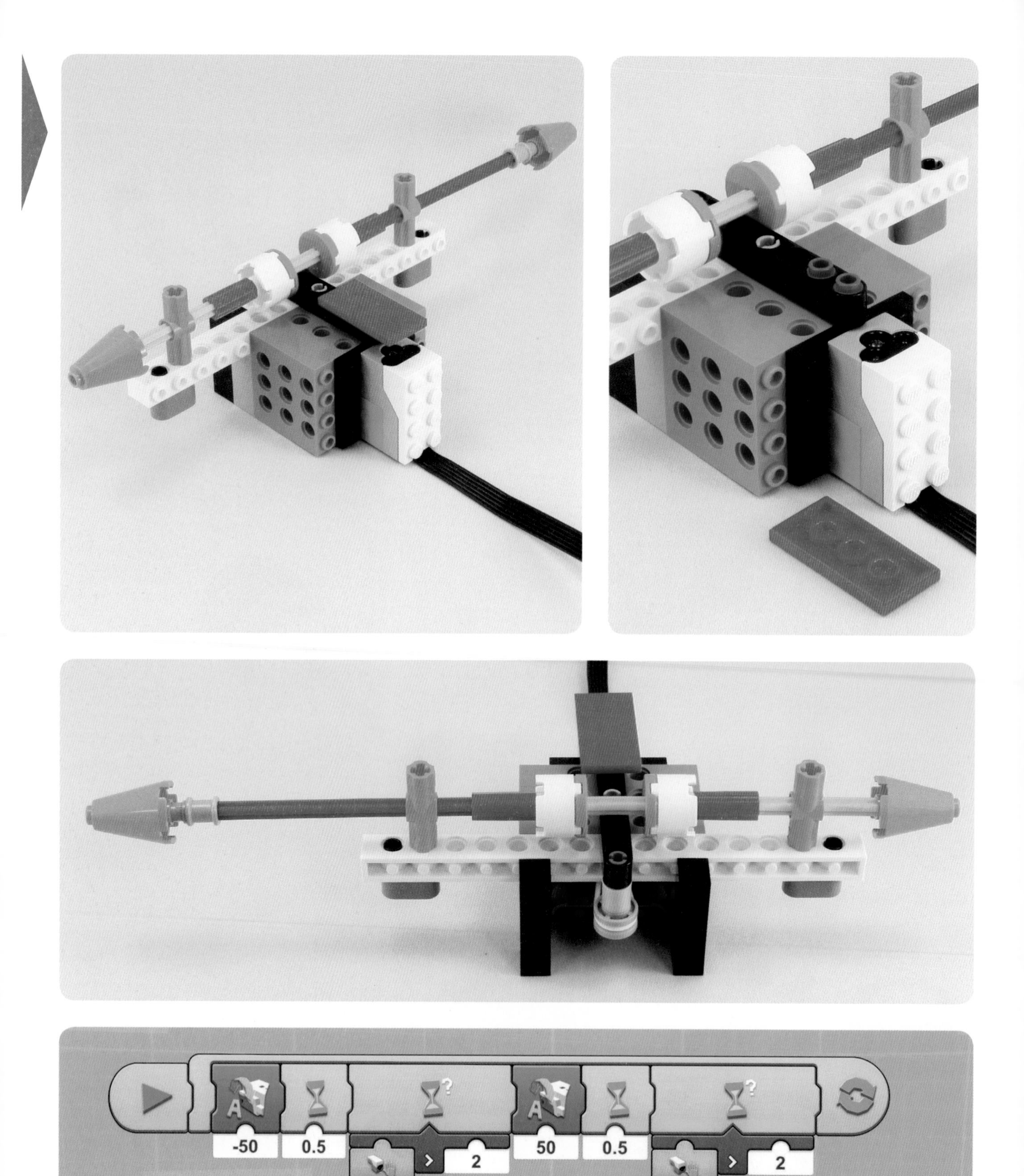
-50
0.5
2
50
0.5
2

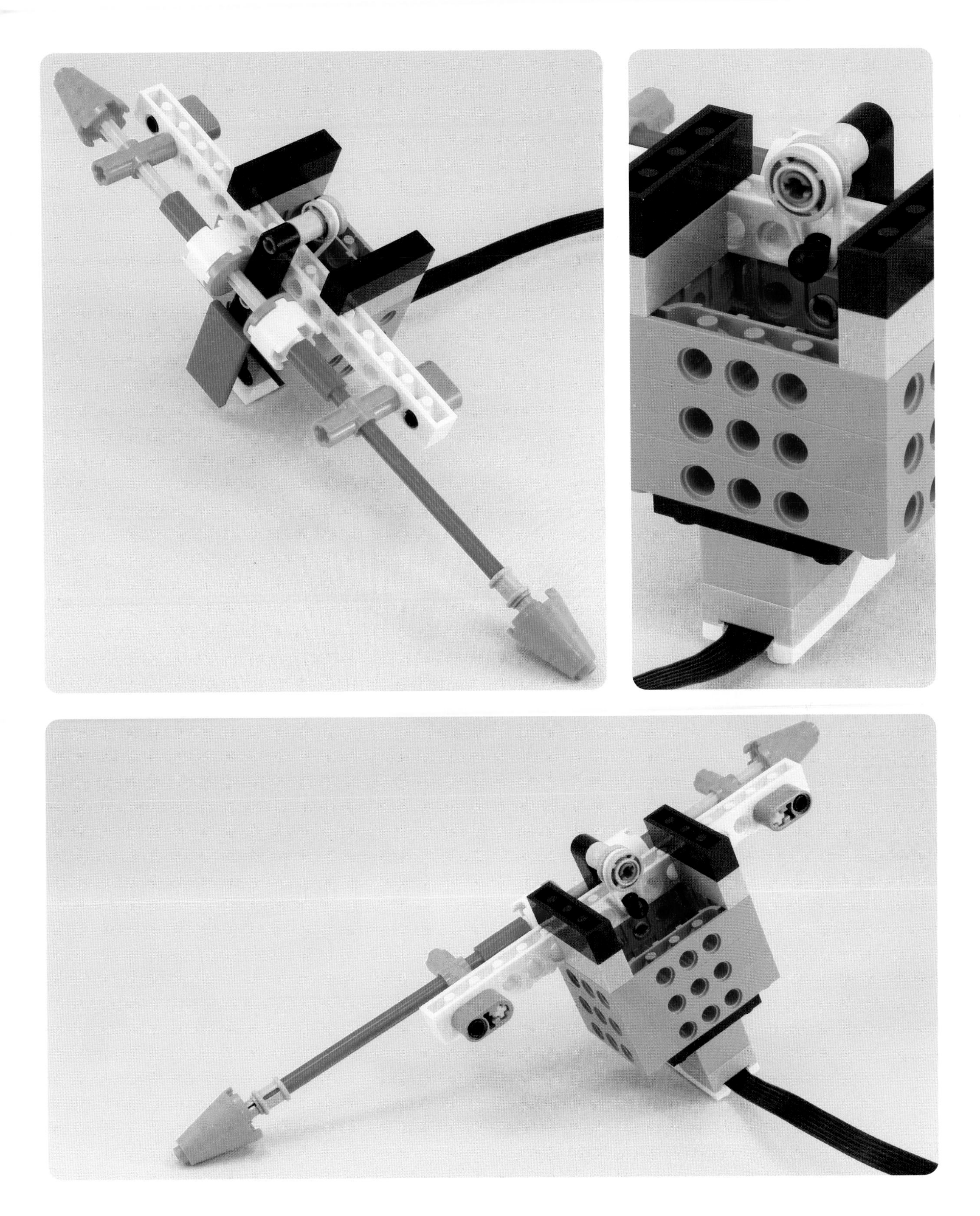

Automatic doors

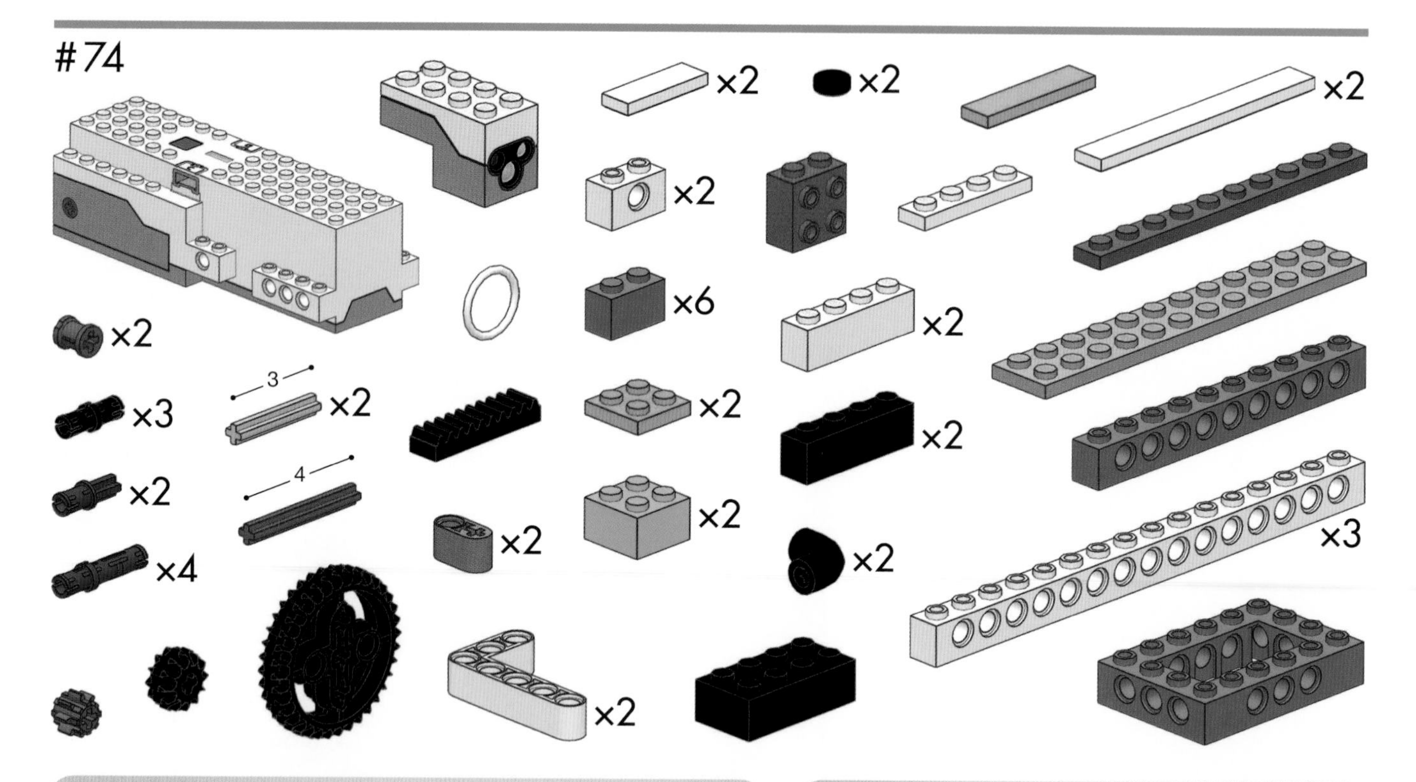

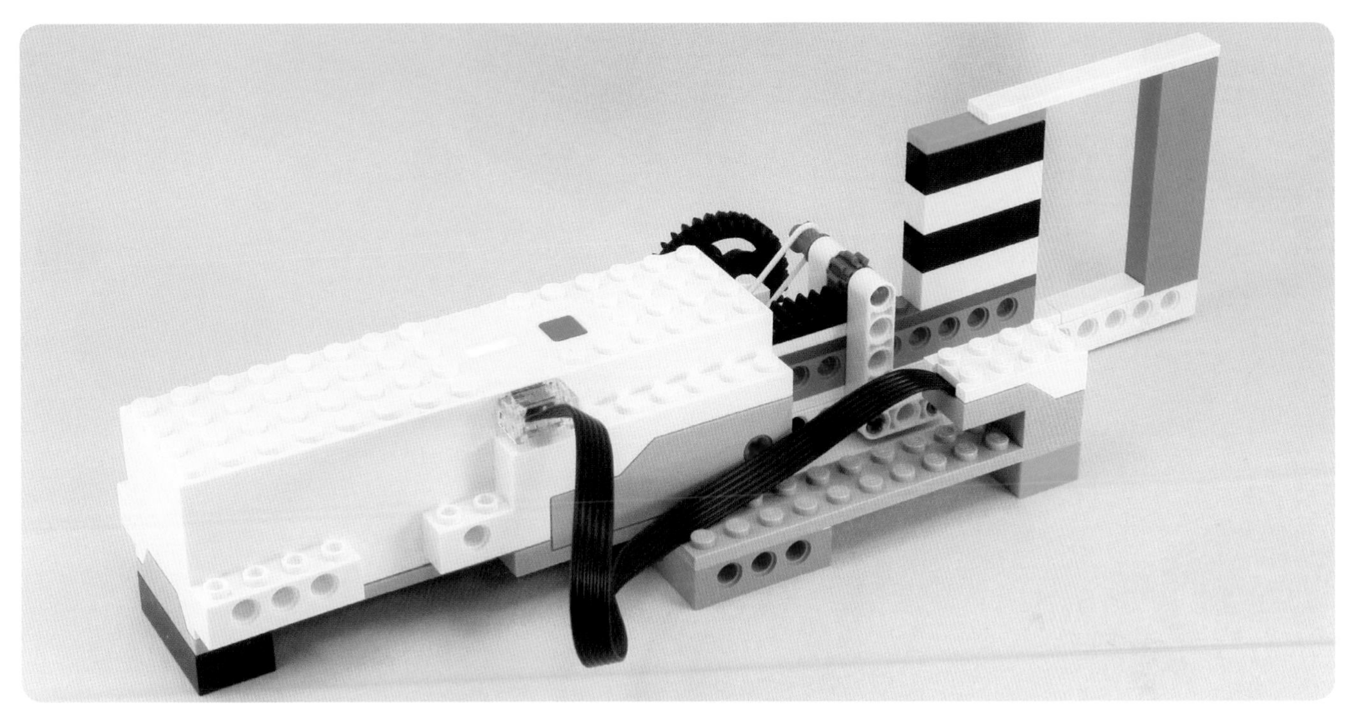

60
2
3
-60
2
2

#75

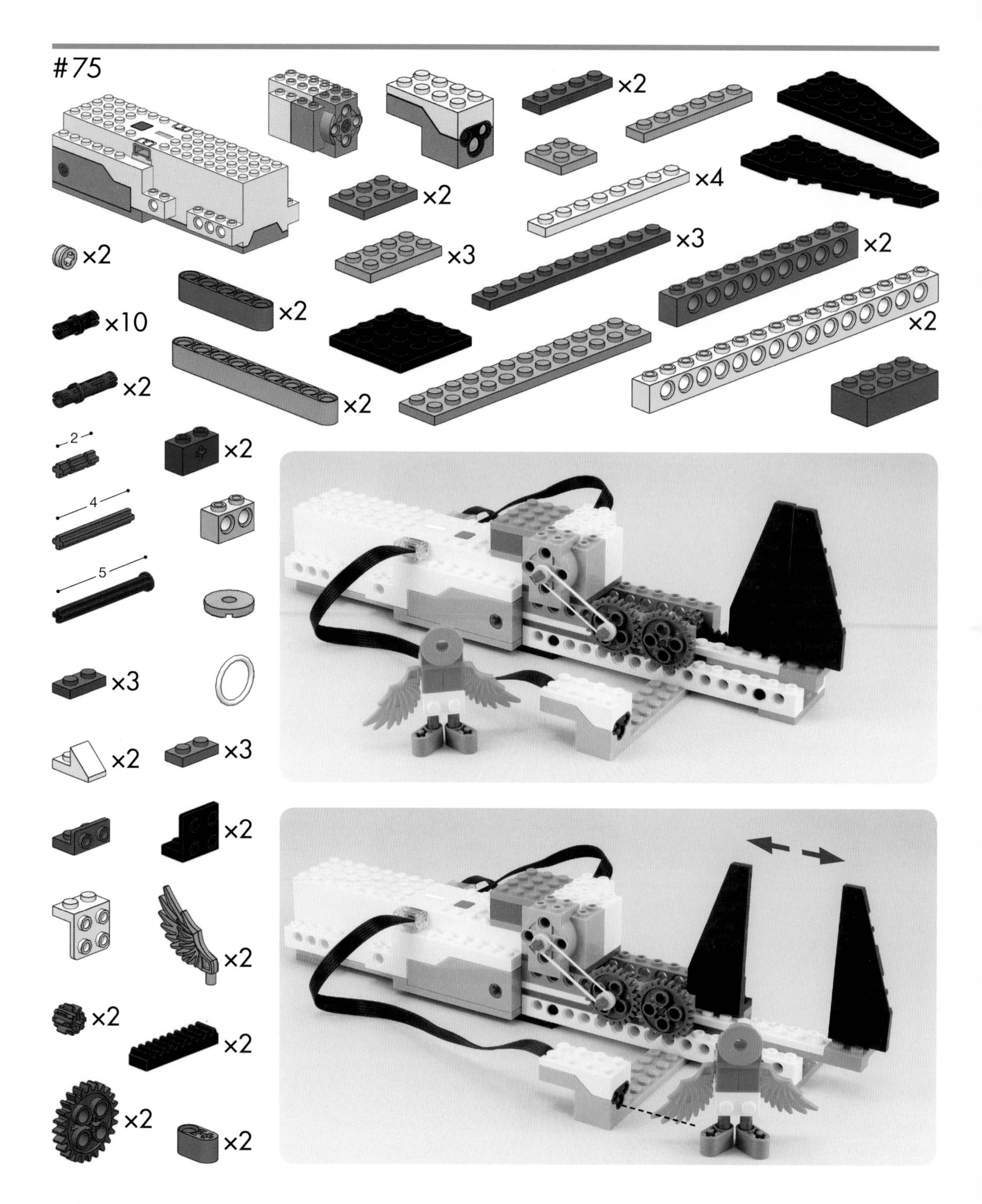

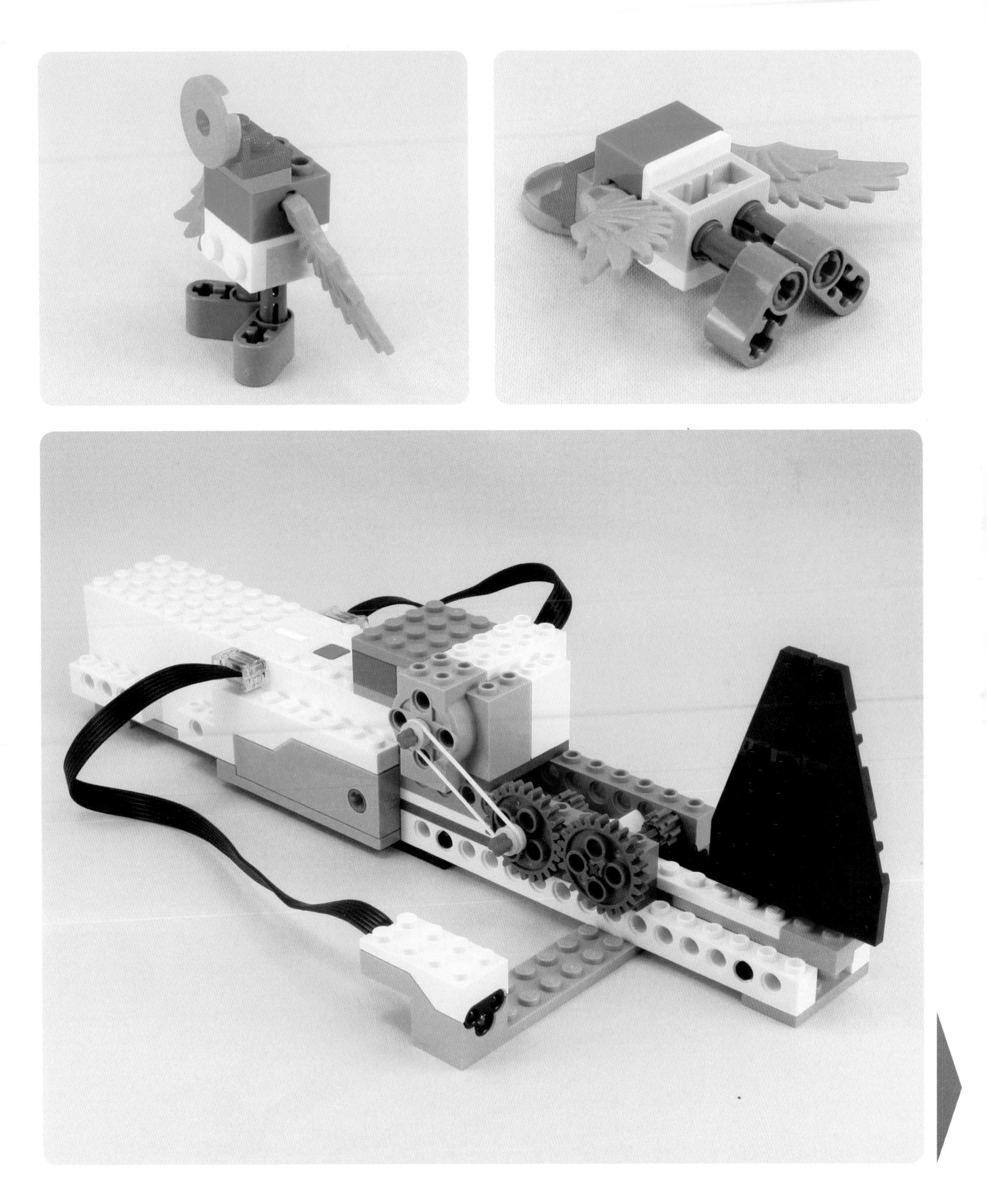

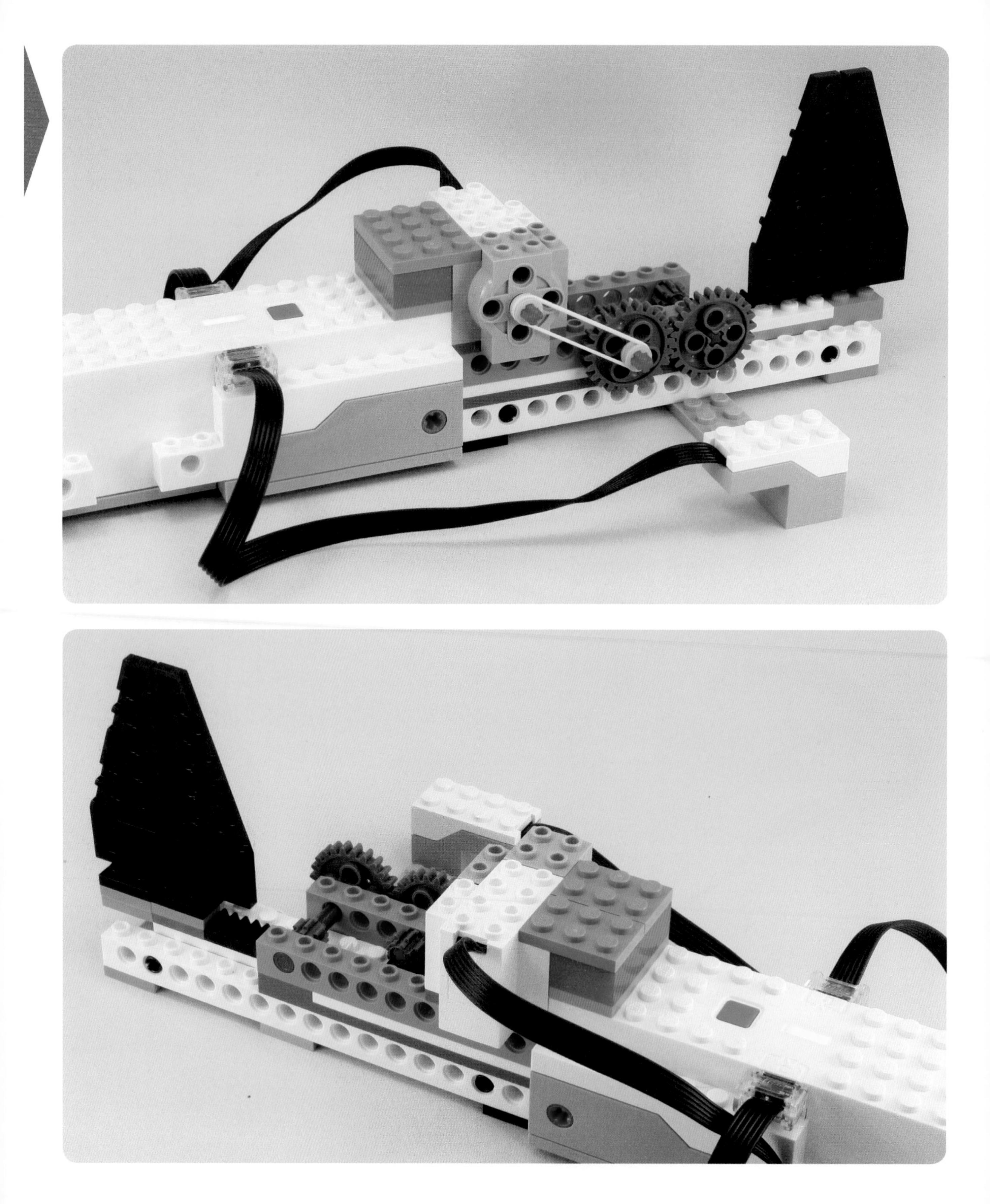

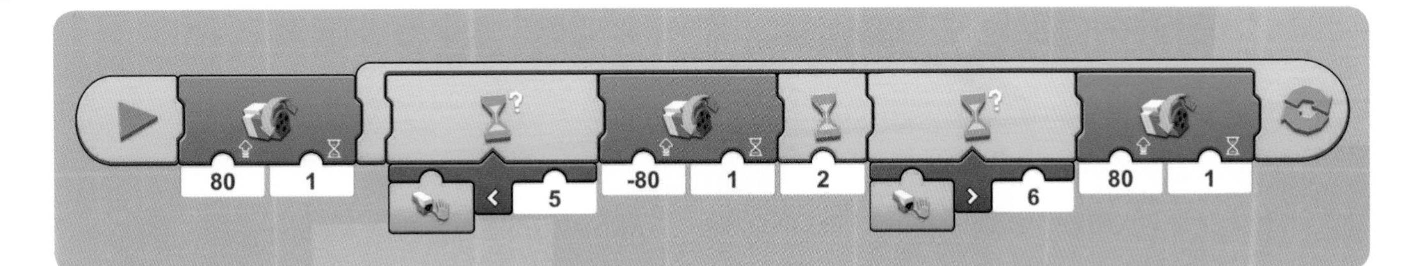
80
1
5
-80
1
2
6
80
1

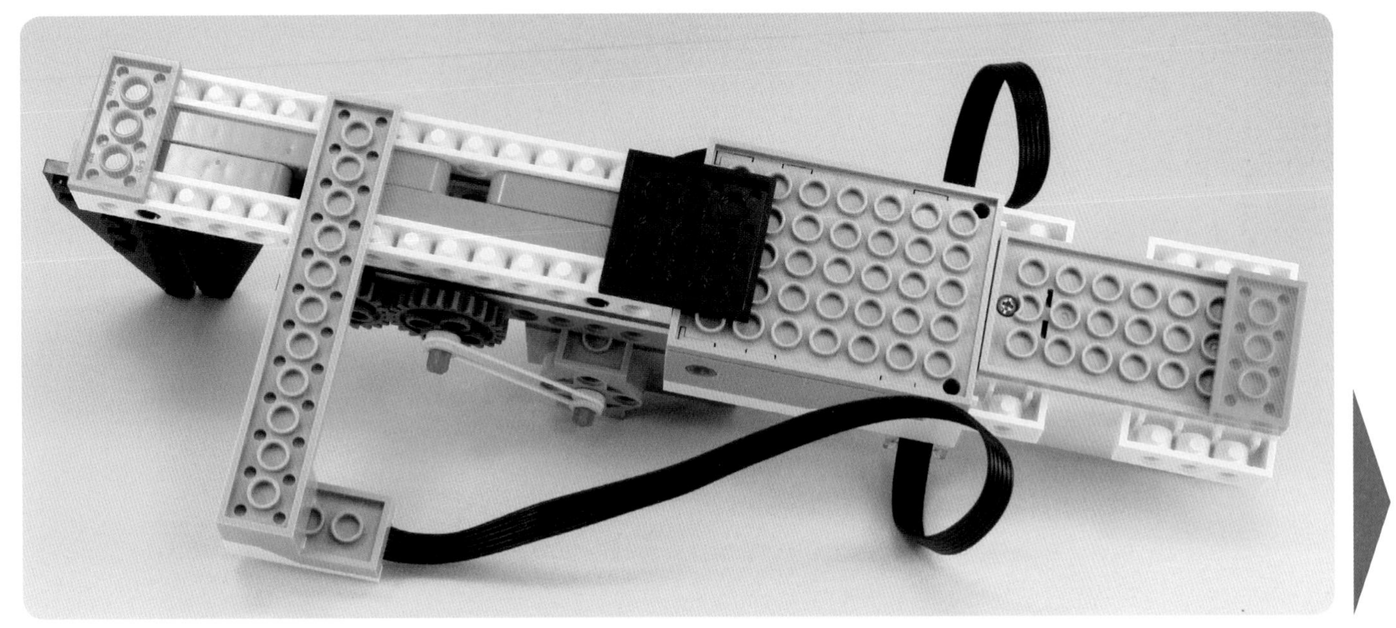

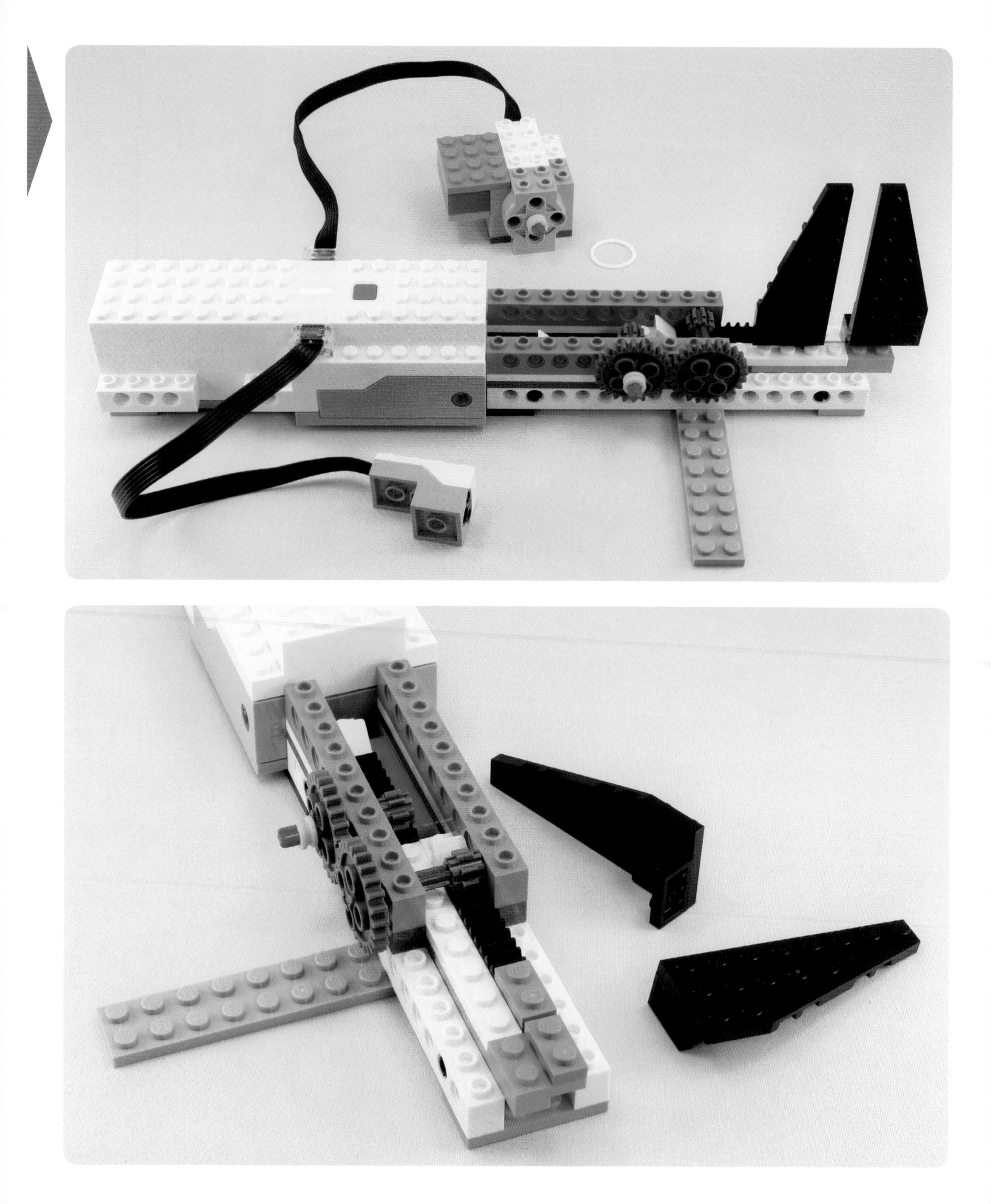

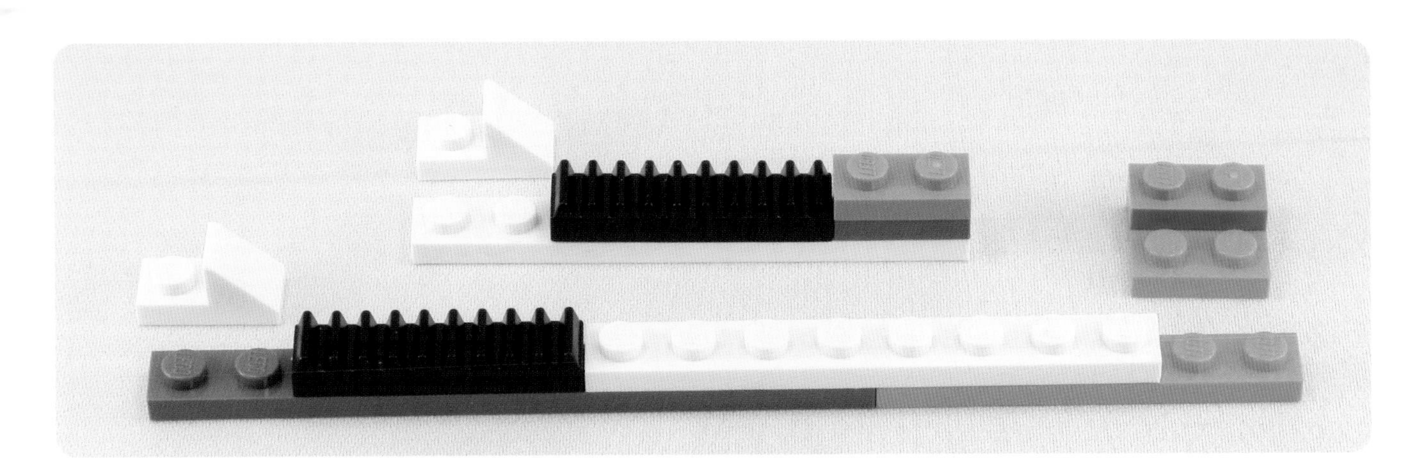

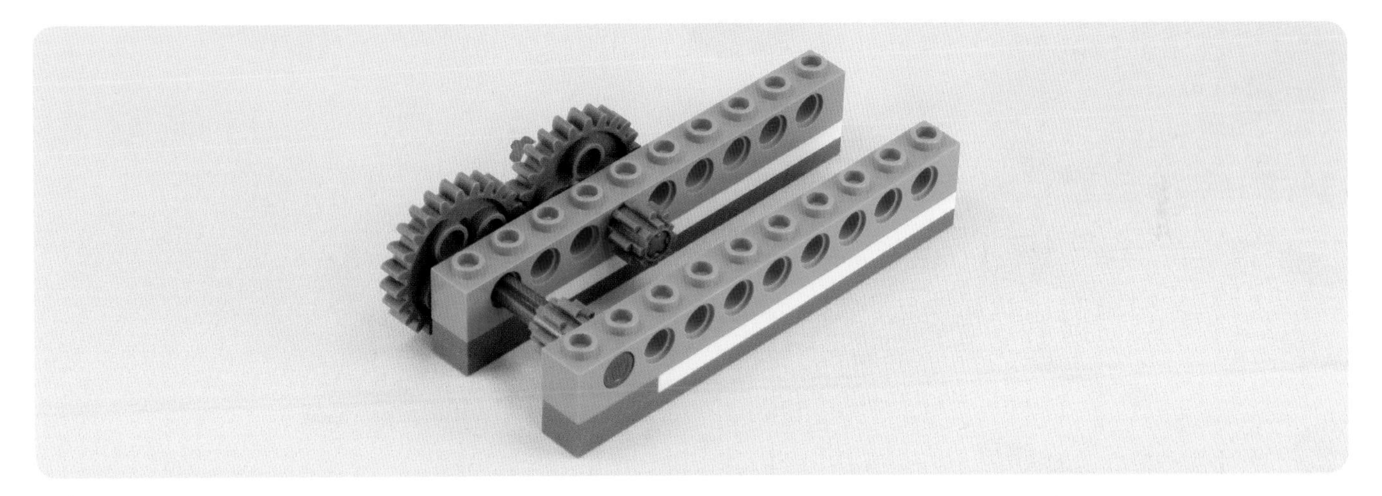

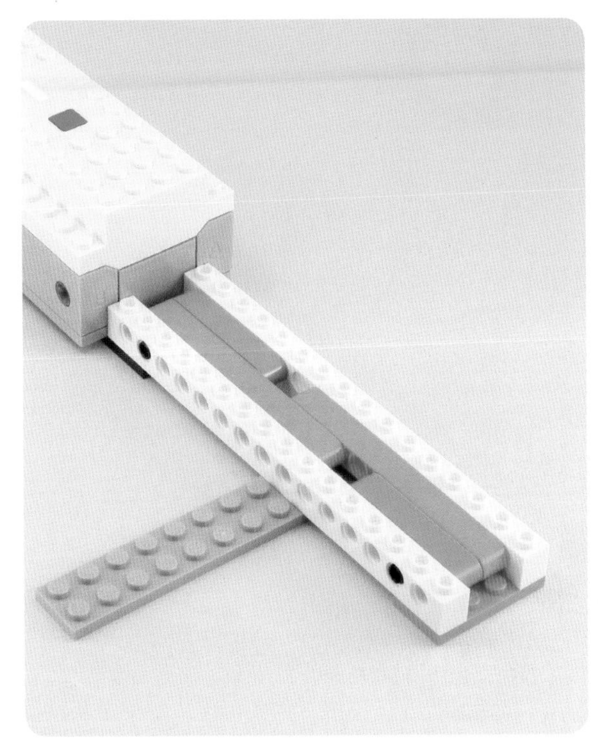

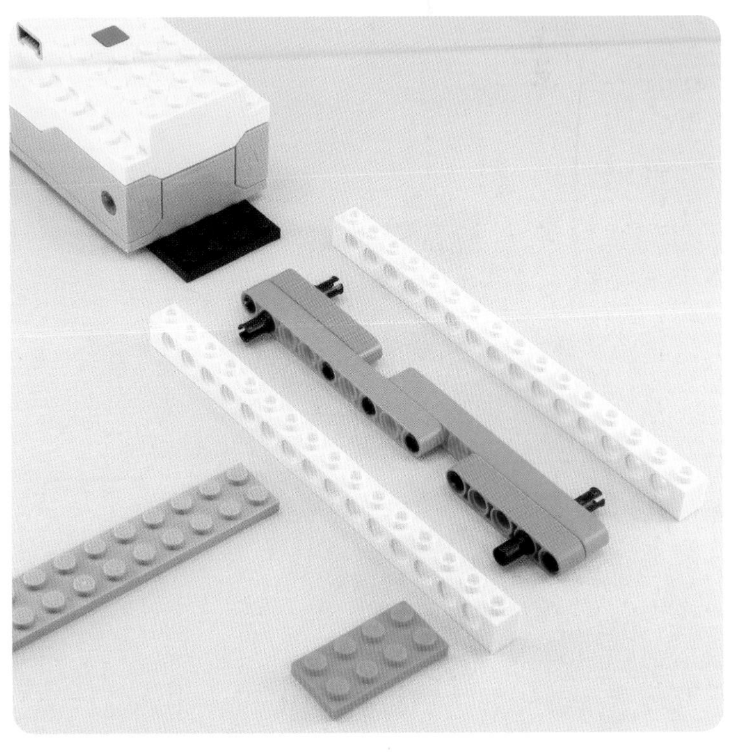

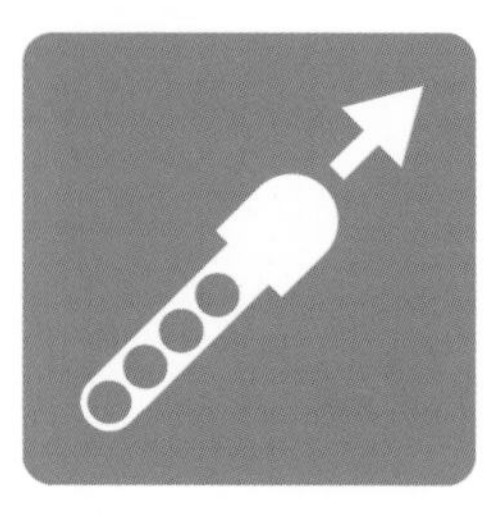

Launching rockets

#76

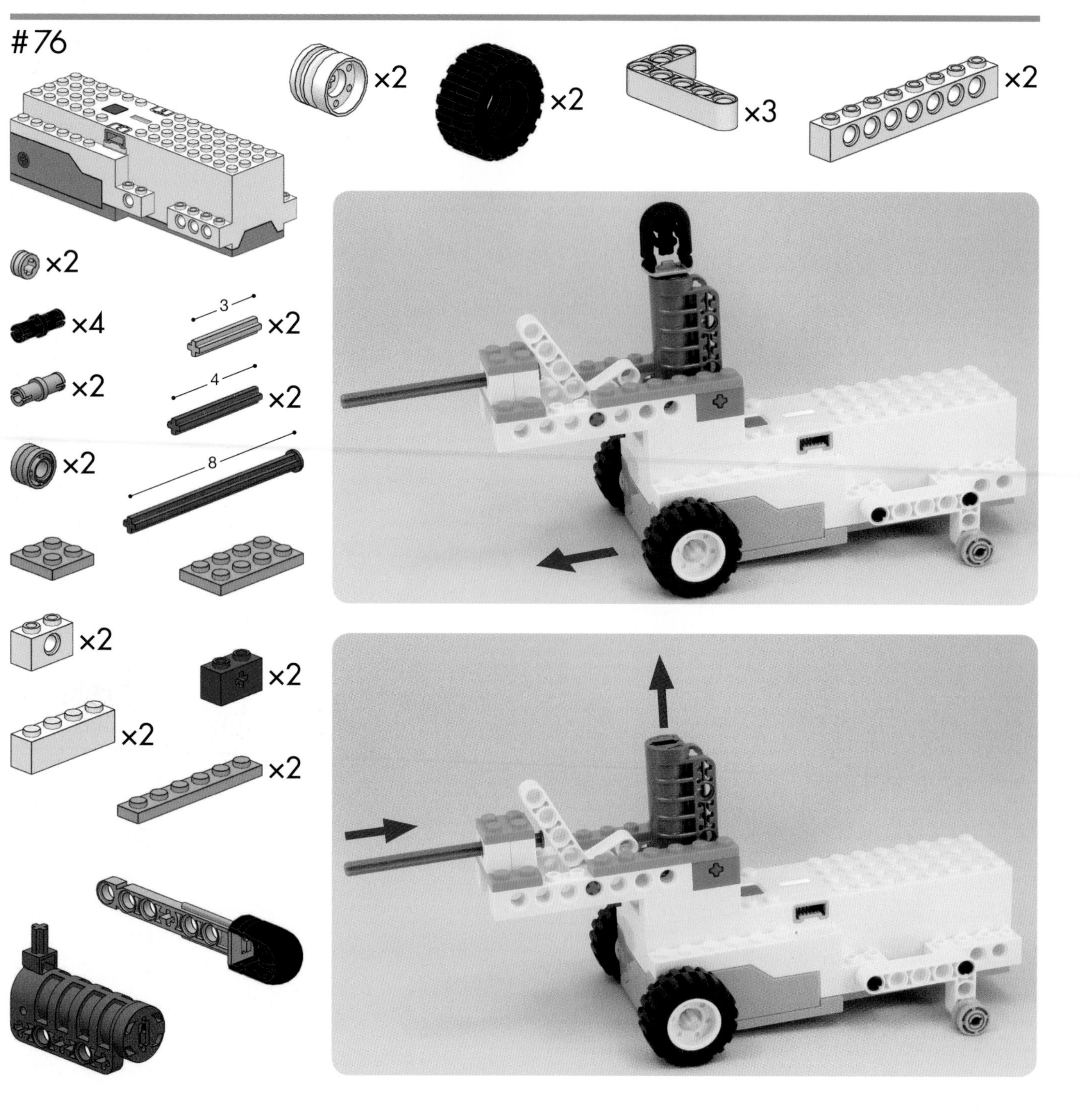

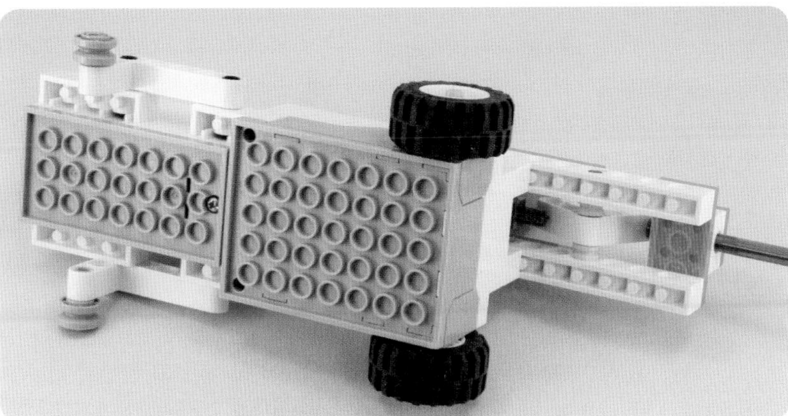

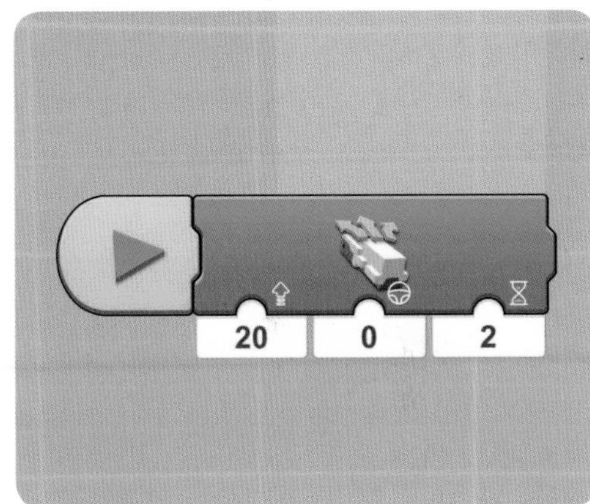
20
0
2

#77

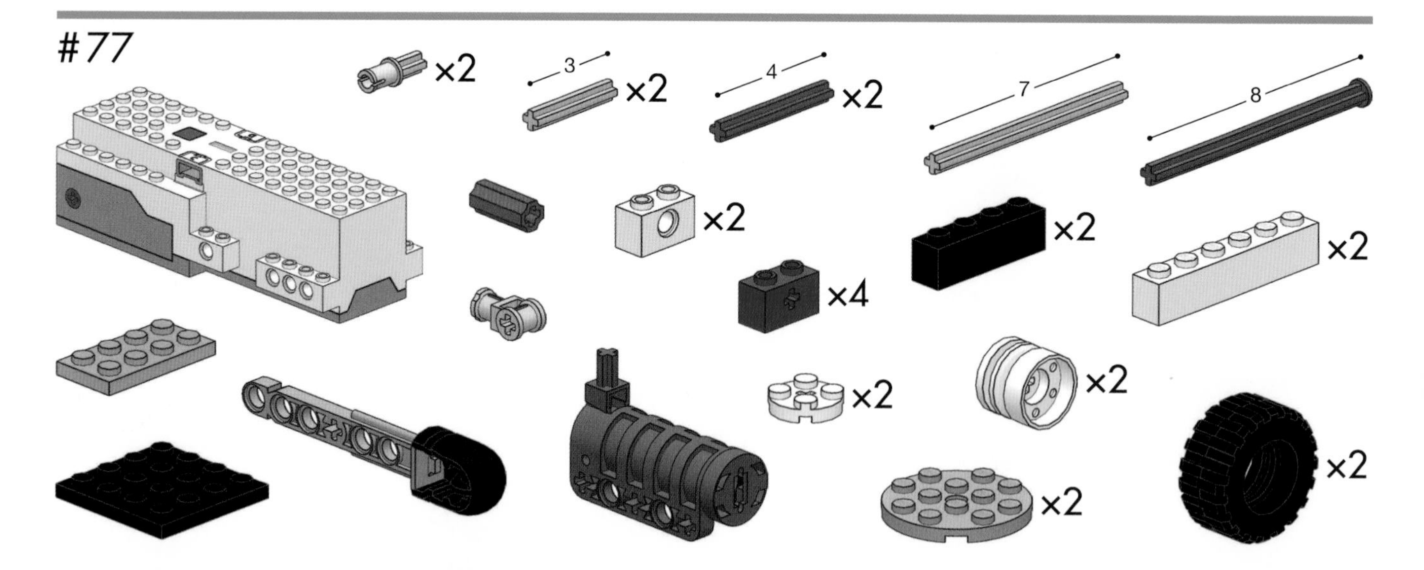

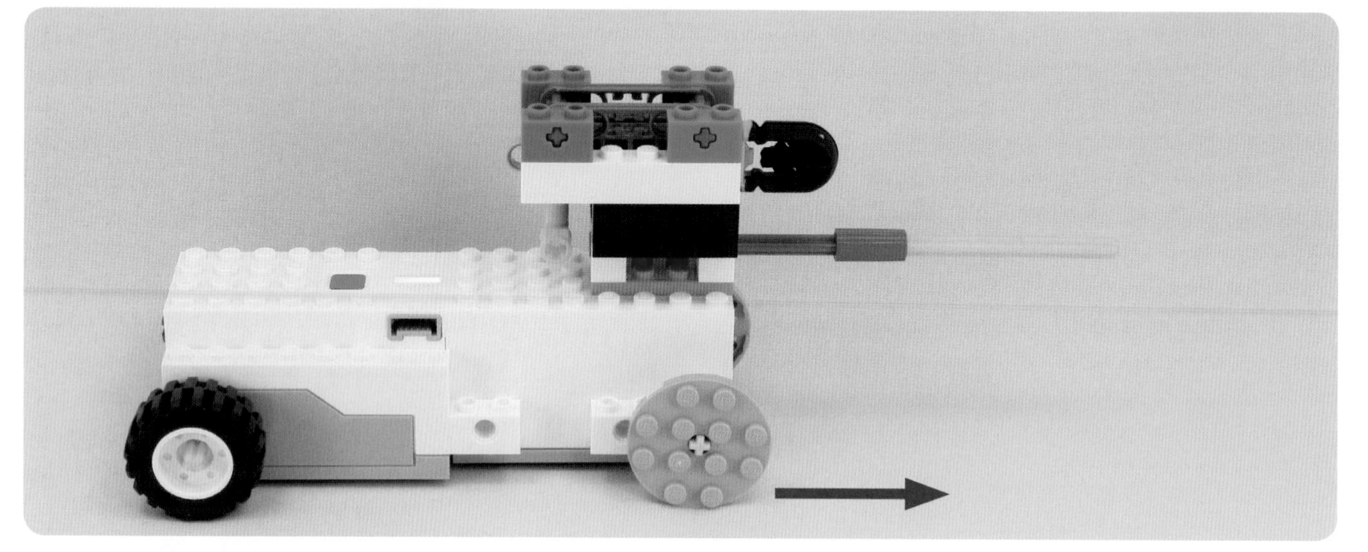

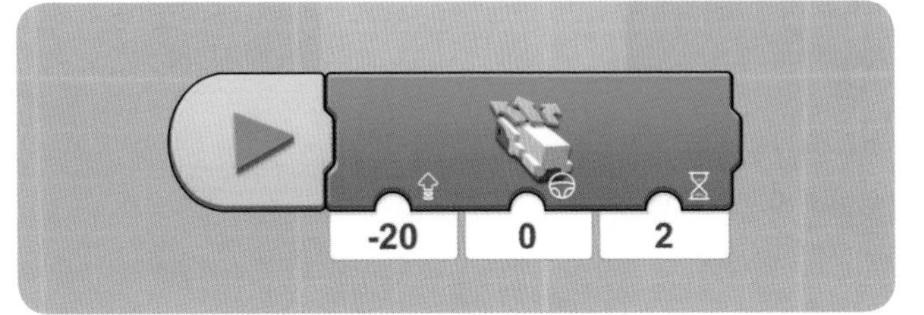
-20
0
2

Drawing with a pen

Use any pen you like.

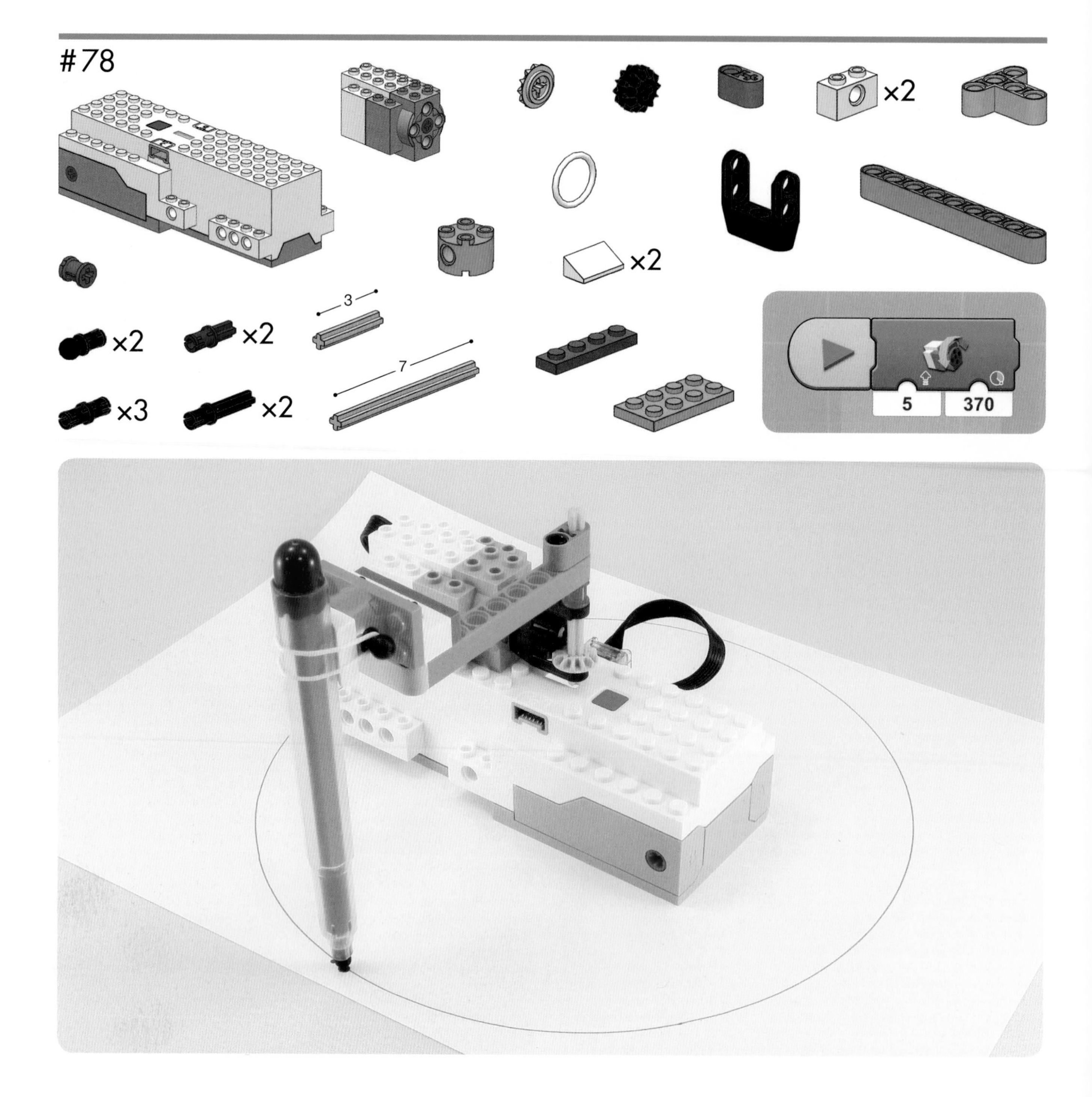

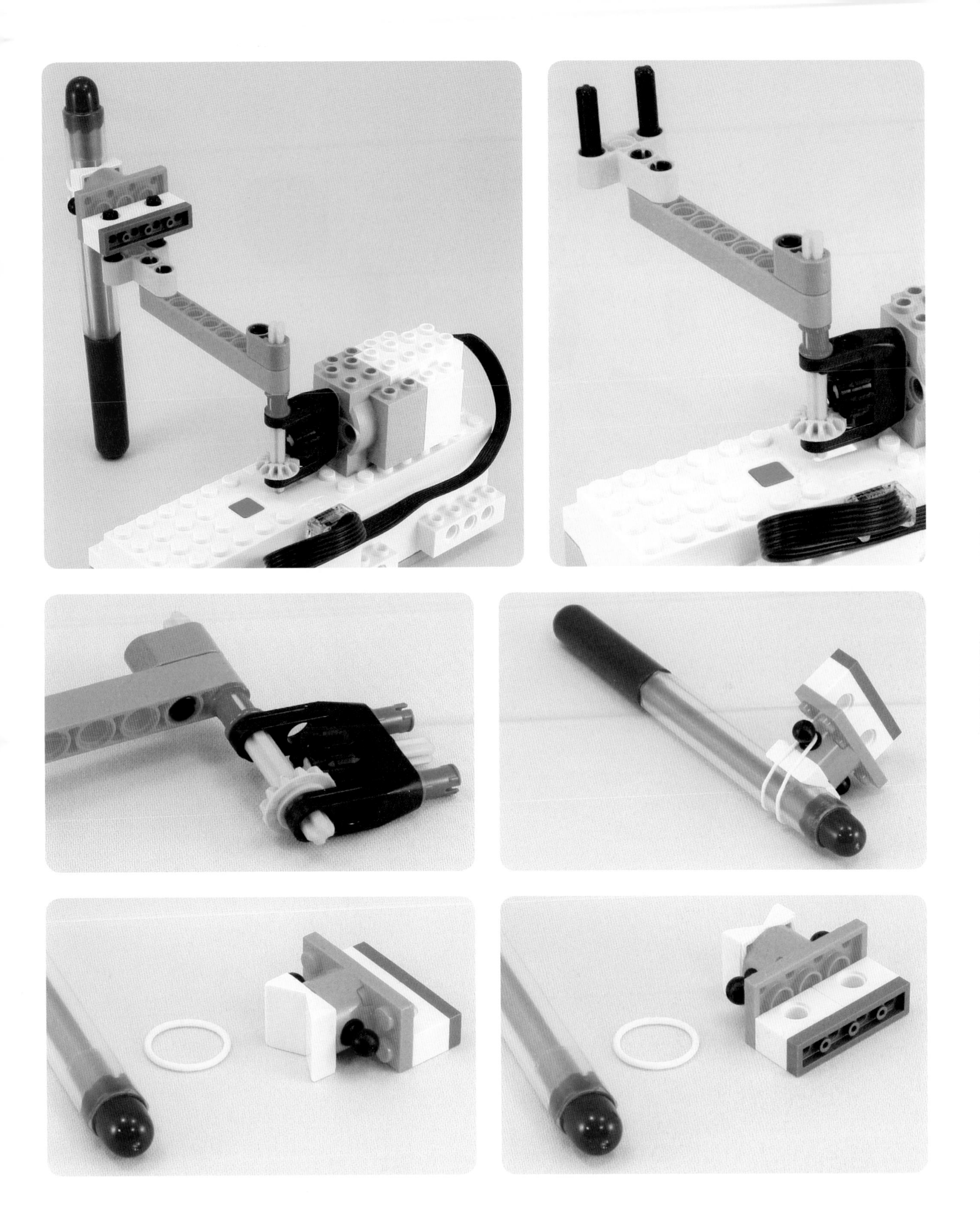

#79

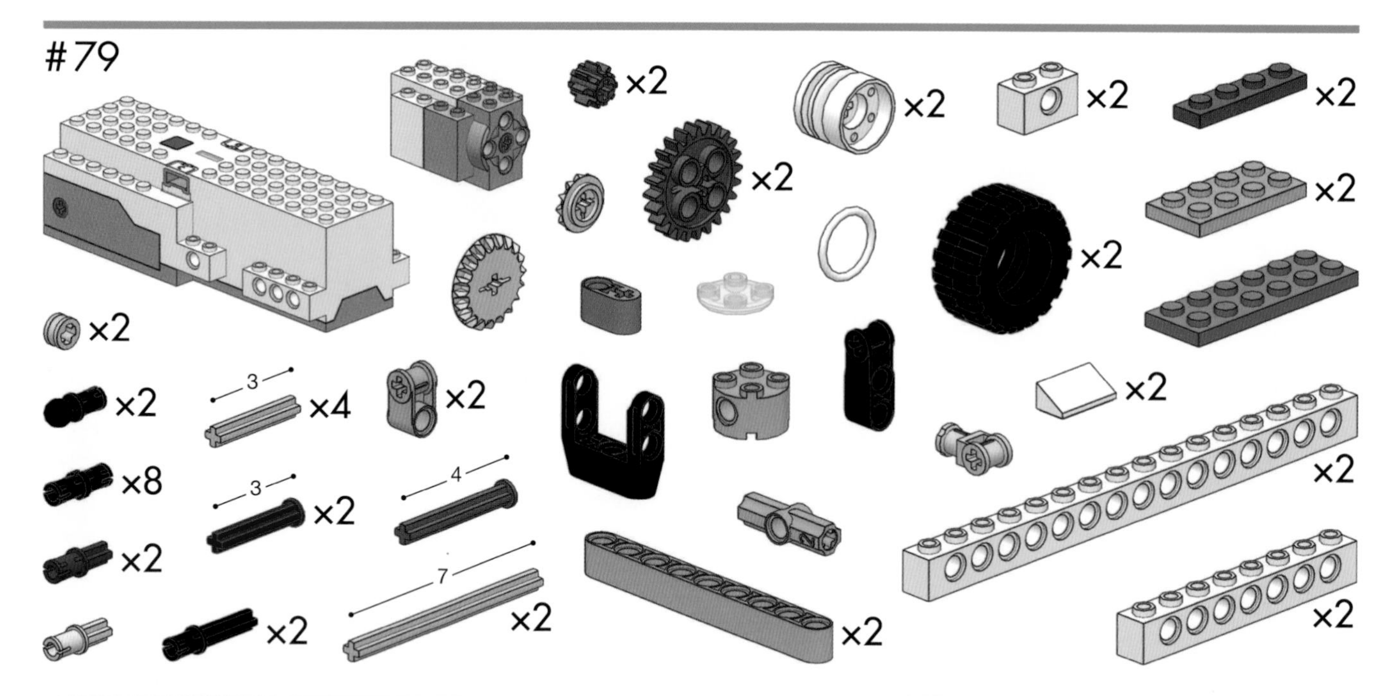

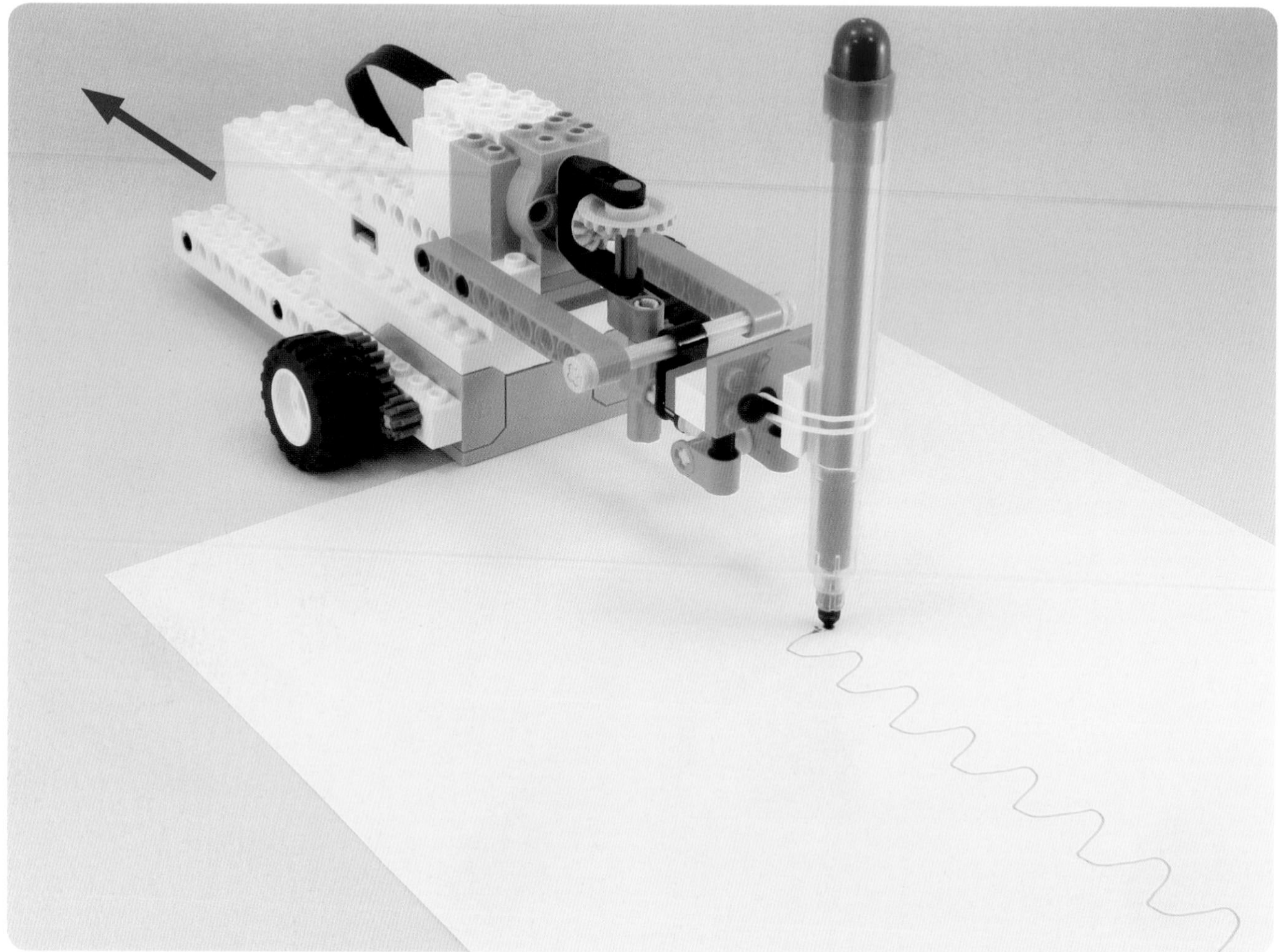

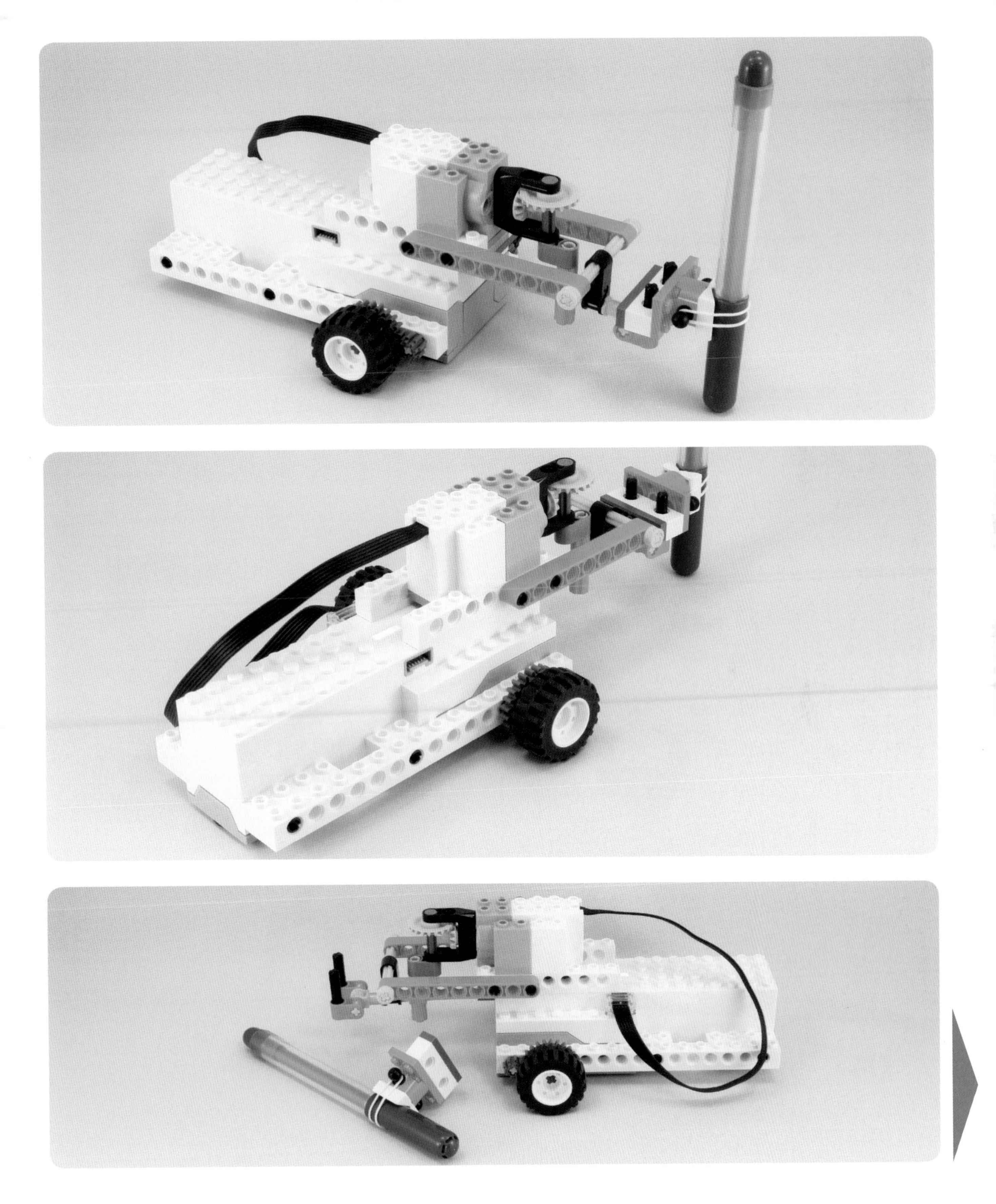

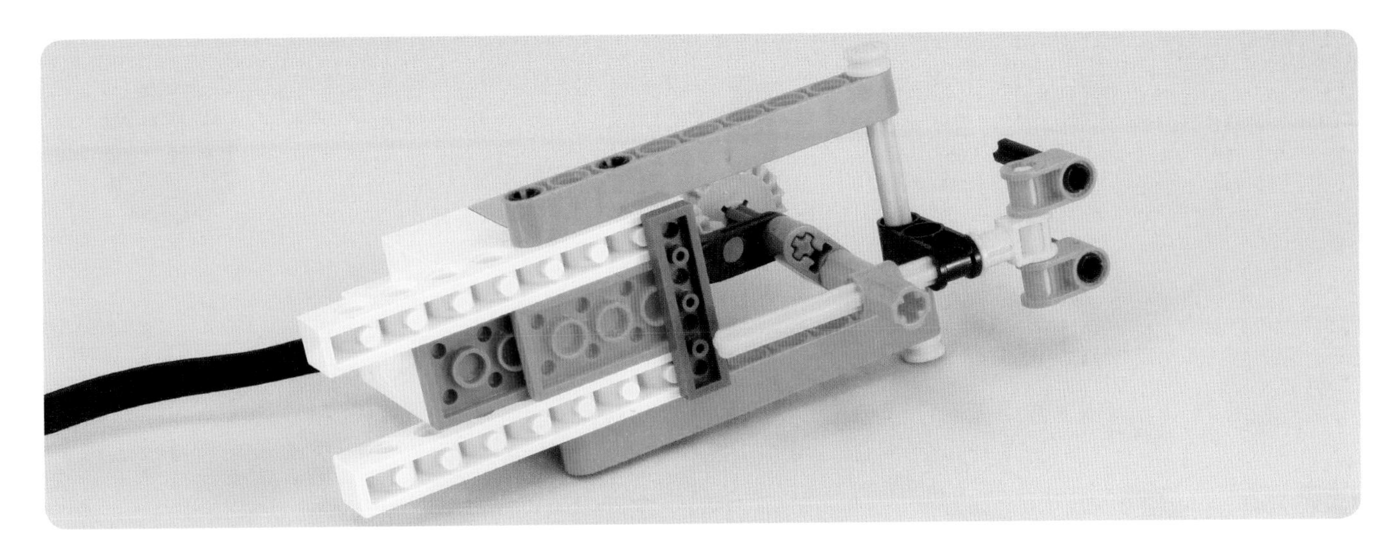

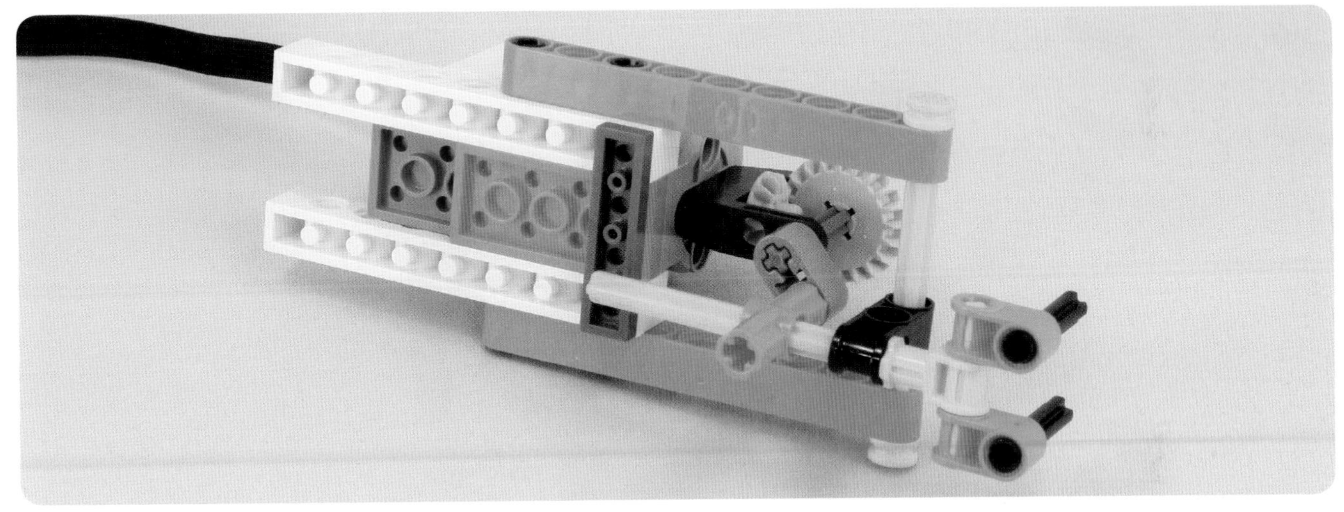

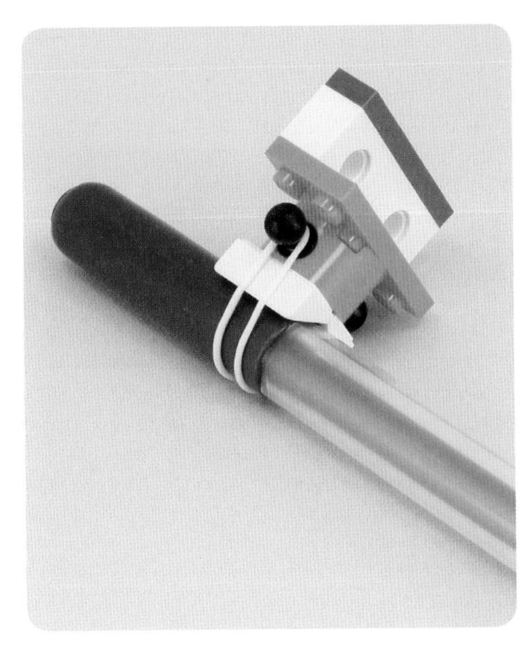

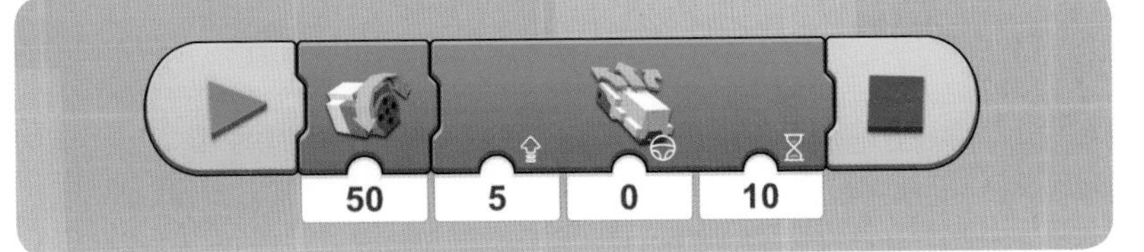
50
5
0
10

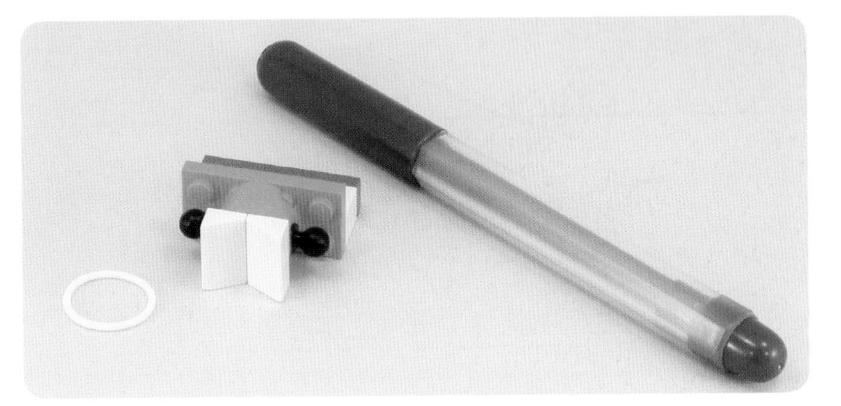

#80

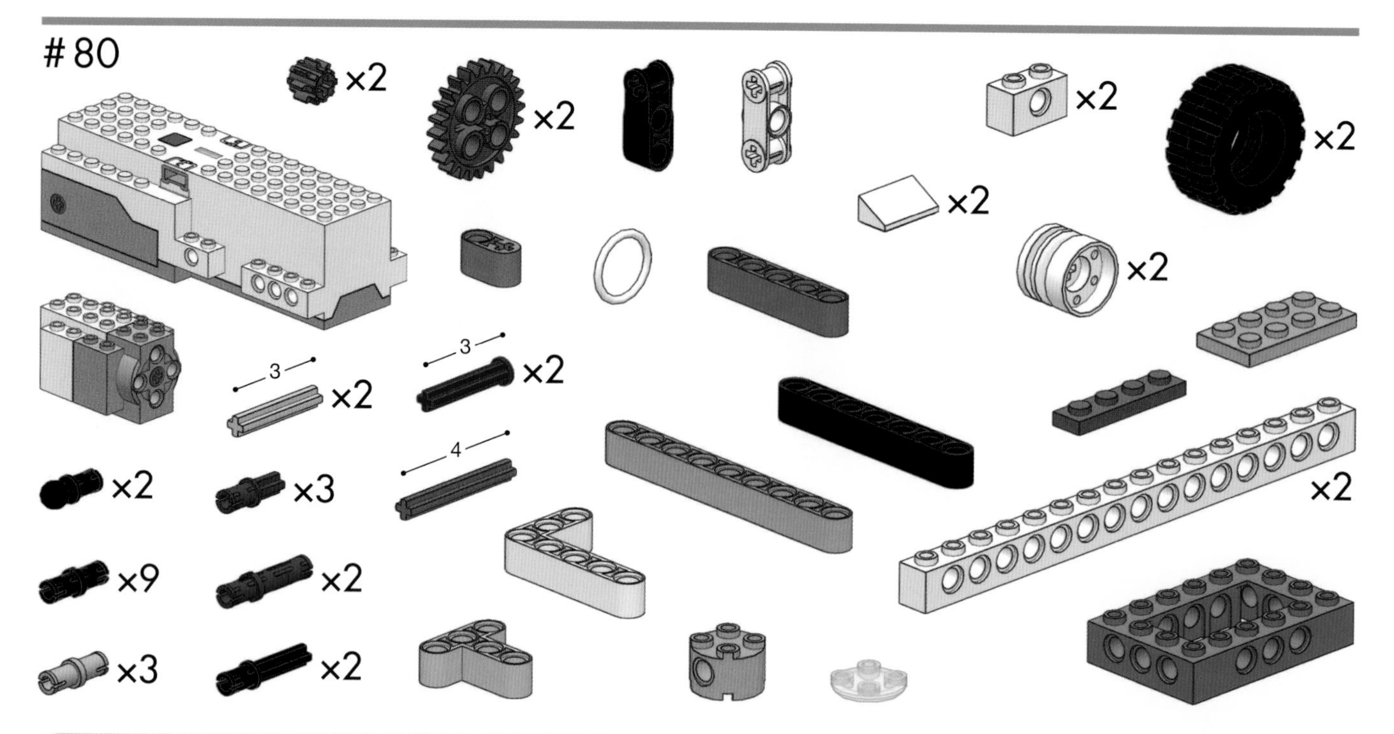

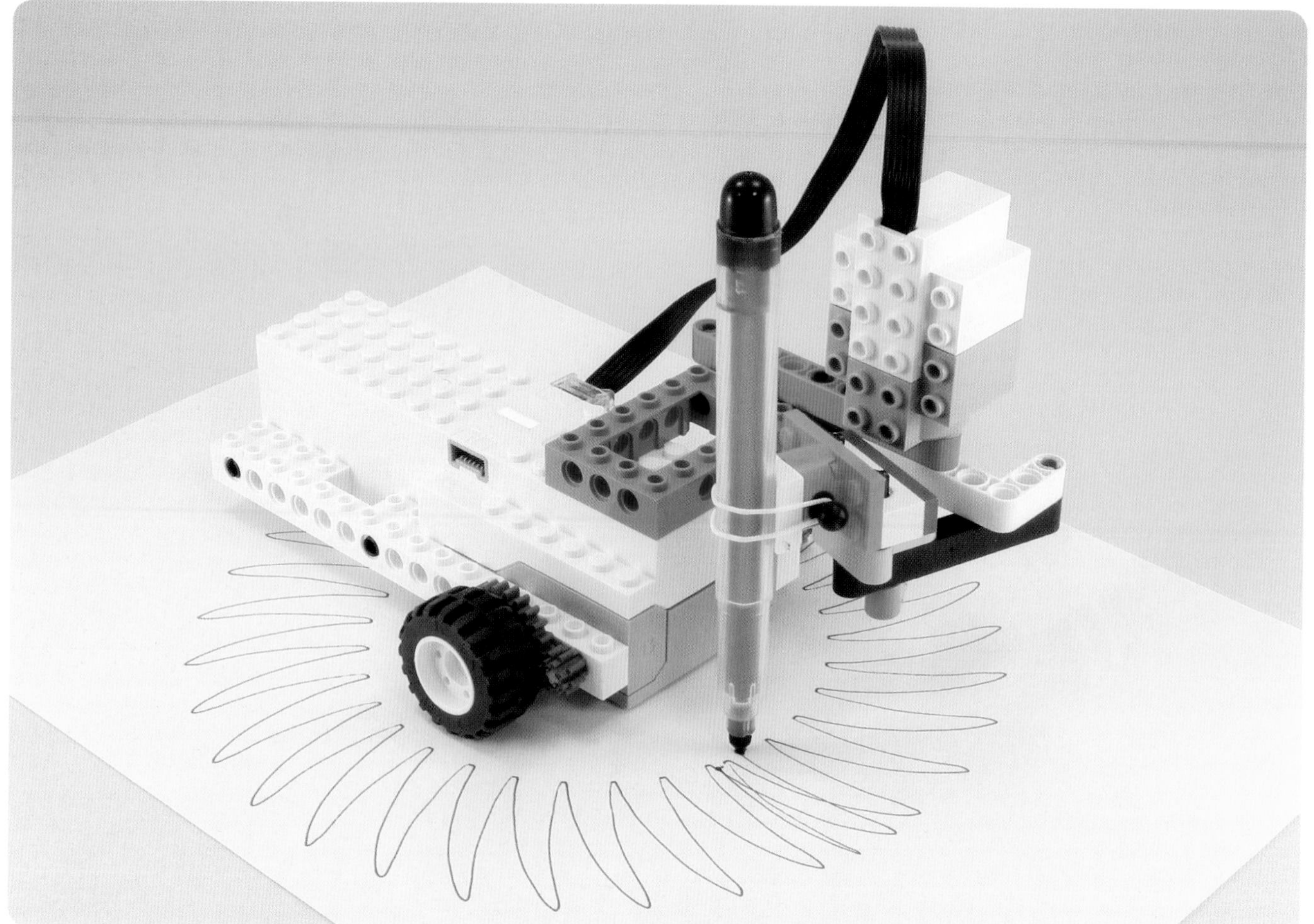

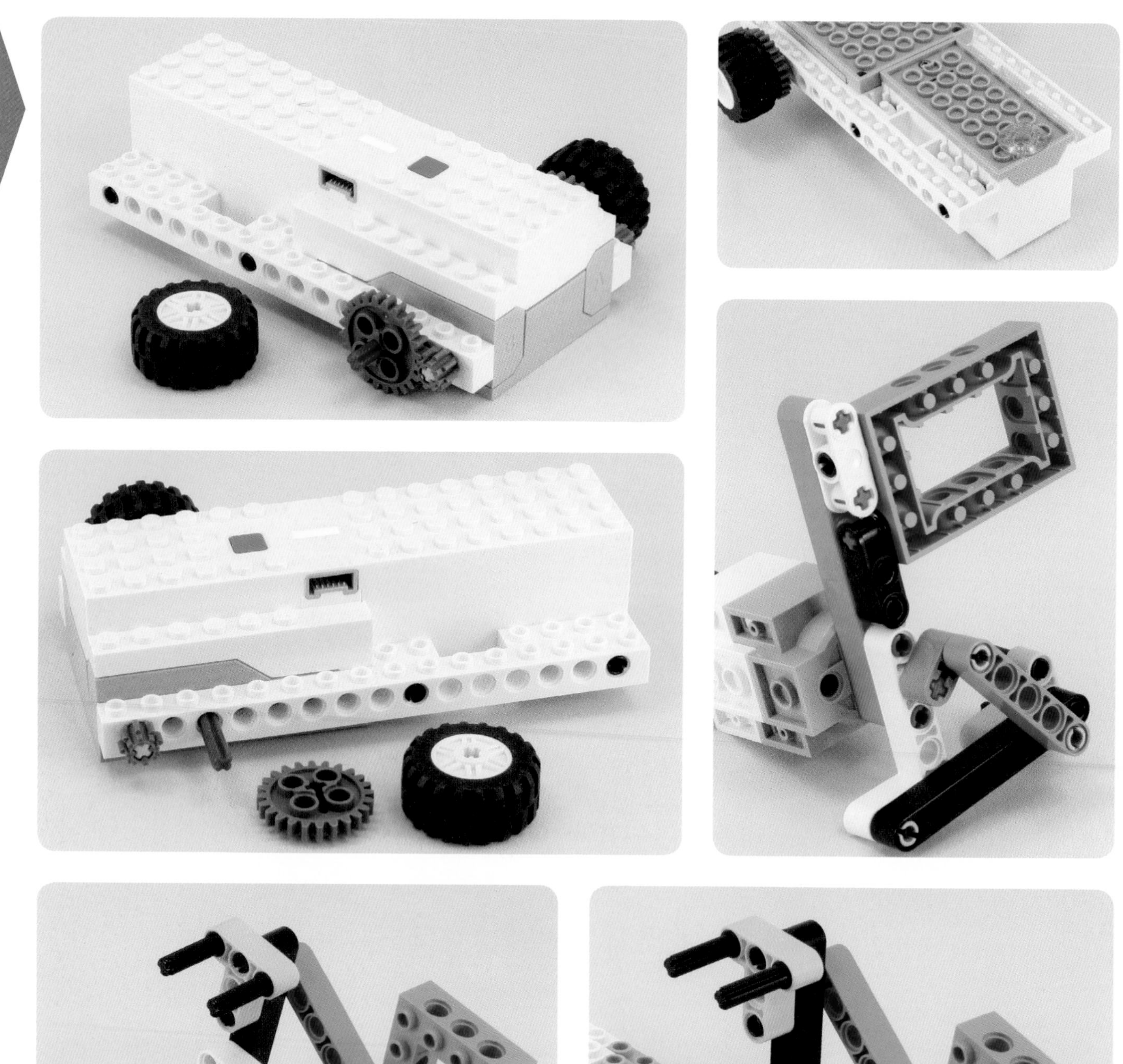

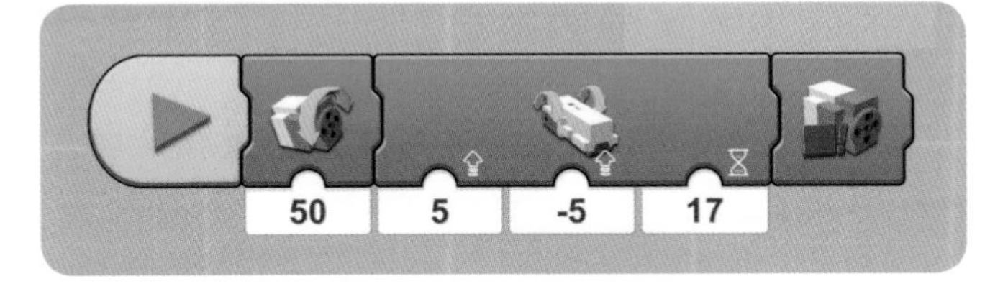
50
5
-5
17

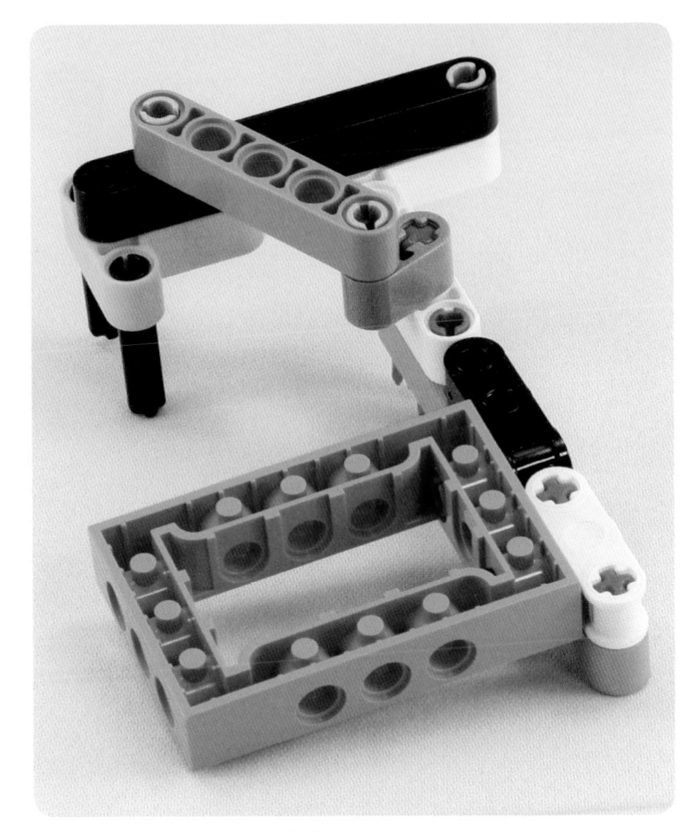

20
5
-5
17

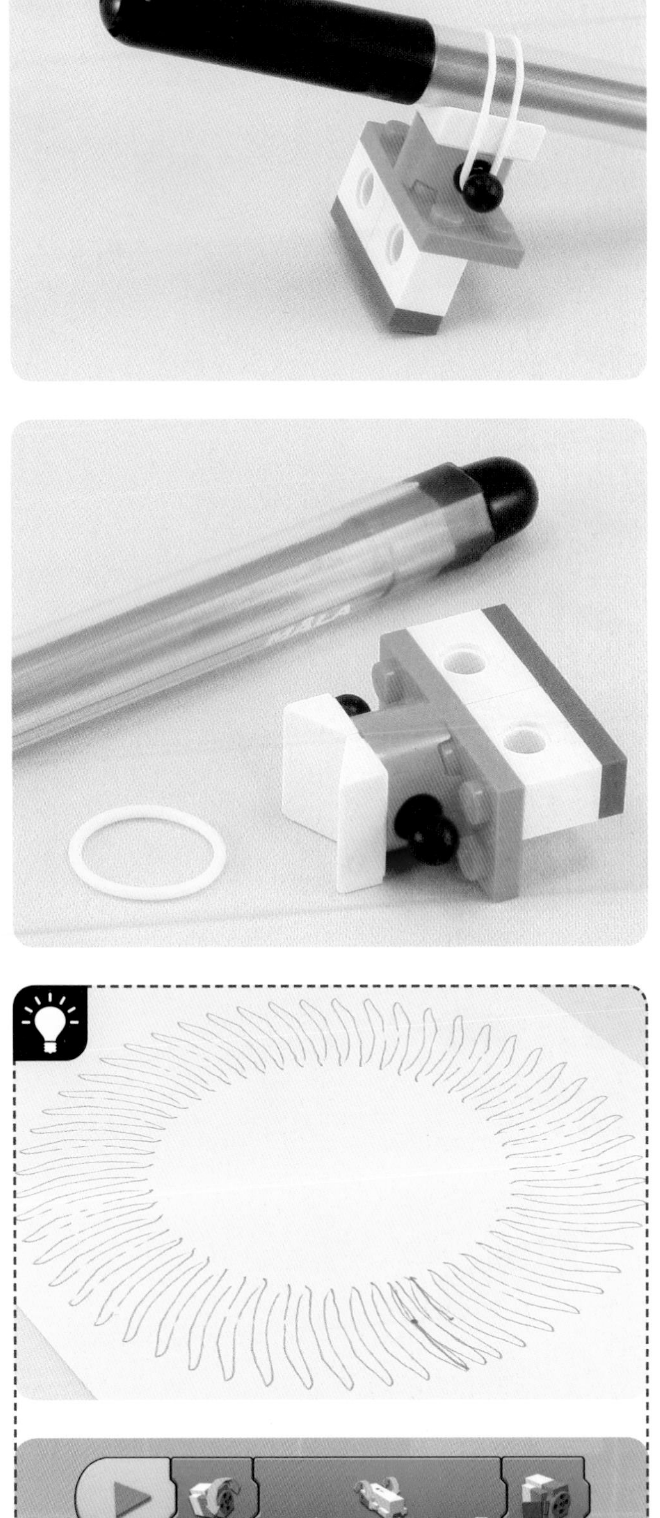
90
5
-5
17

81

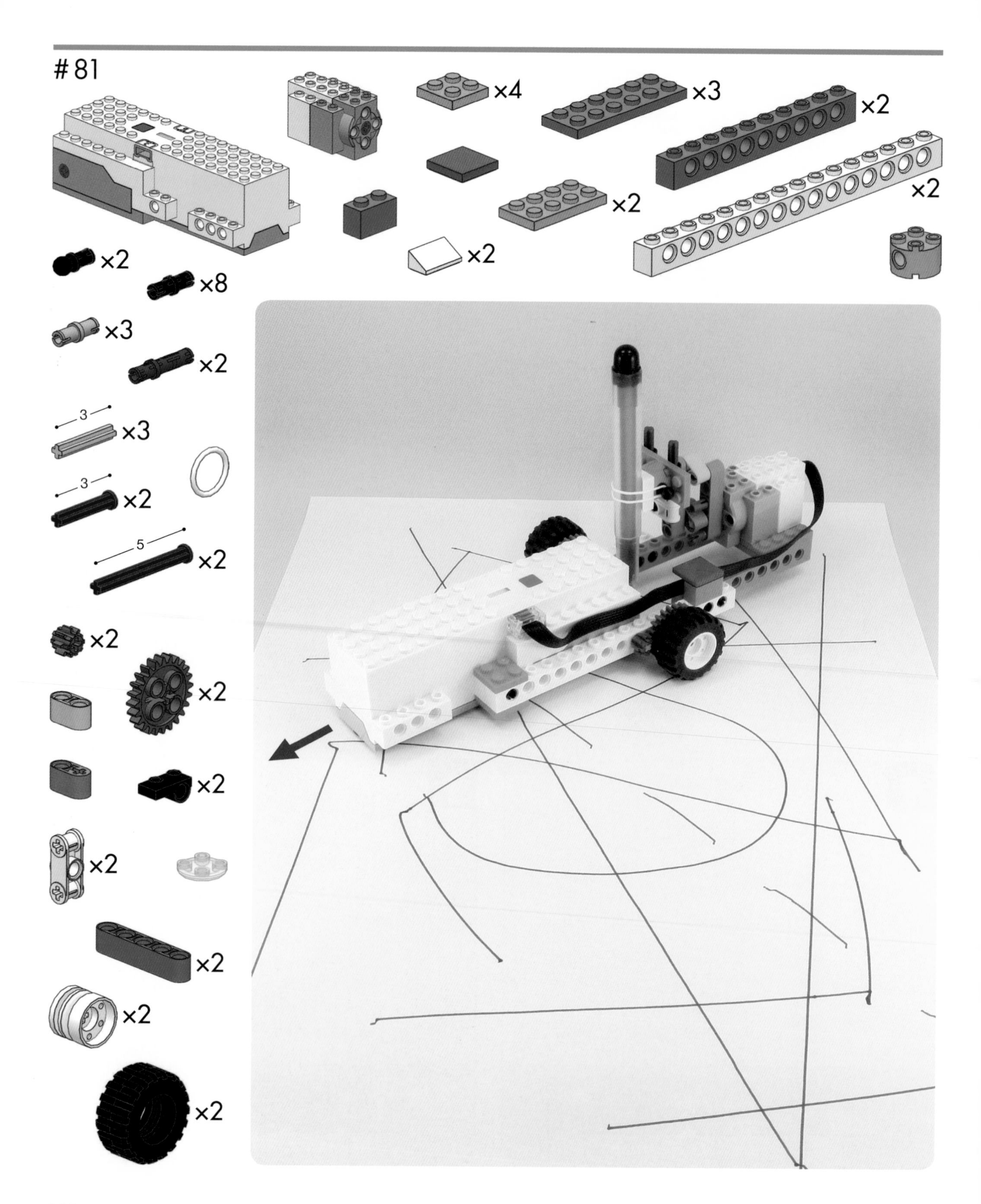

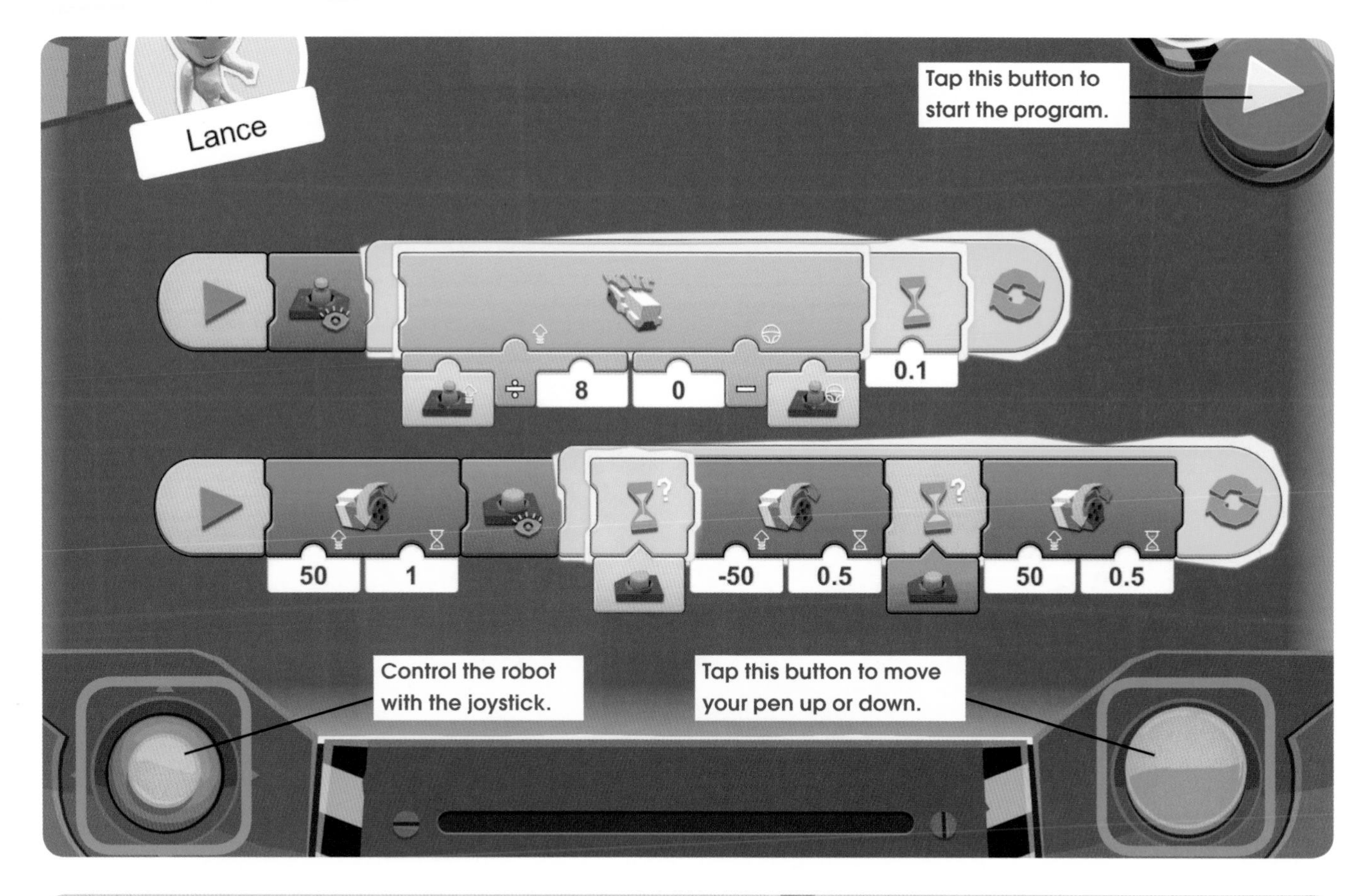
Lance
Tap this button to start the program.
8
0
0.1
50
1
-50
0.5
50
0.5
Control the robot with the joystick.
Tap this button to move your pen up or down.

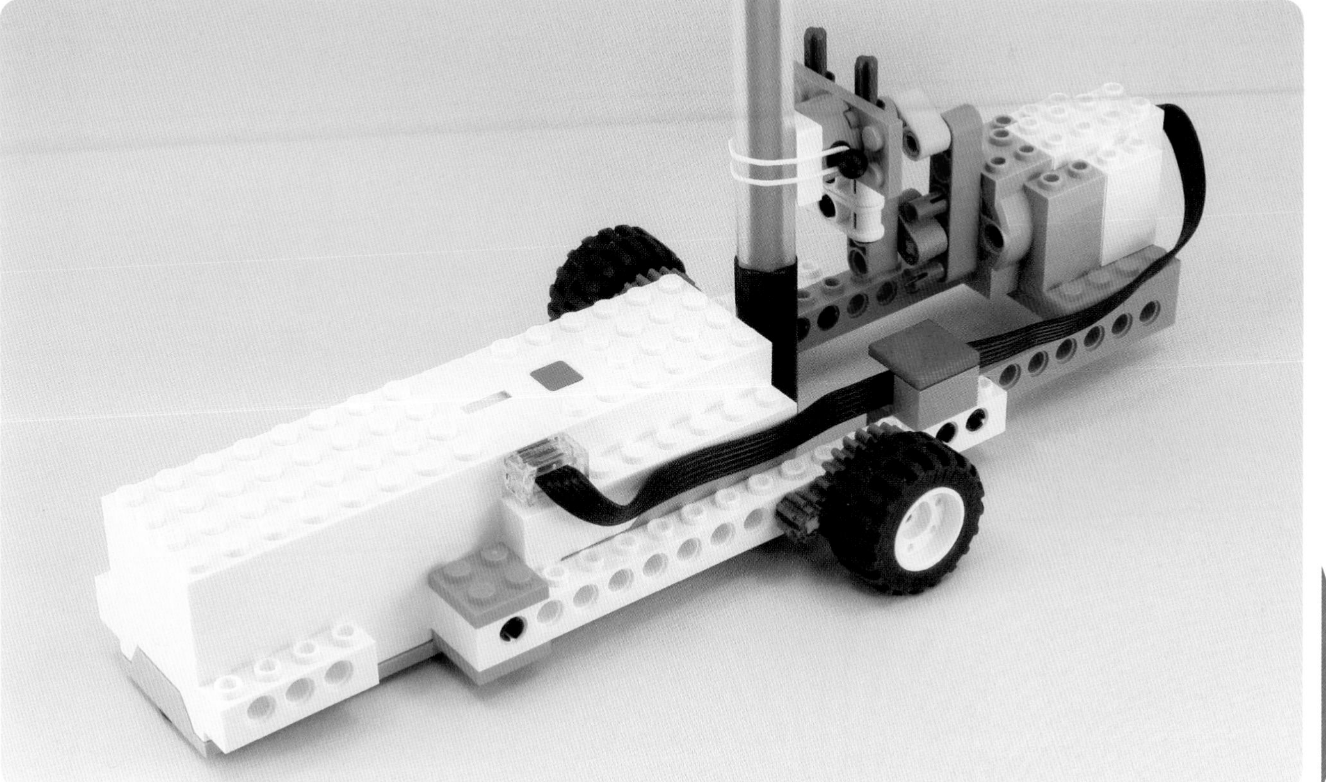

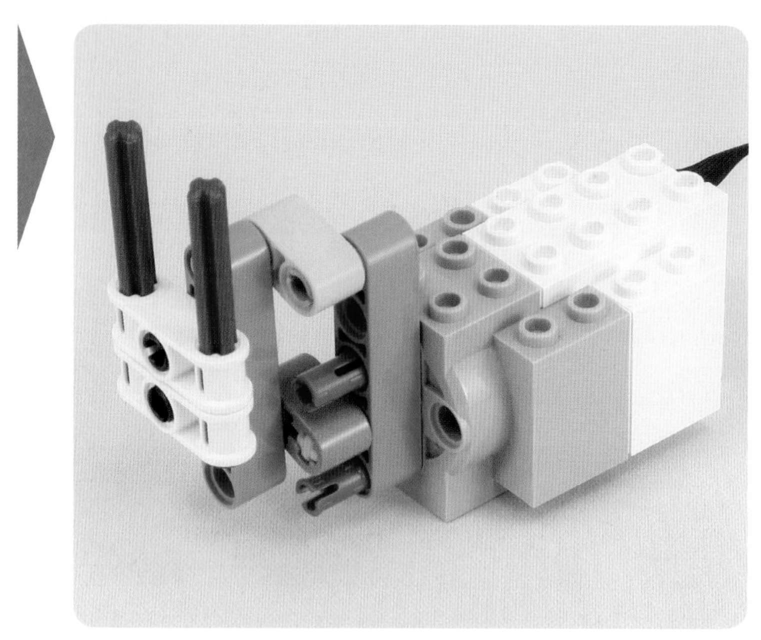

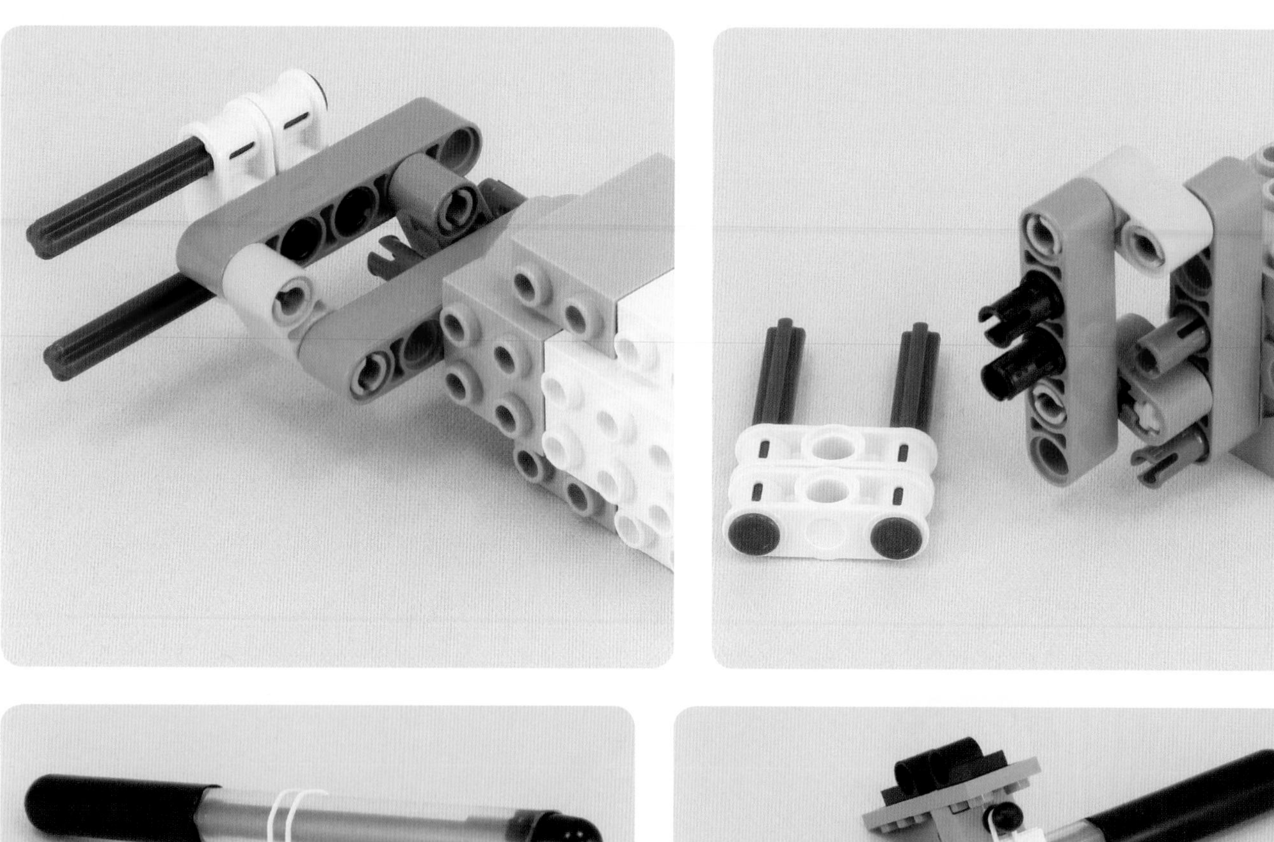

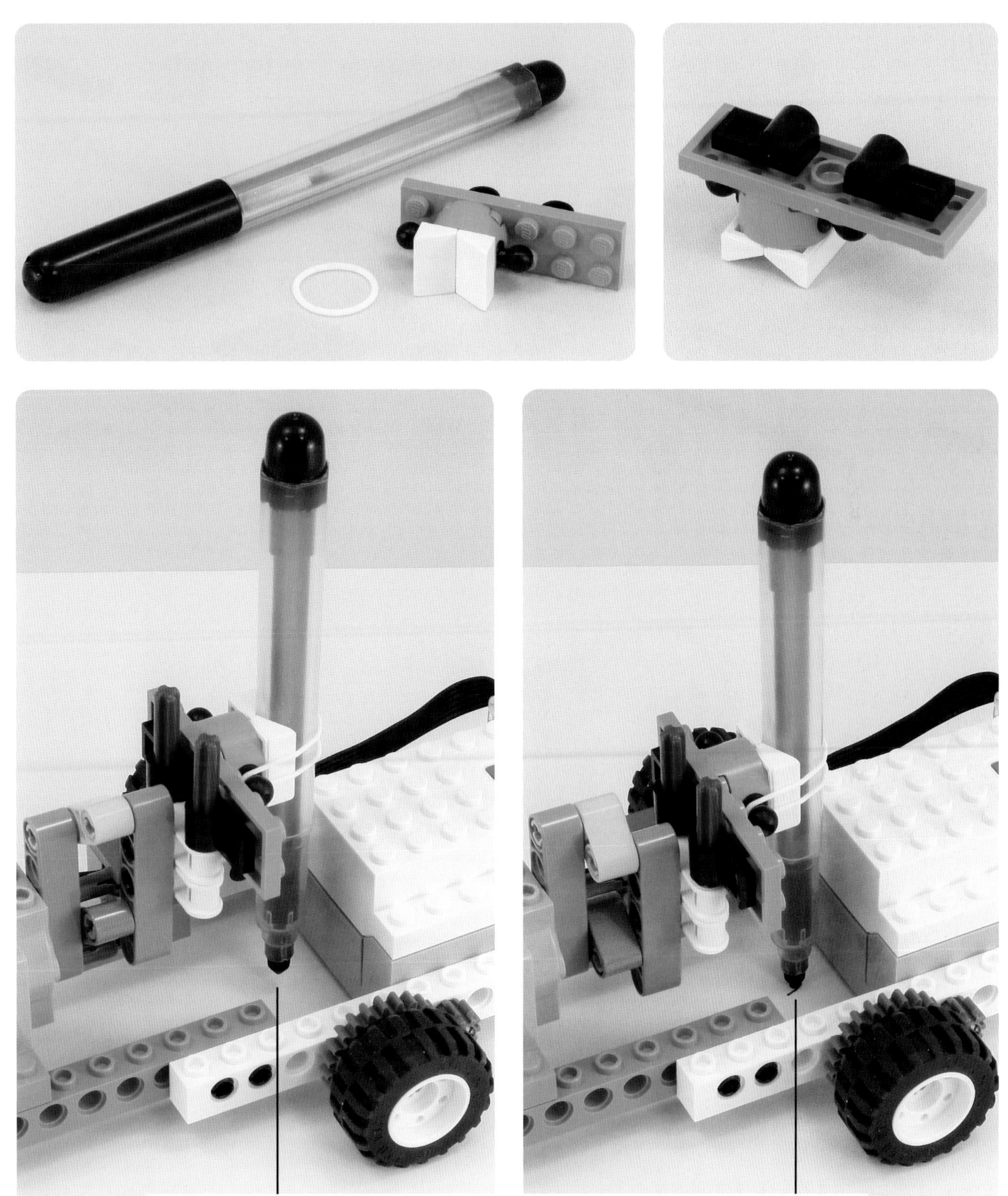
Adjust the position of the pen so that the tip doesn't touch the surface of the paper when the pen holder is up and does touch the surface of the paper when the holder is down.

#82

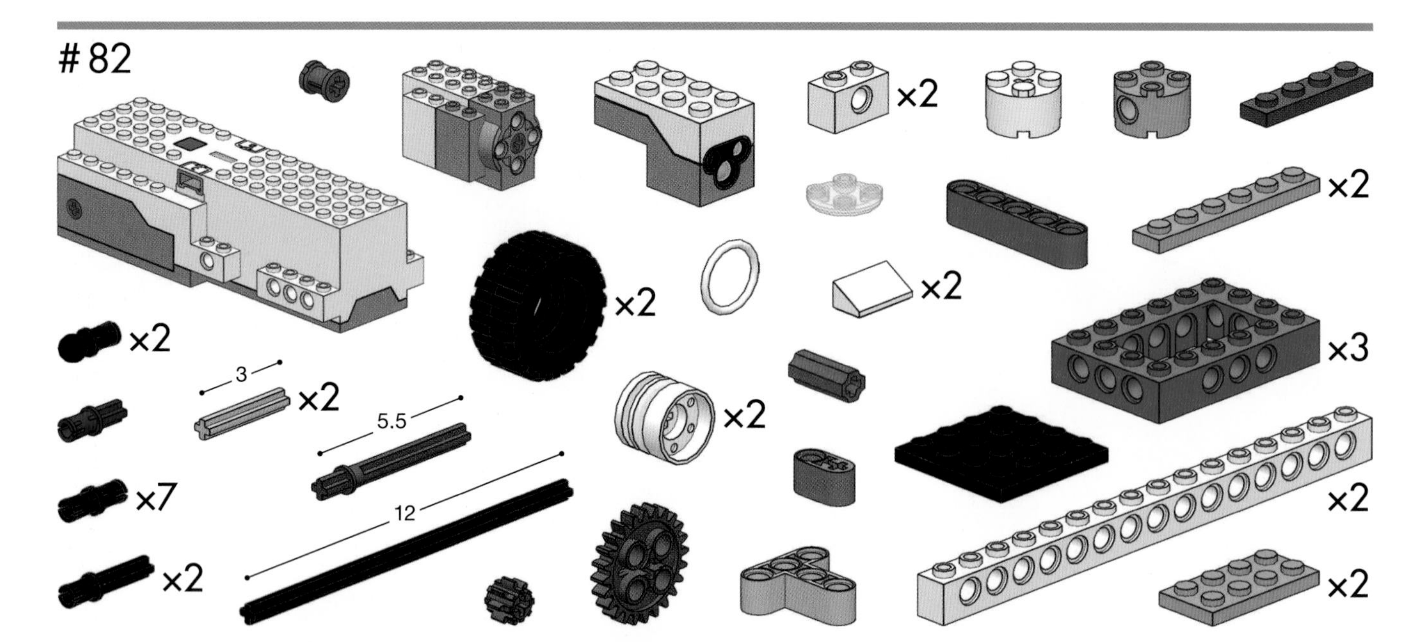
×2
×2
×2
×2
×2
3
×2
5.5
×2
×7
12
×2
×3
×2
×2
×2

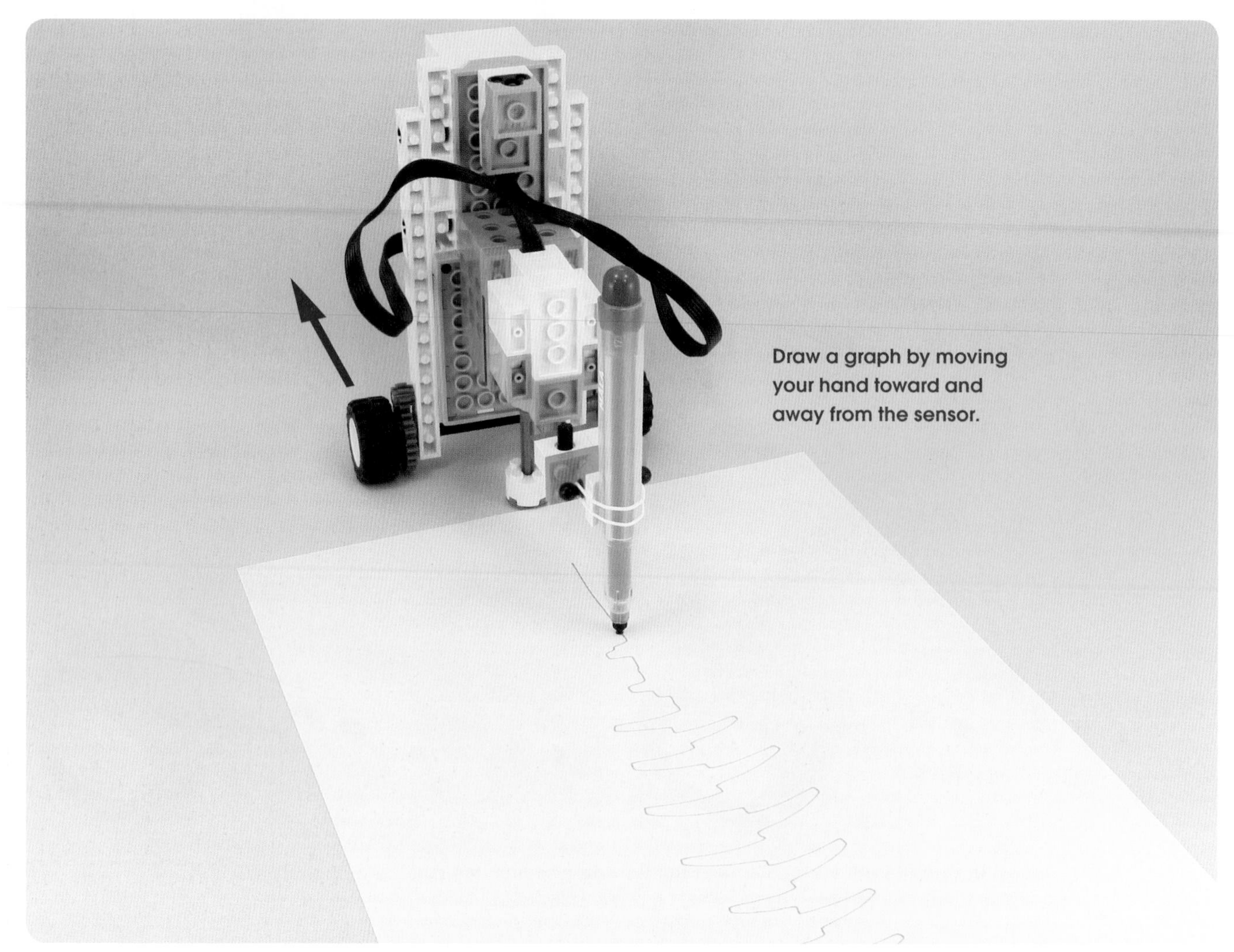
Draw a graph by moving your hand toward and away from the sensor.

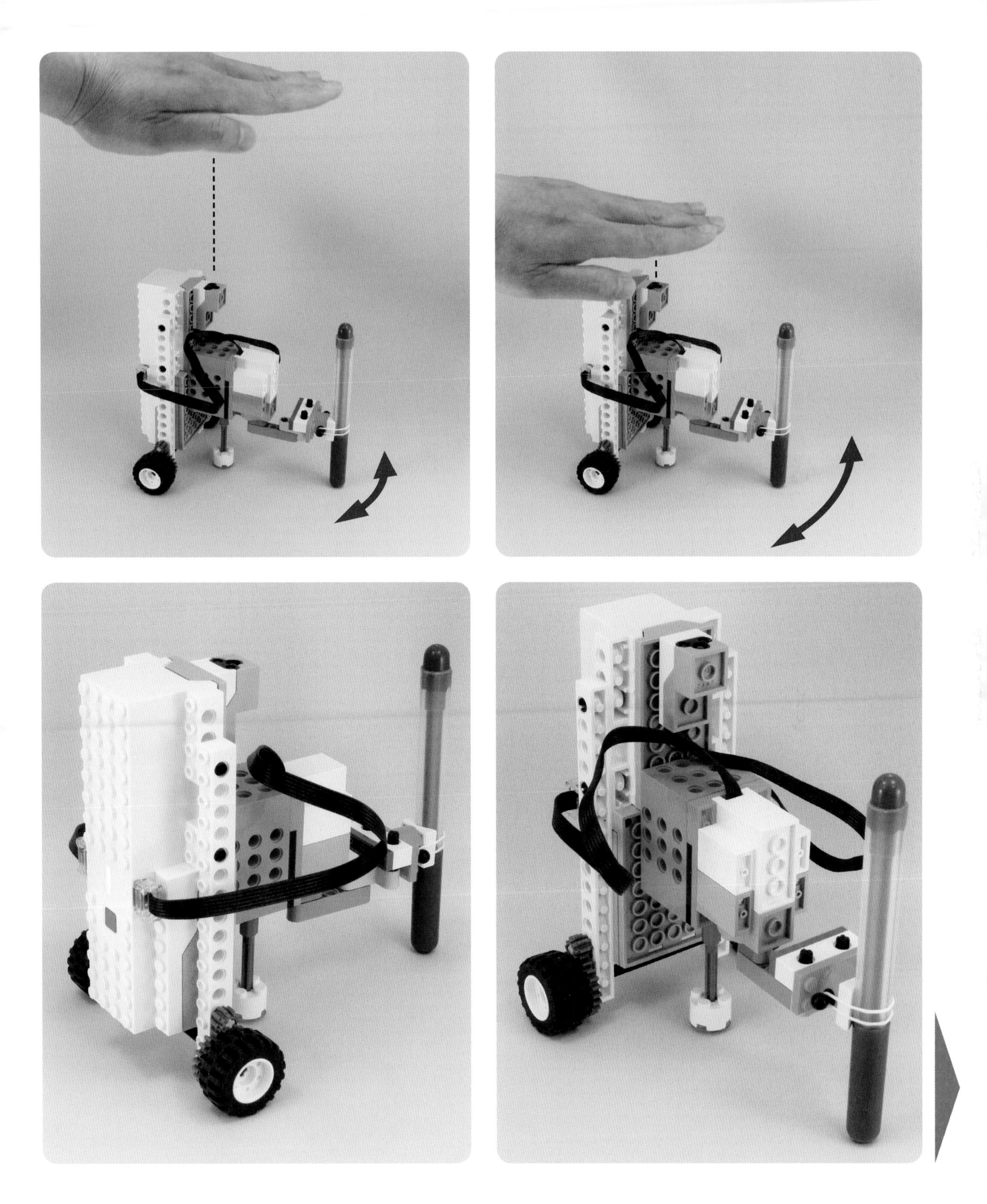

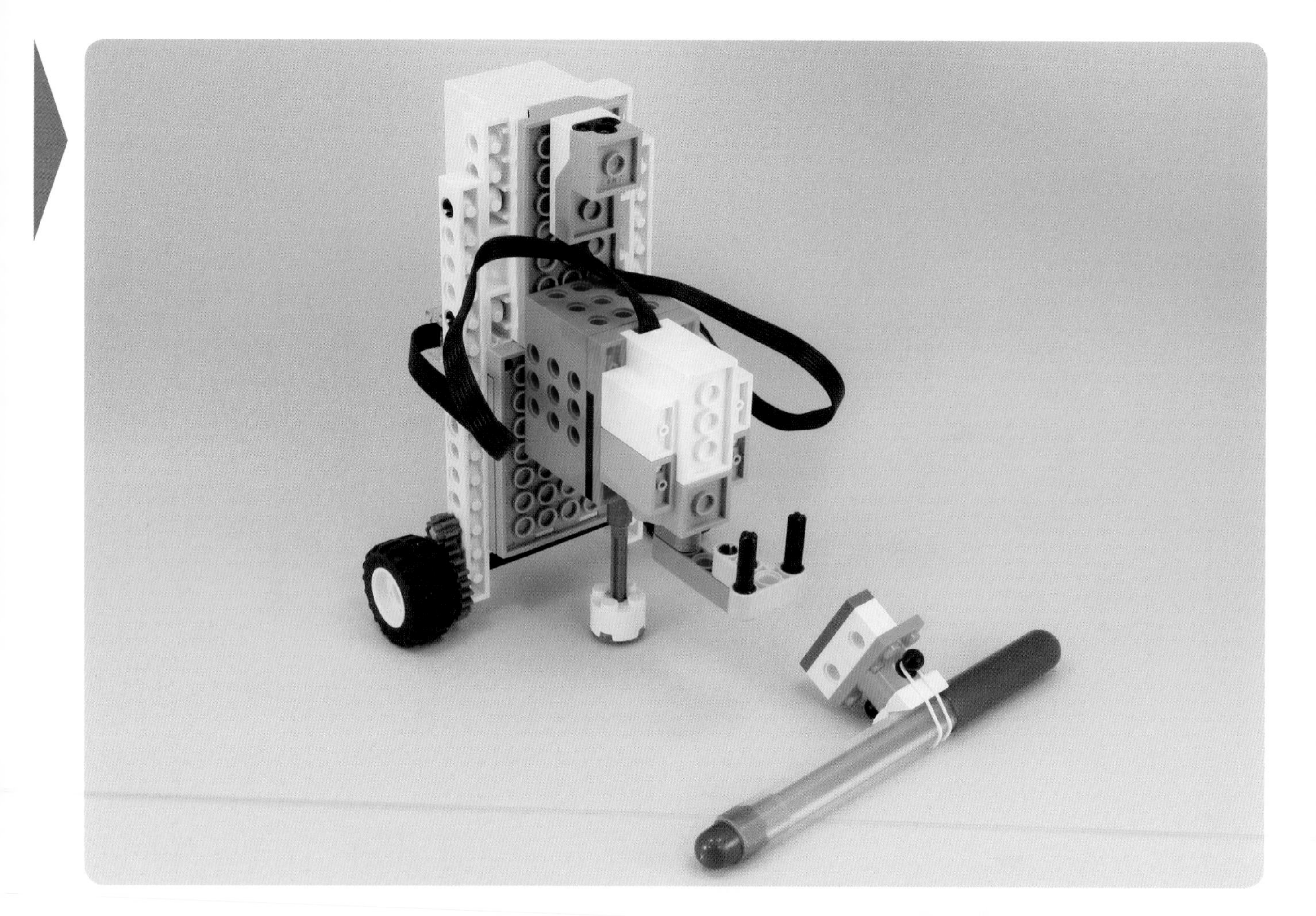

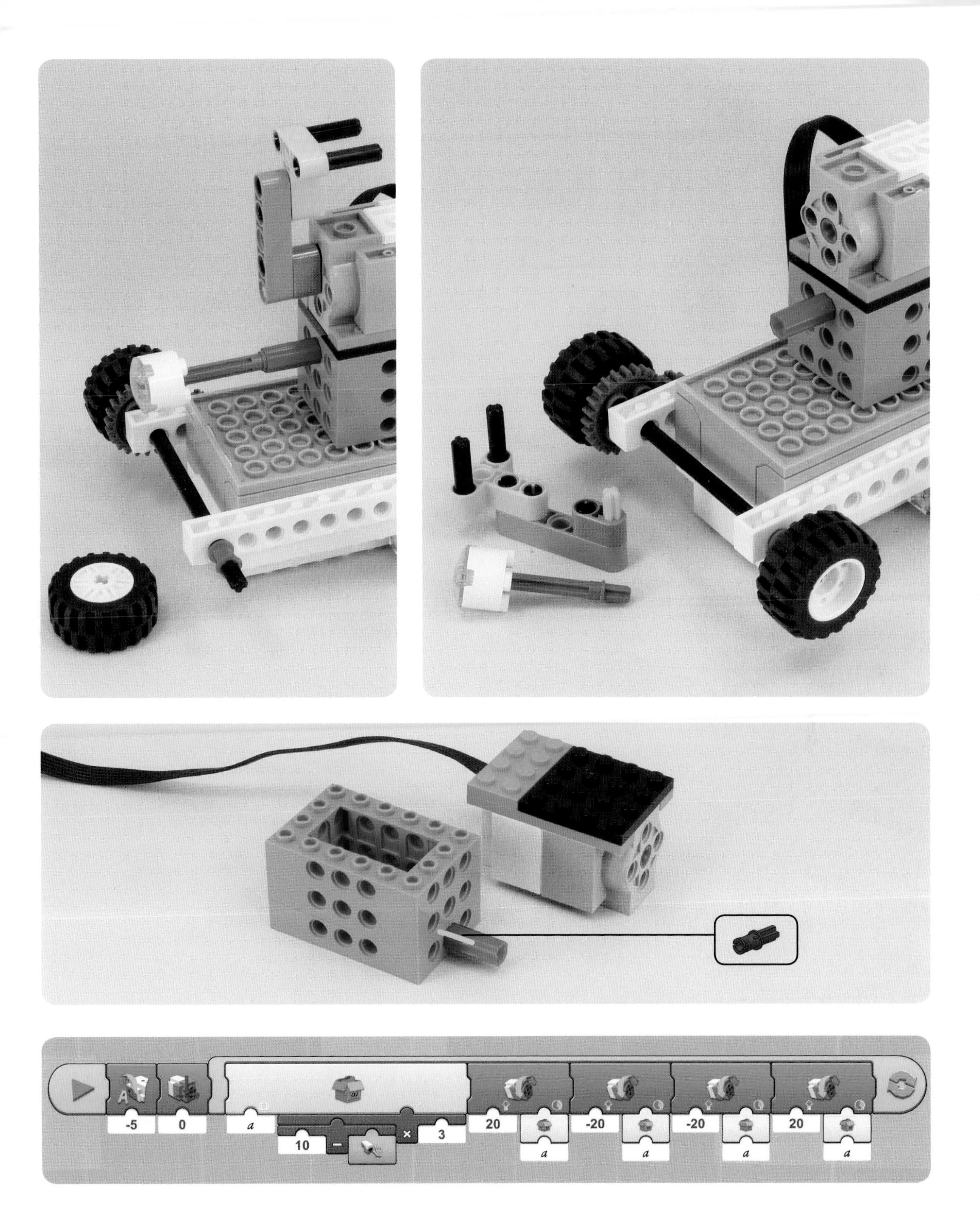

-5
0
a
10
–
×
3
20
a
-20
a
-20
a
20
a

Using turntables

#83

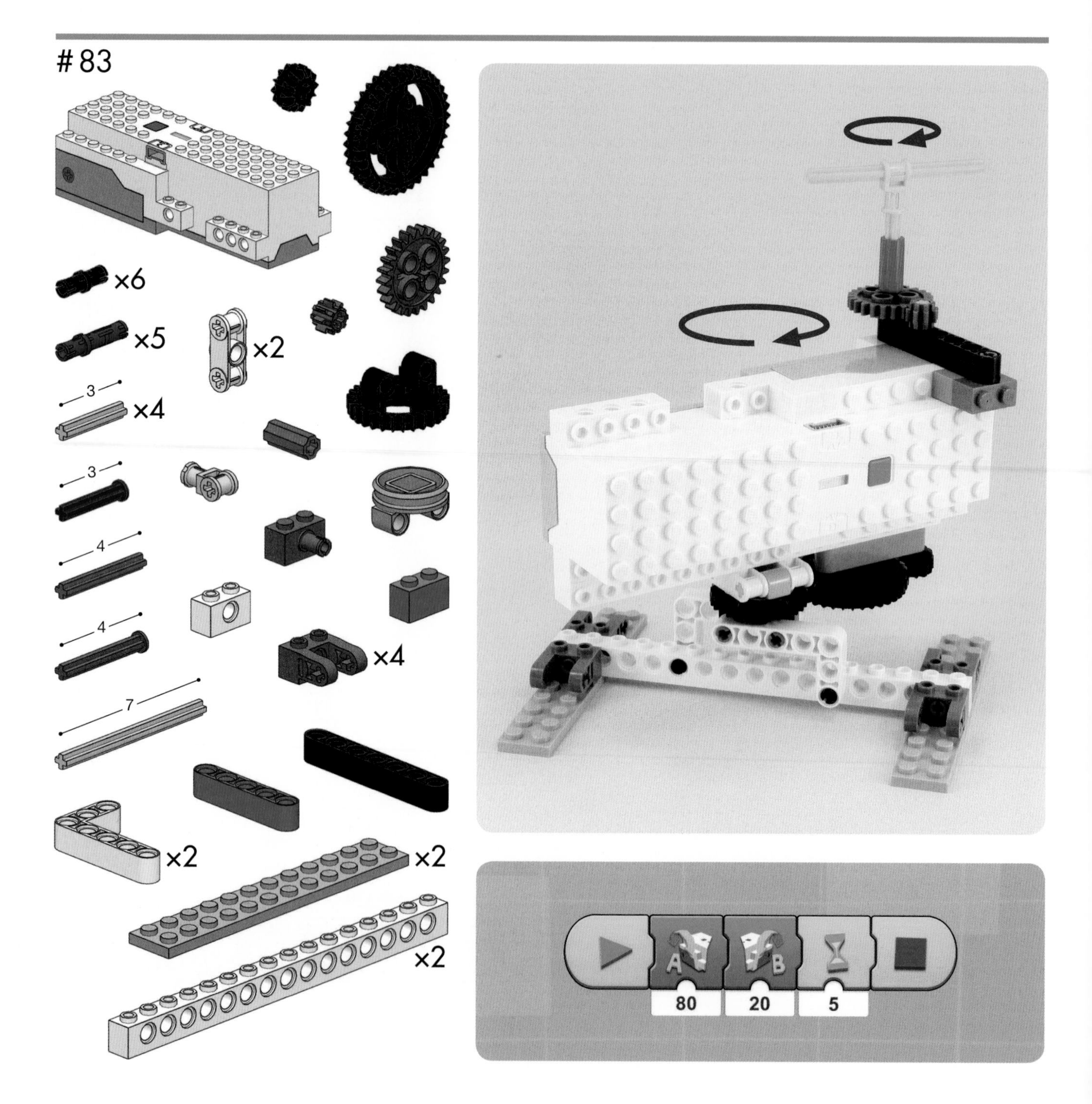

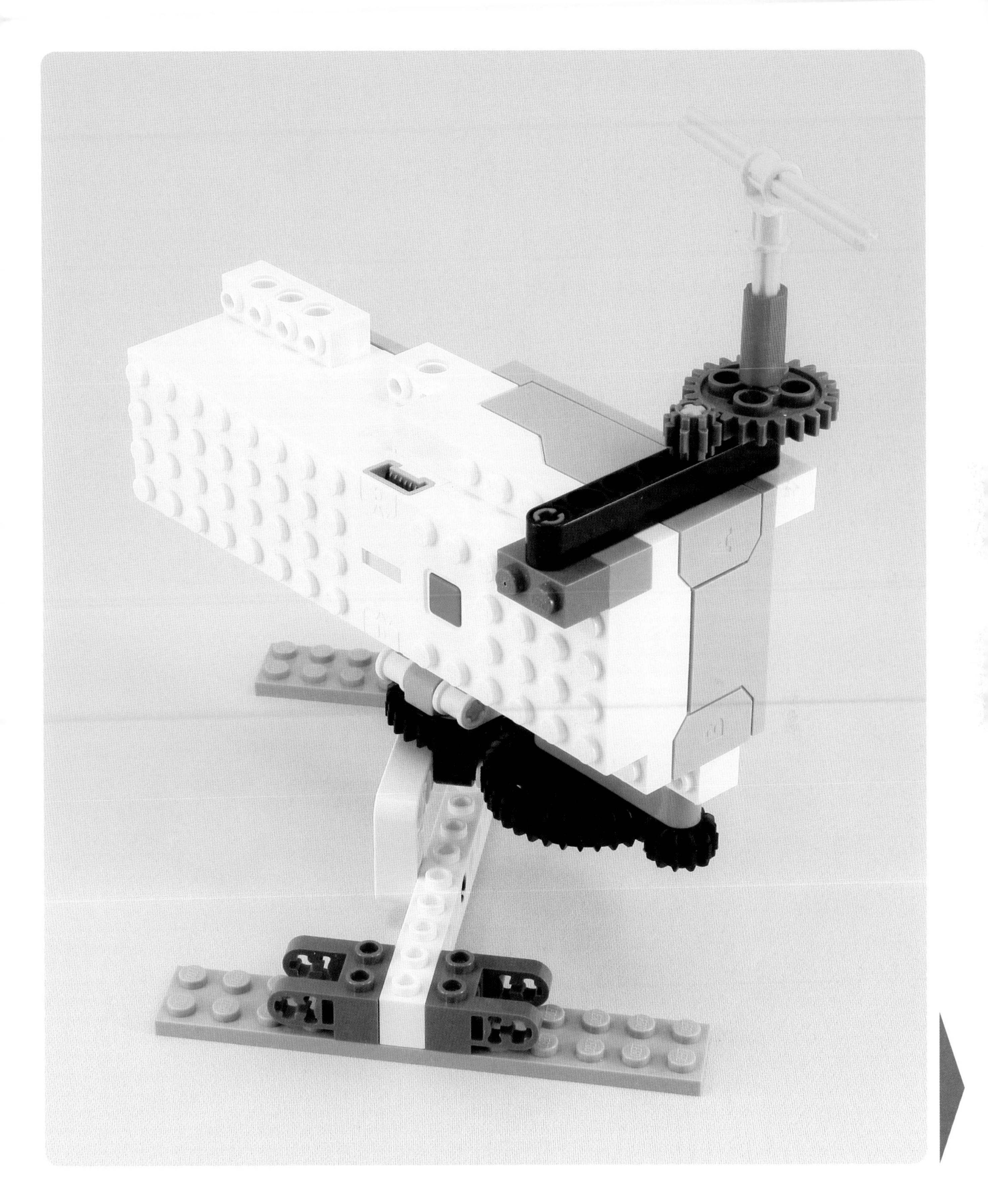

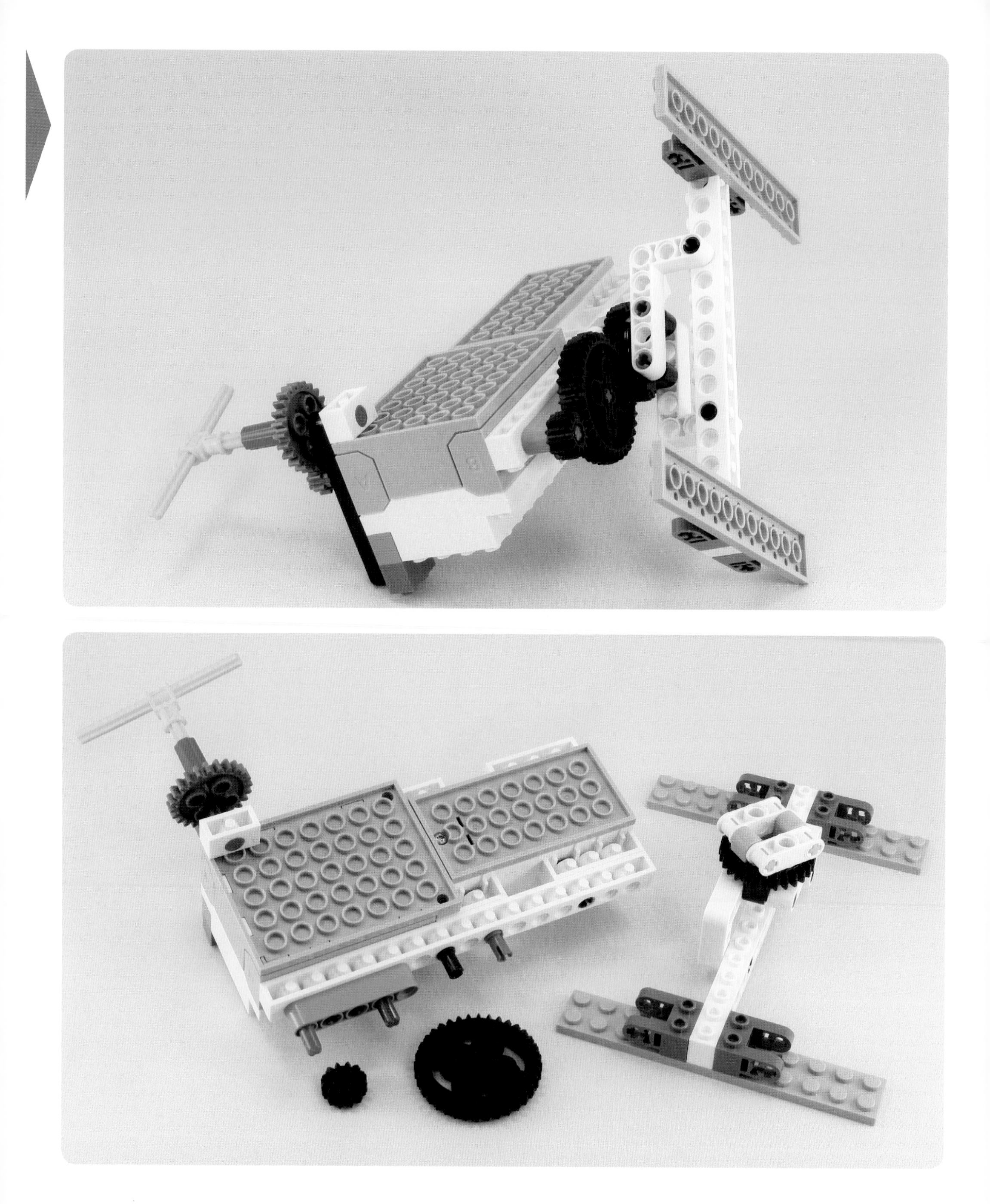

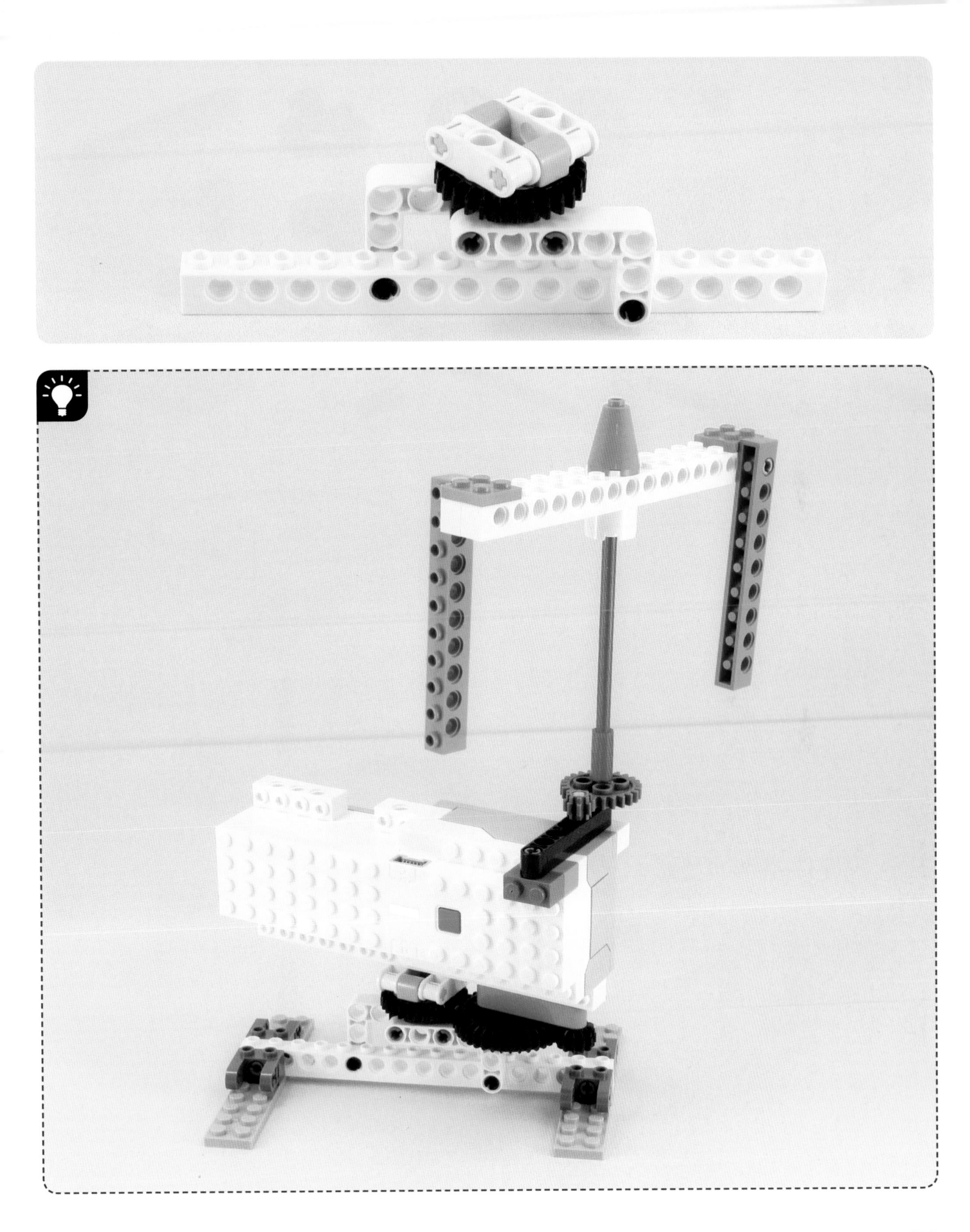

#84

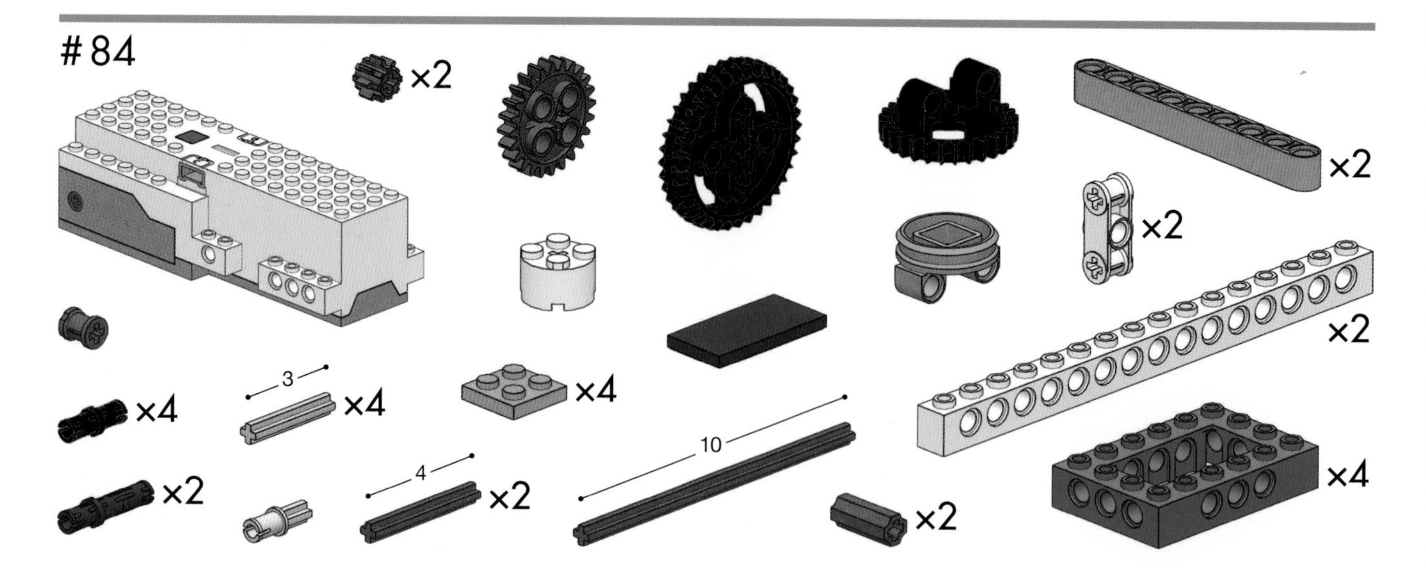

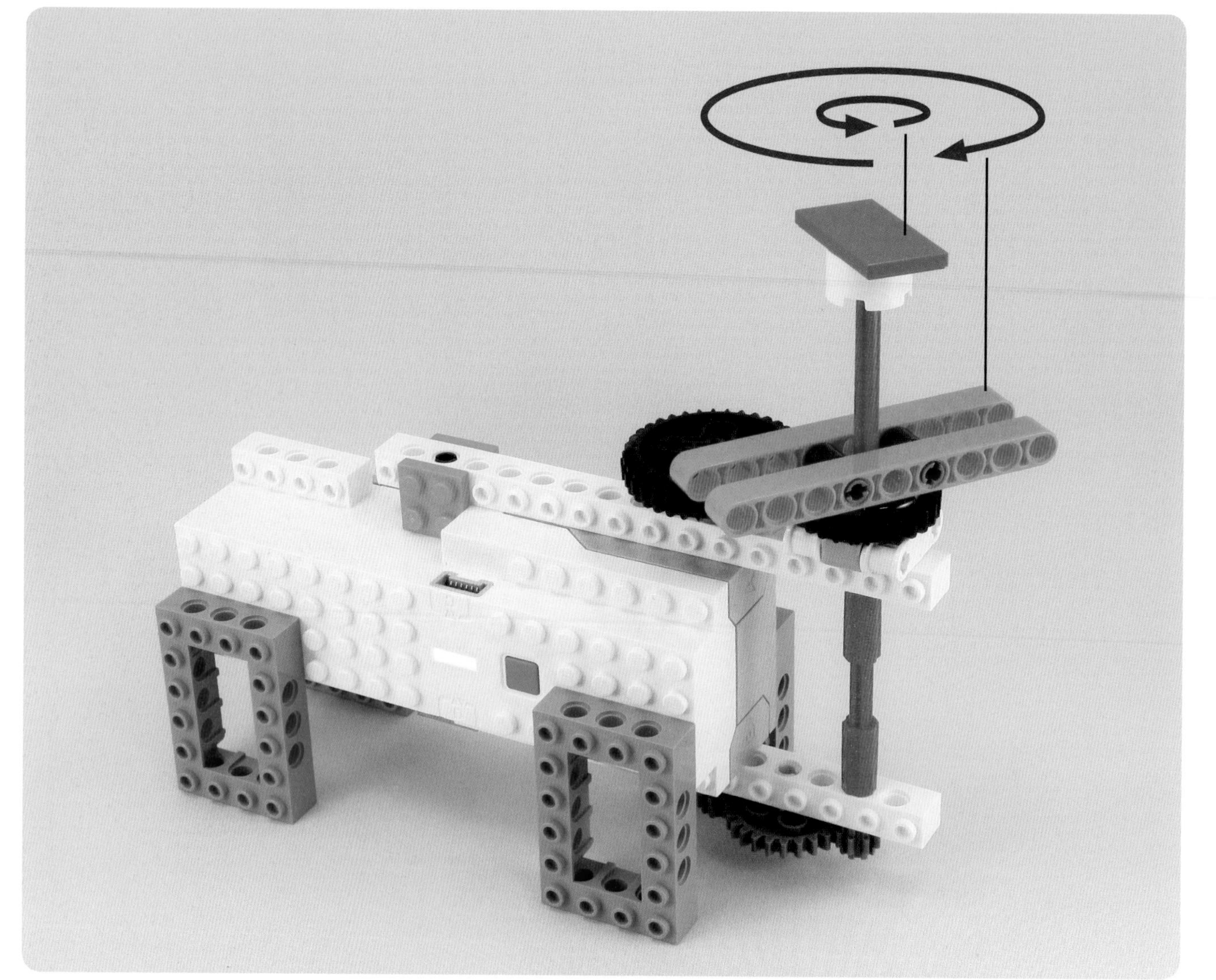

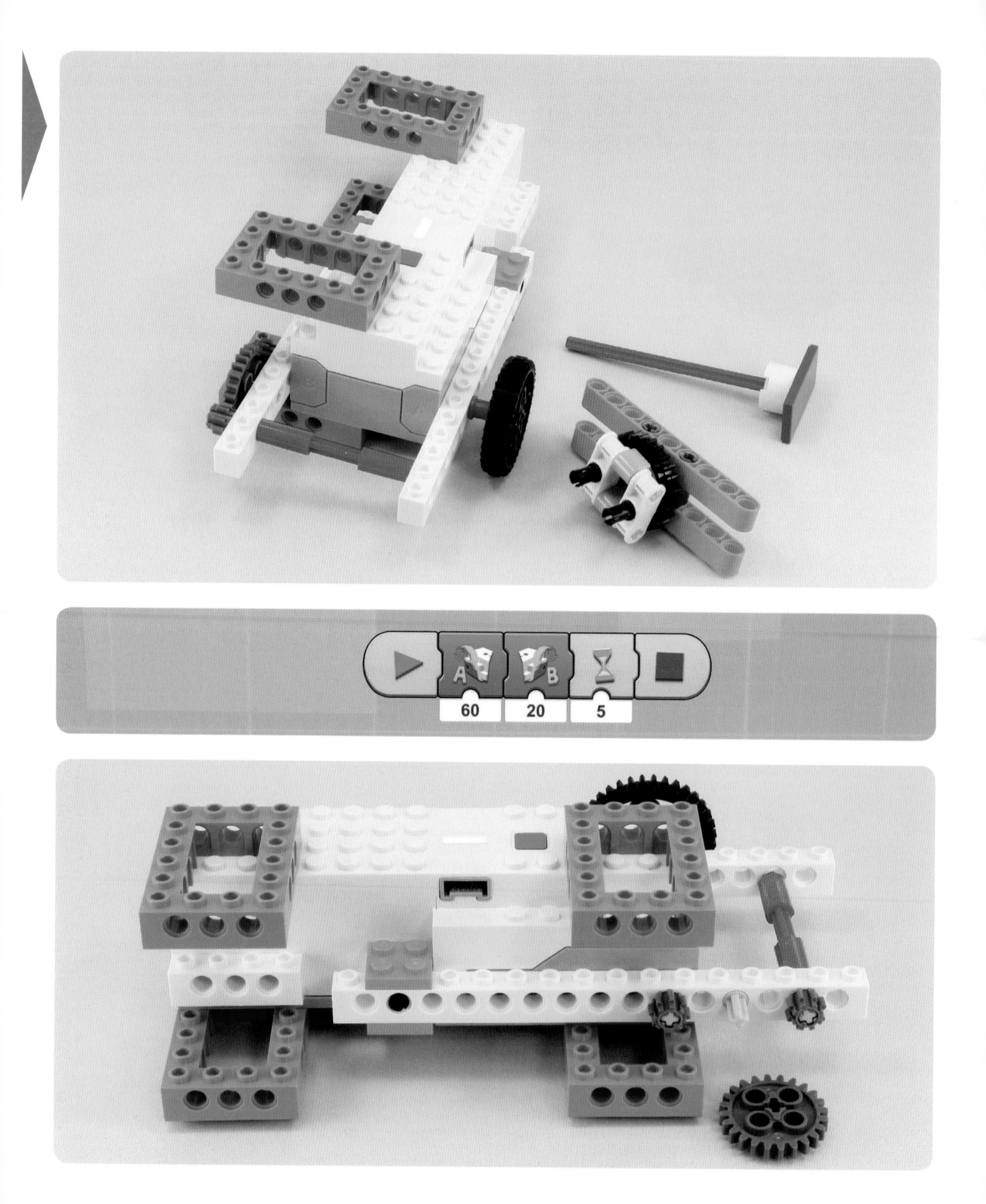
60
20
5

Changing direction by steering

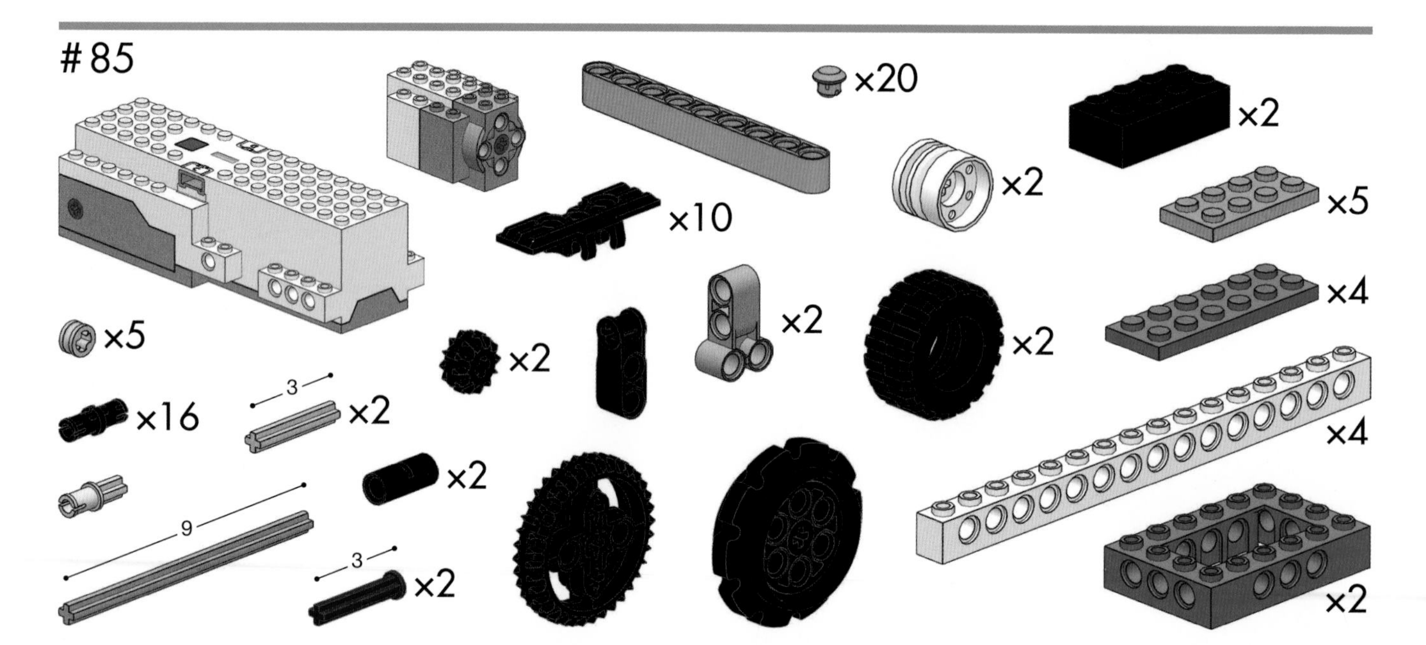

-10
1
-10
2
50
75
2
50
-75

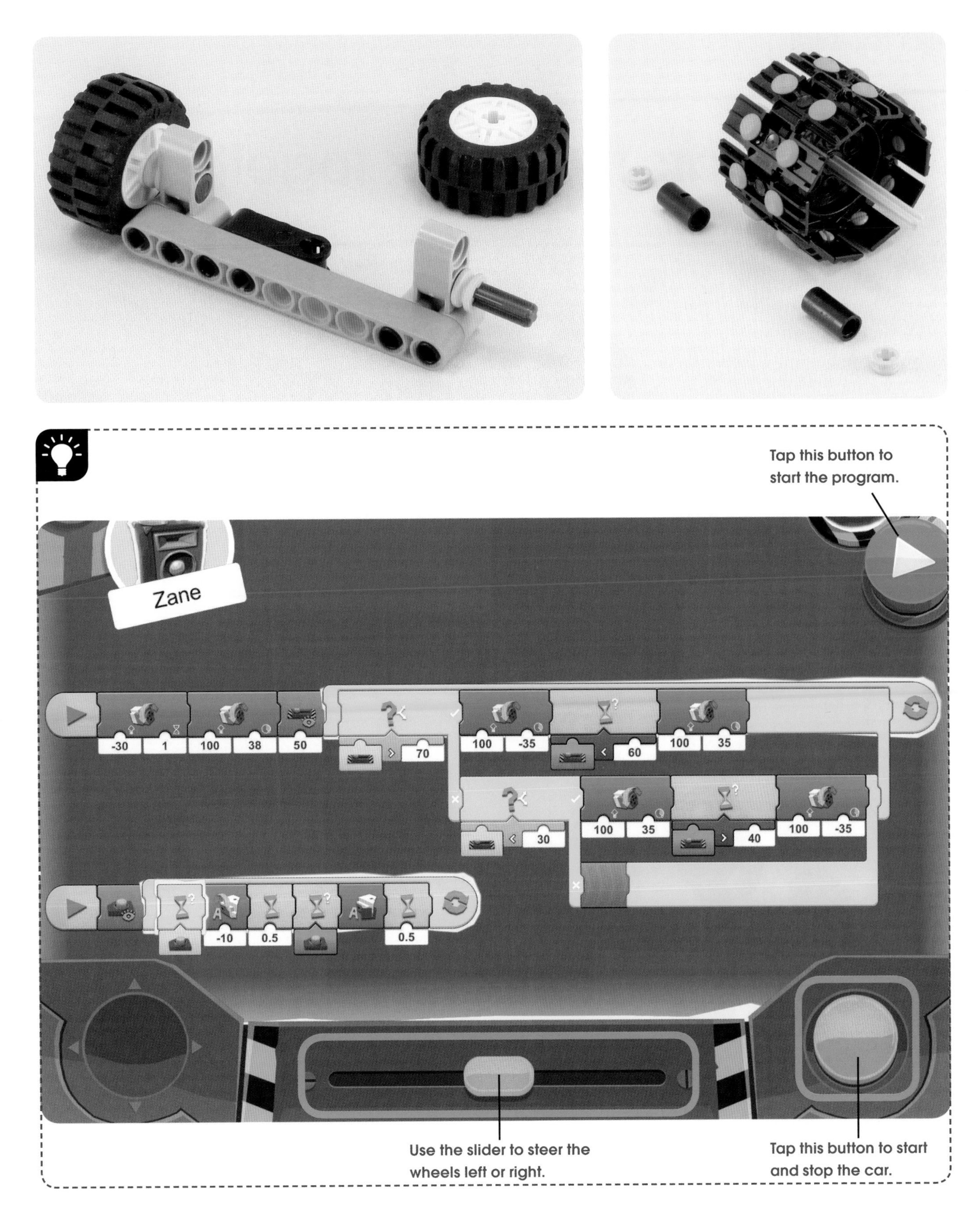
Tap this button to start the program.
Zane
-30
1
100
38
50
70
100
-35
60
100
35
30
100
35
40
100
-35
-10
0.5
0.5
Use the slider to steer the wheels left or right.
Tap this button to start and stop the car.

Cars that work together

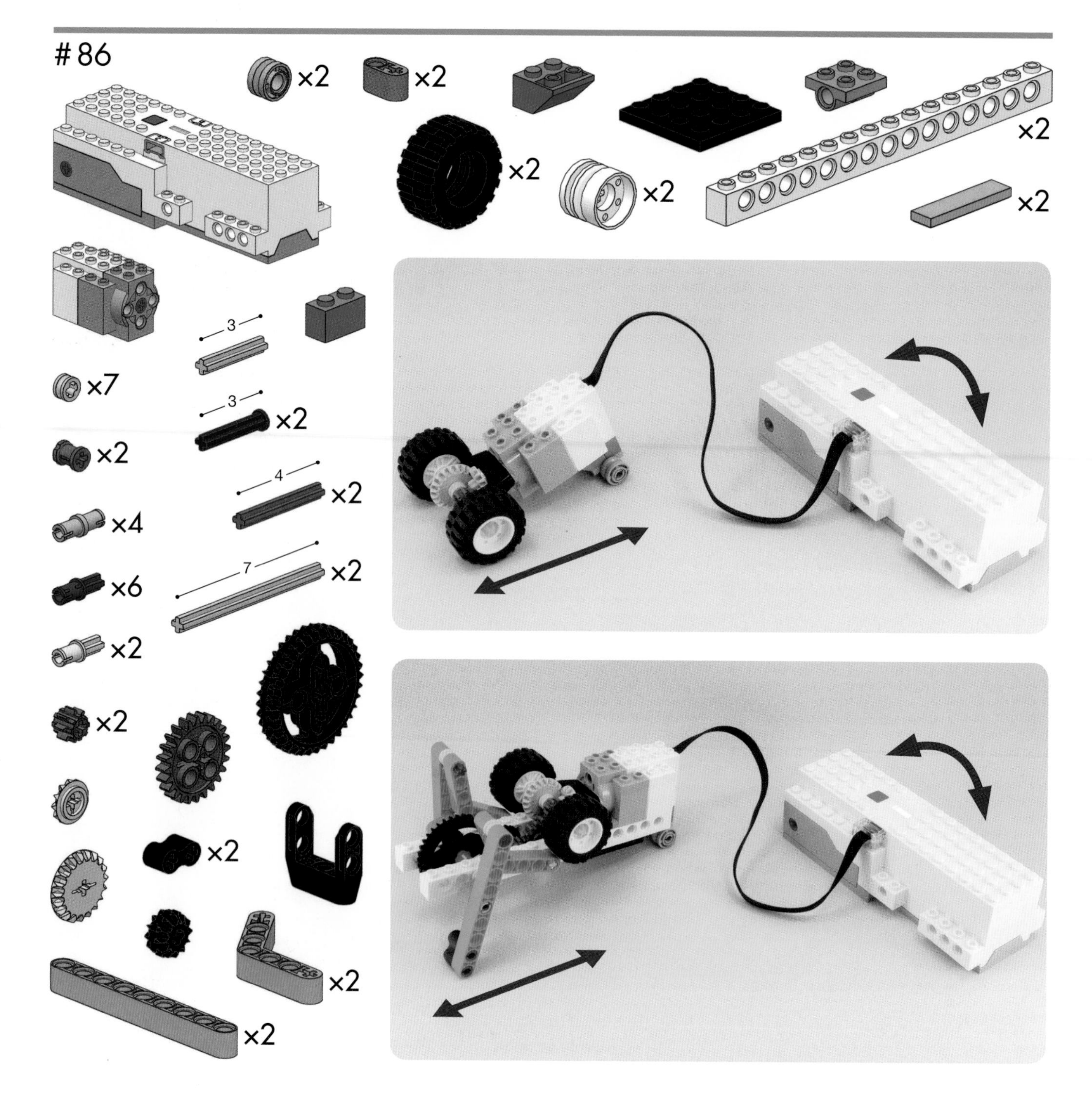

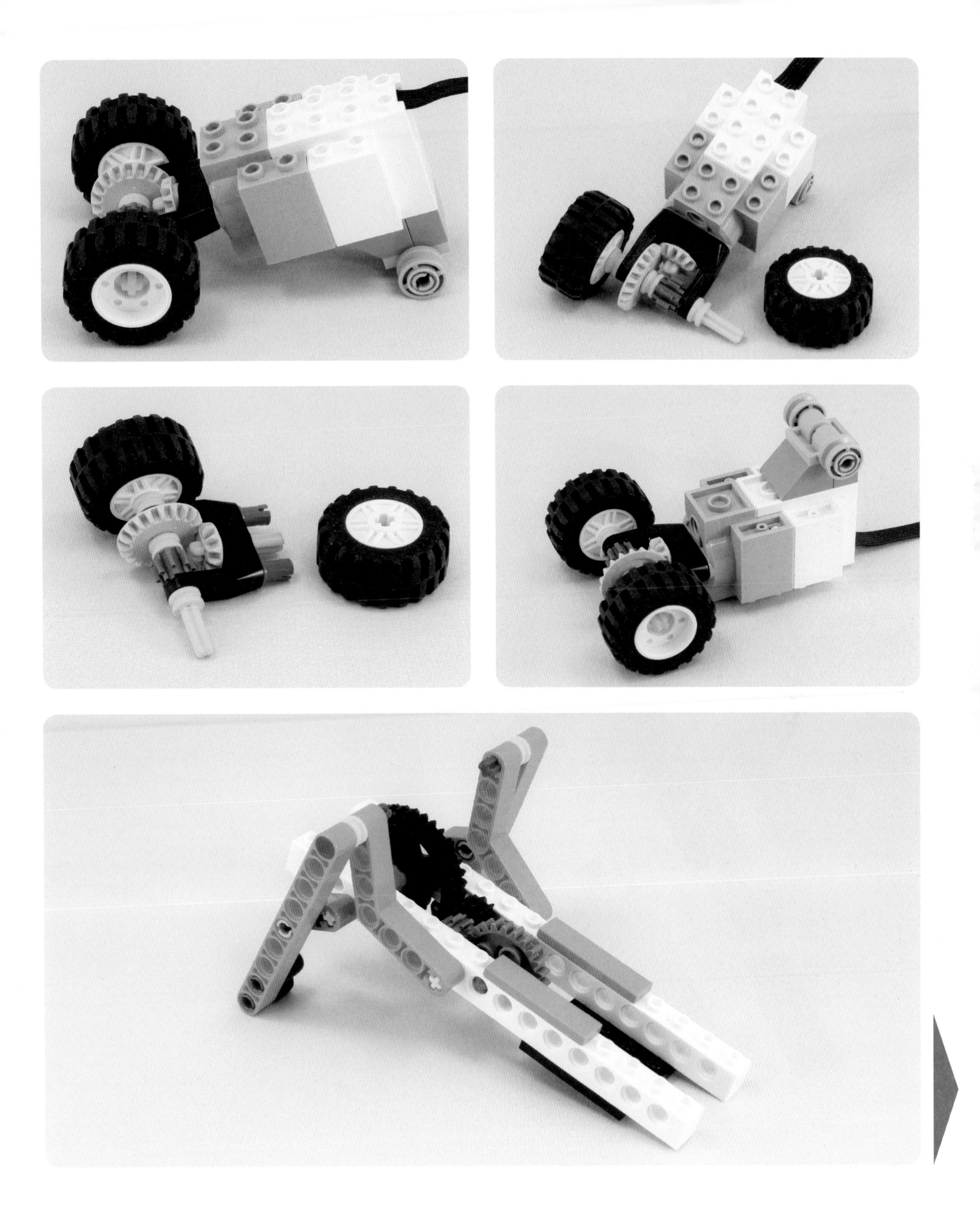

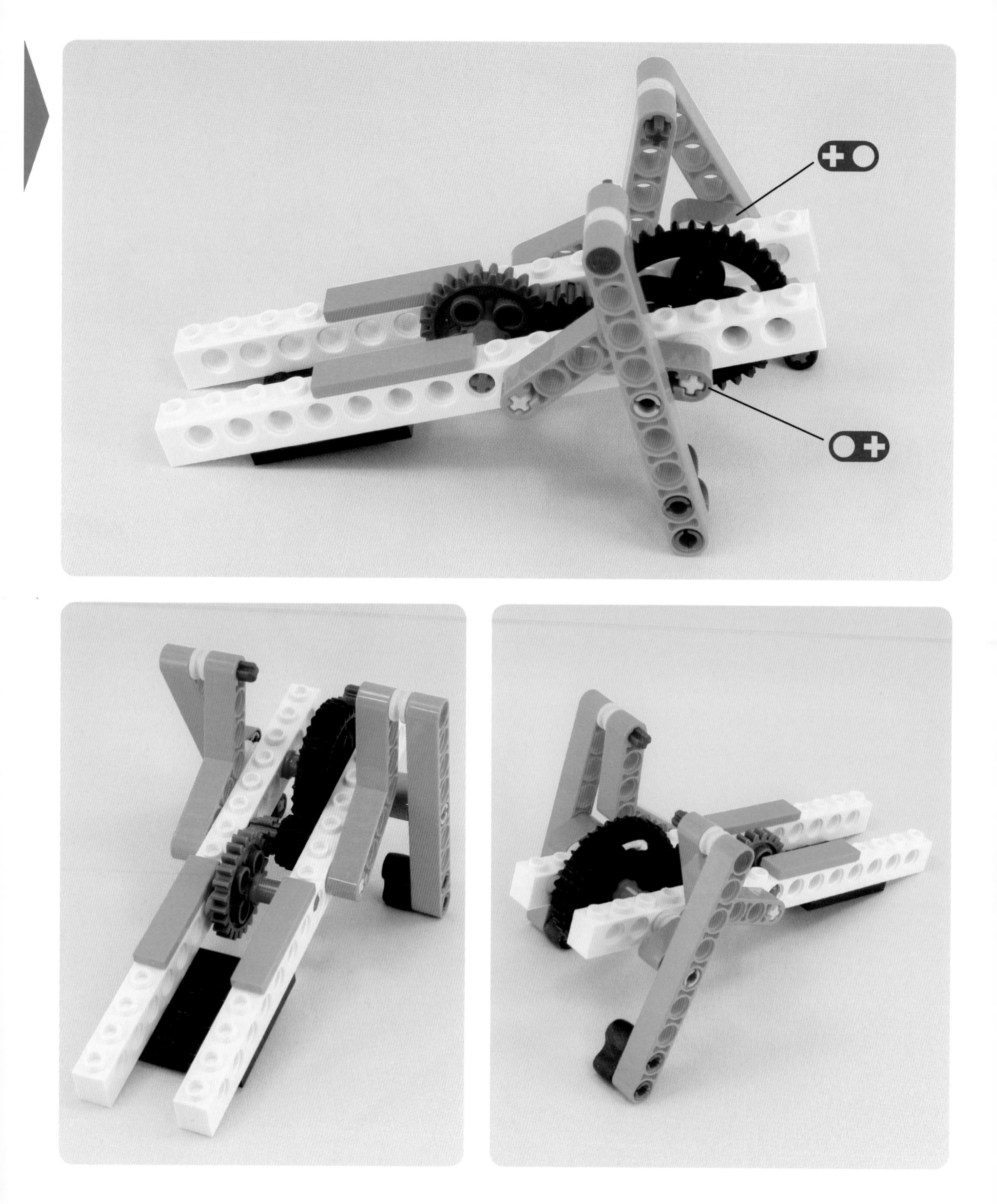

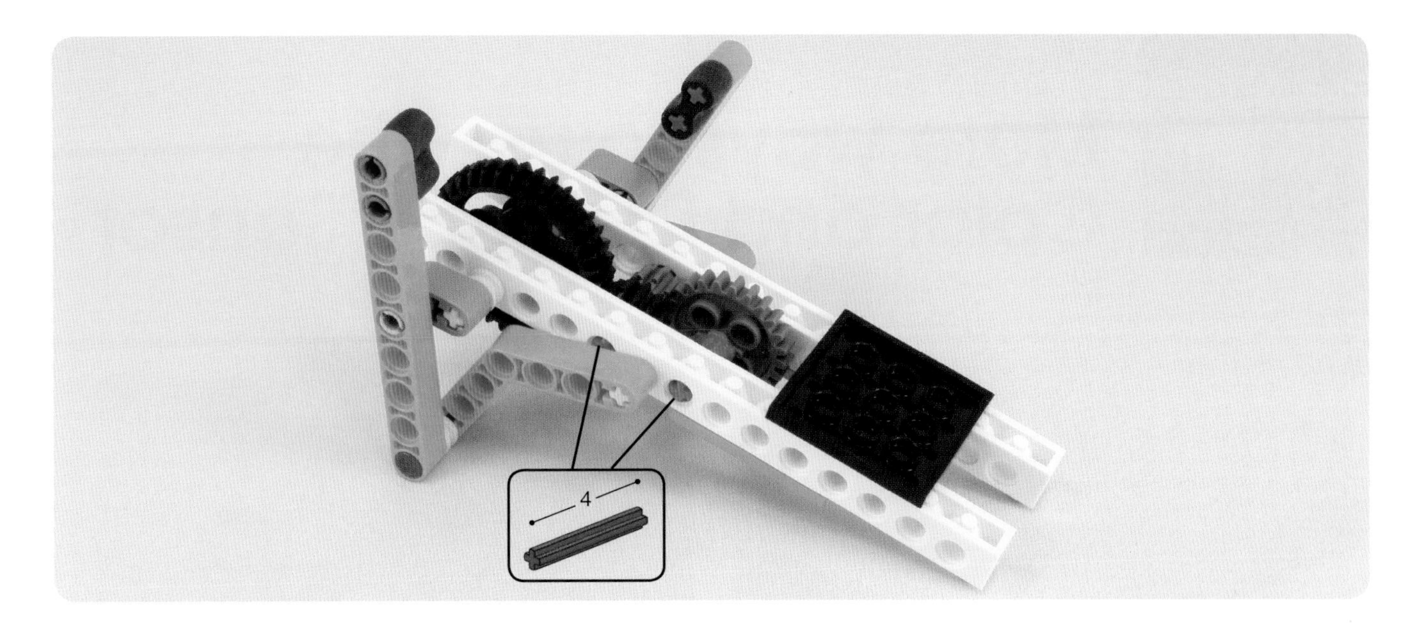
4

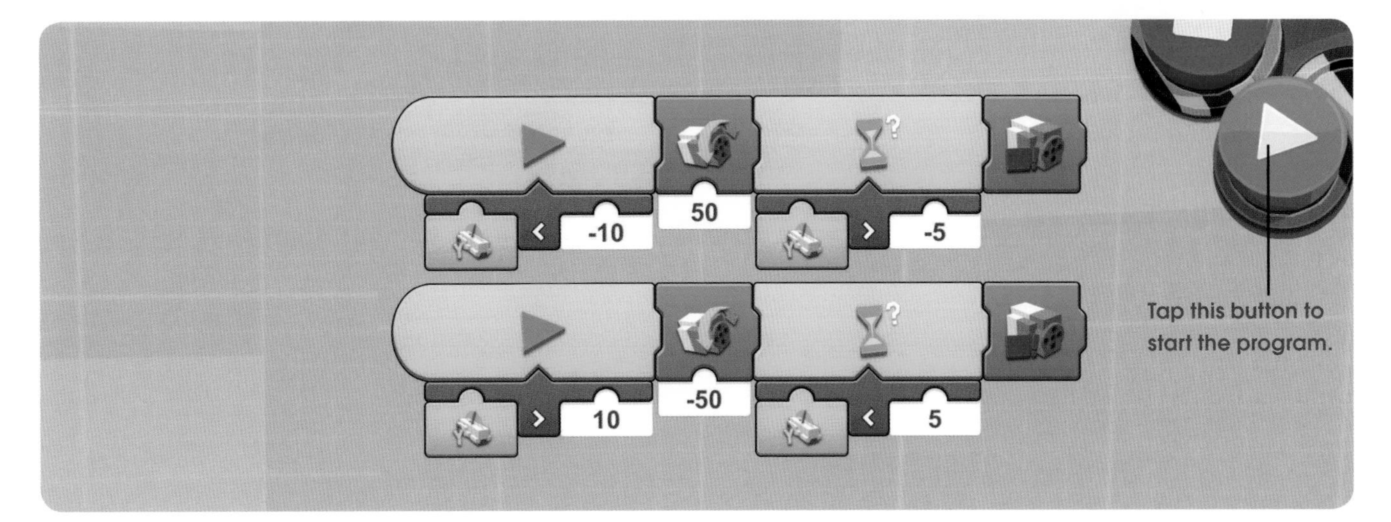
50
-10
-5
-50
10
5
Tap this button to start the program.

More ways to use the Color and Distance Sensor

87

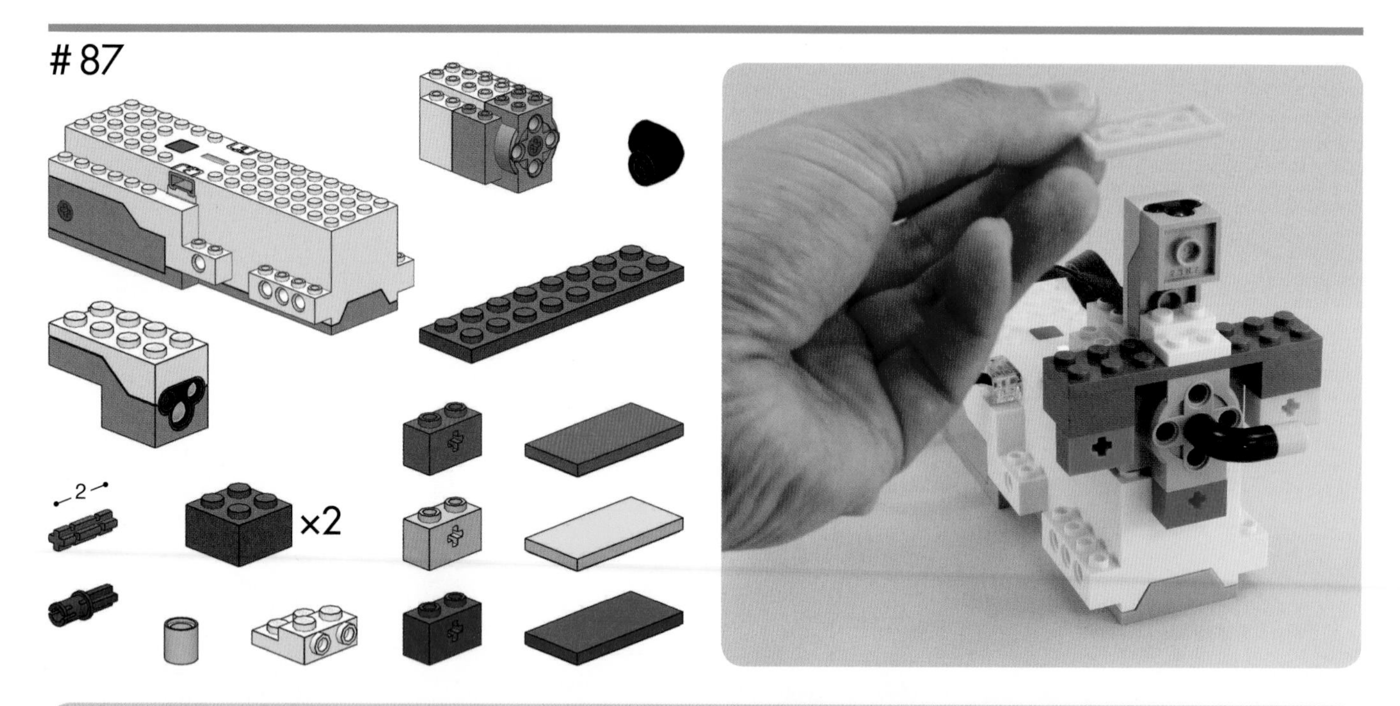

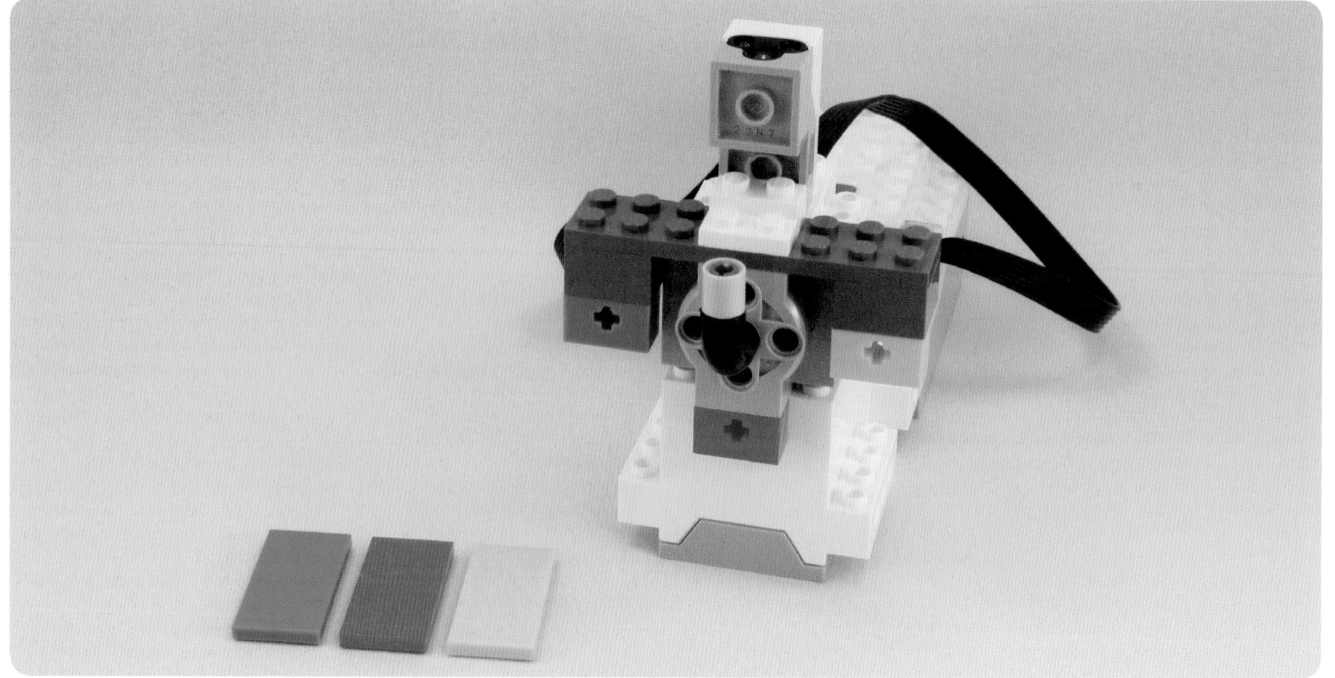

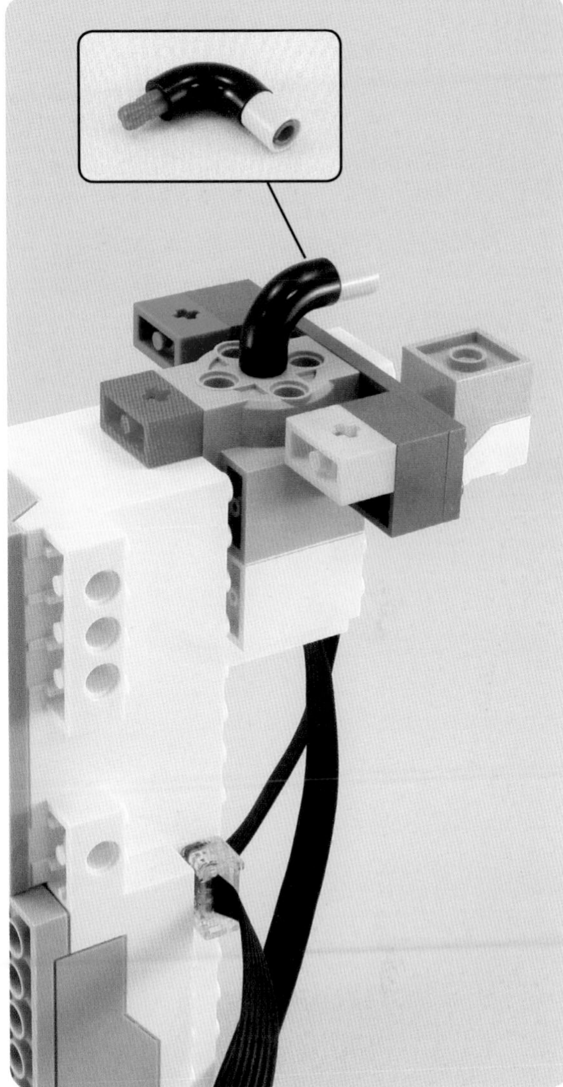

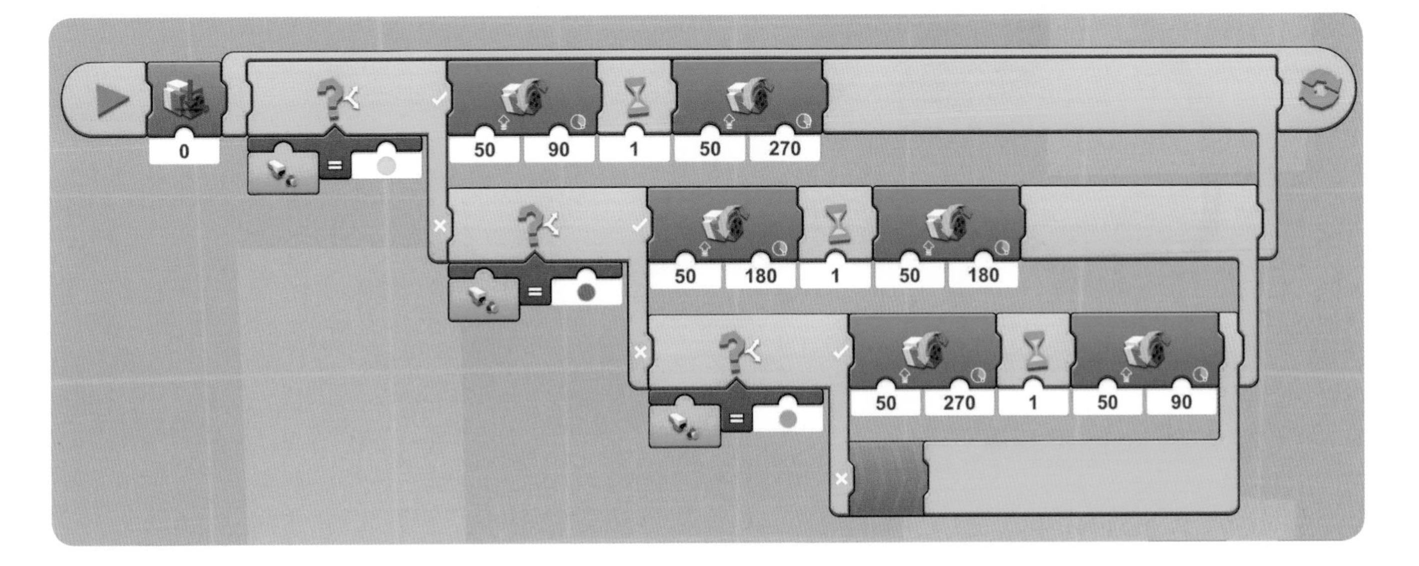
0
50
90
1
50
270
50
180
1
50
180
50
270
1
50
90

#88

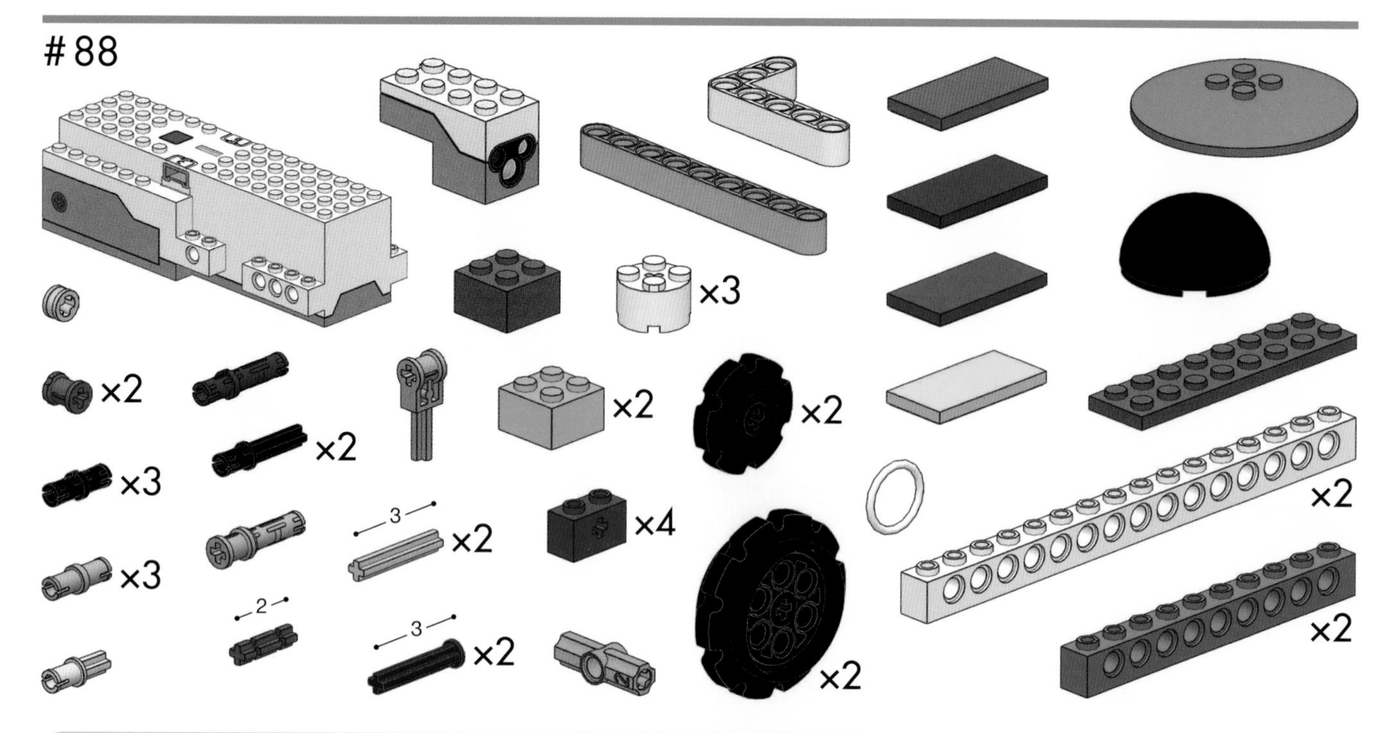

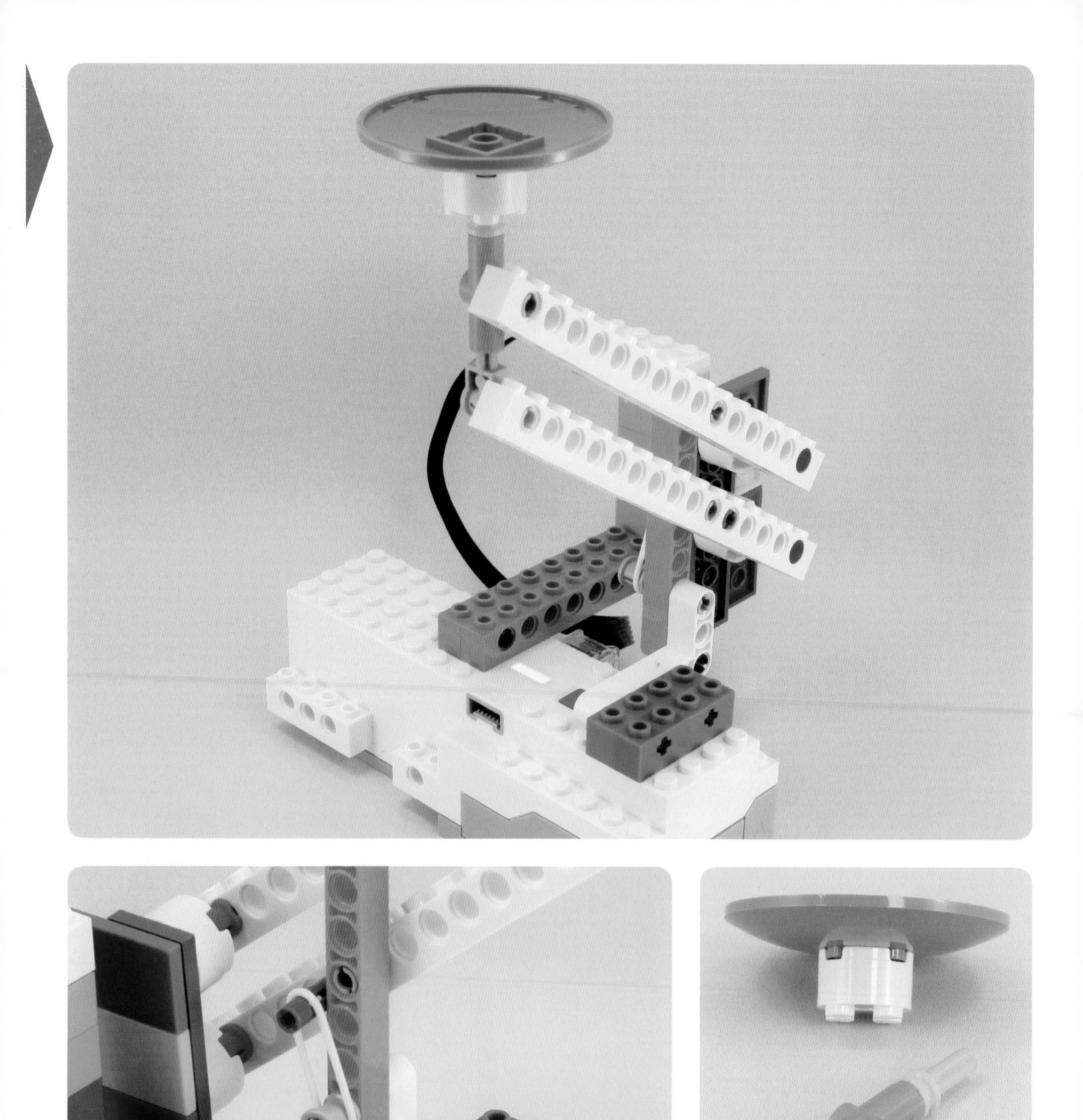

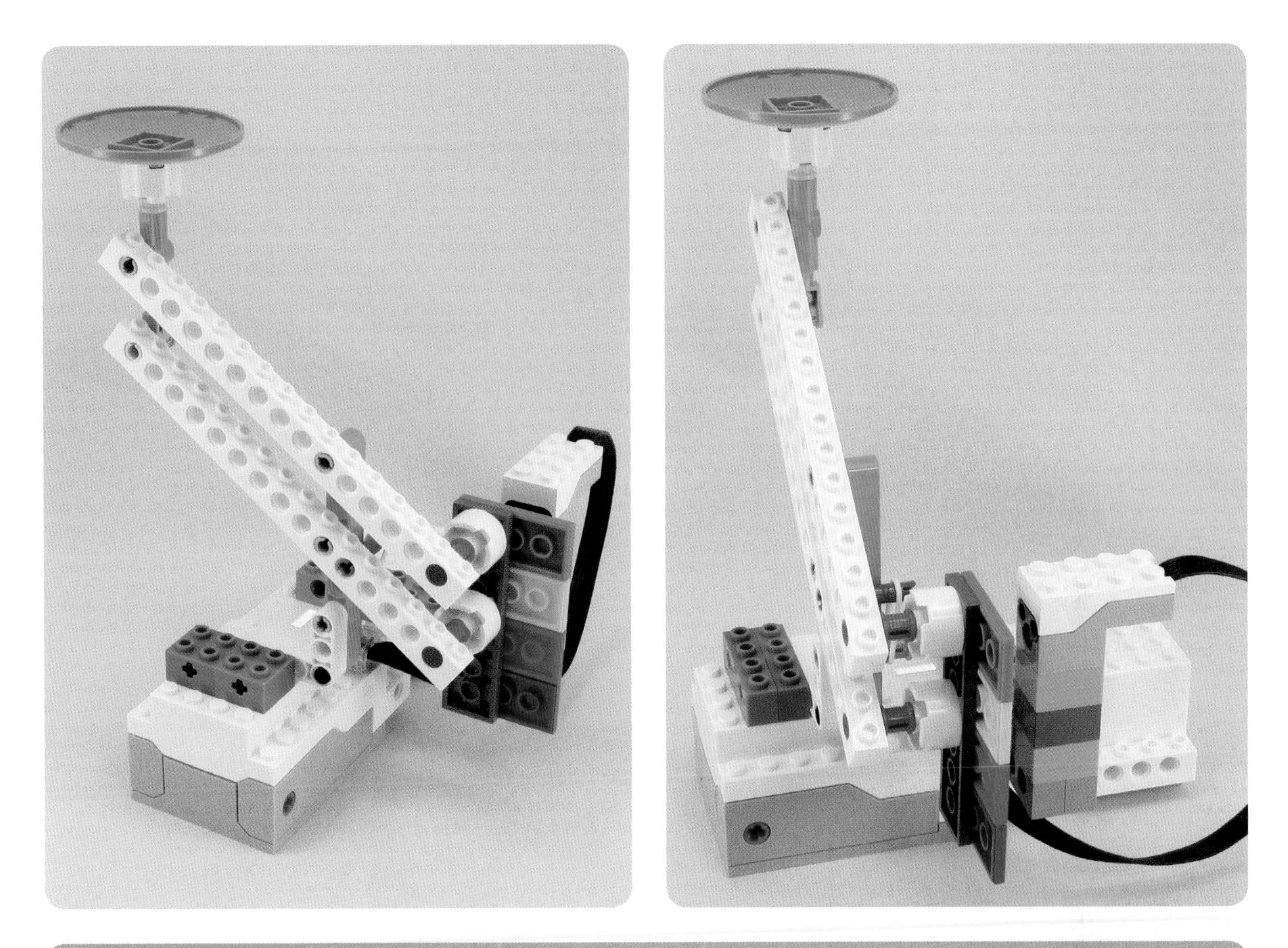

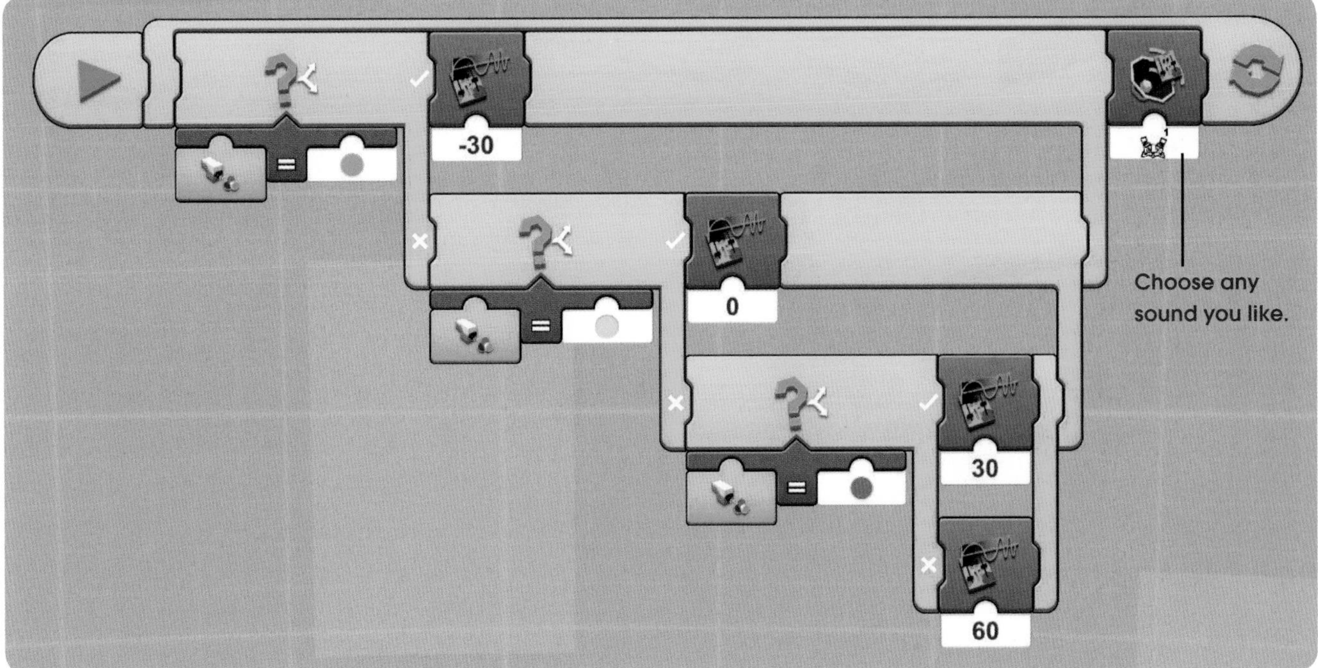

-30
0
30
60
Choose any sound you like.

89

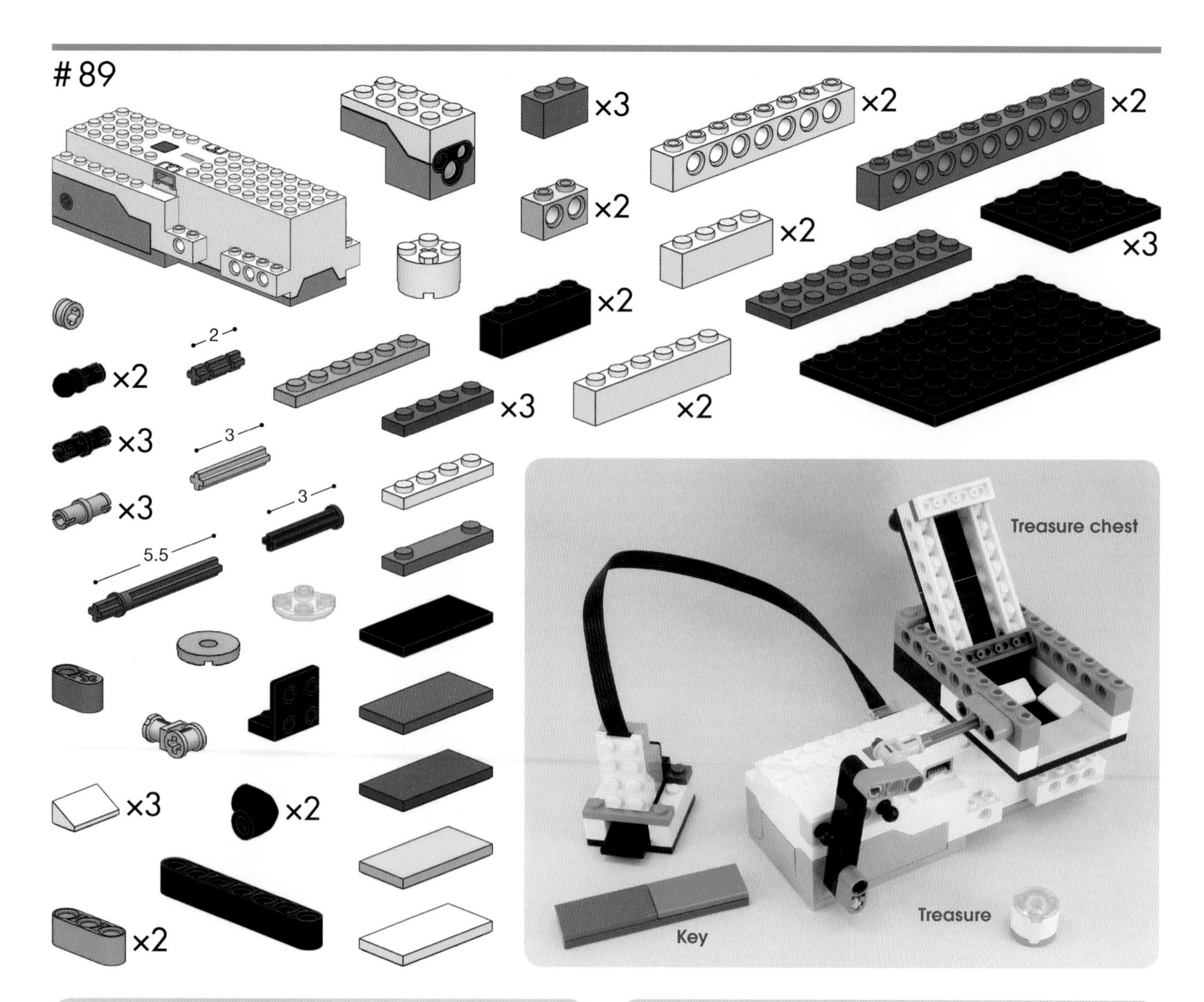
×3
×2
×2
×2
×2
×3
×2
×2
×3
×3
×2
2
3
3
5.5
×3
×2
×2
Treasure chest
Key
Treasure

Unlocked.
You can open and close
the lid manually.
Locked

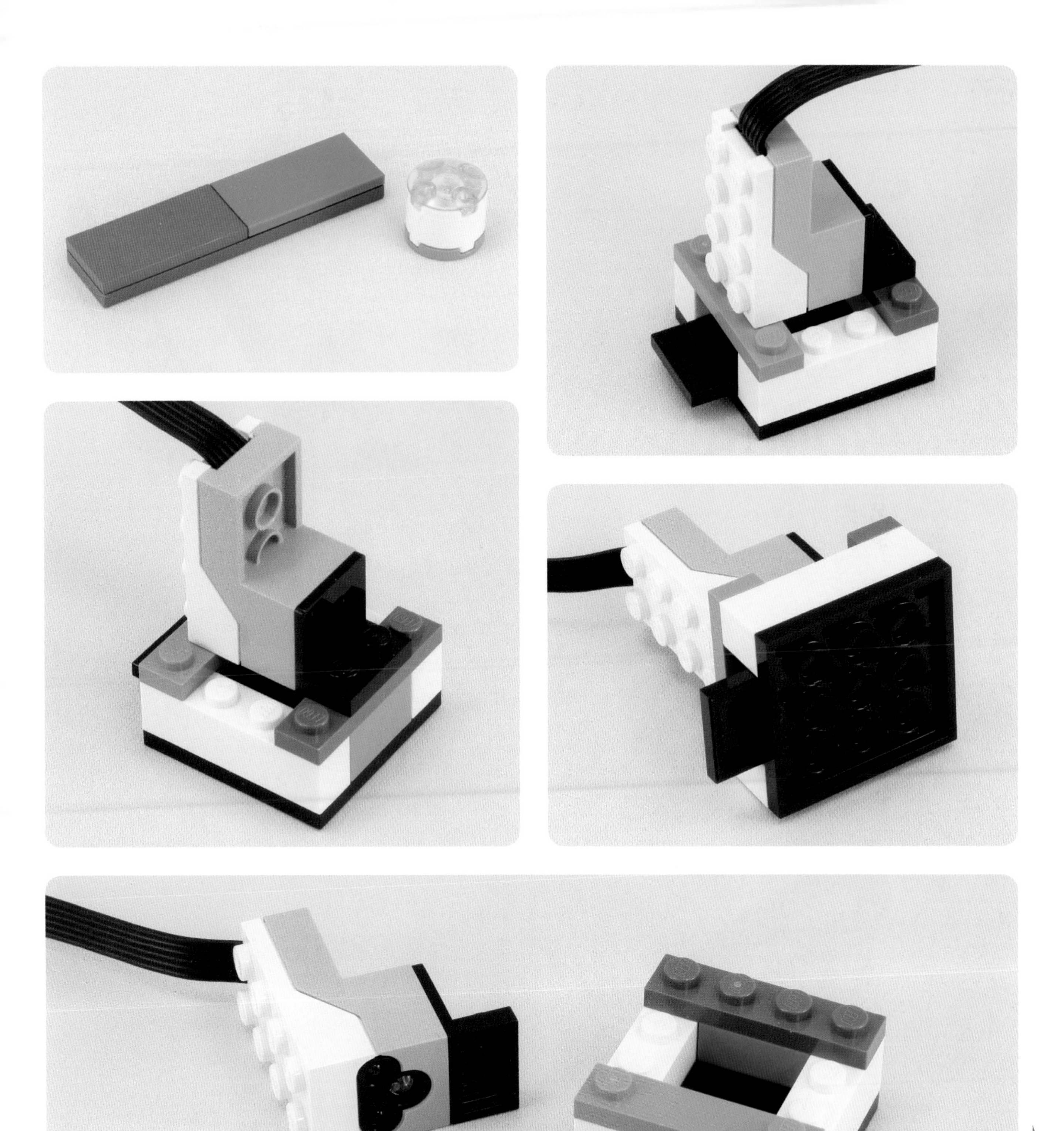

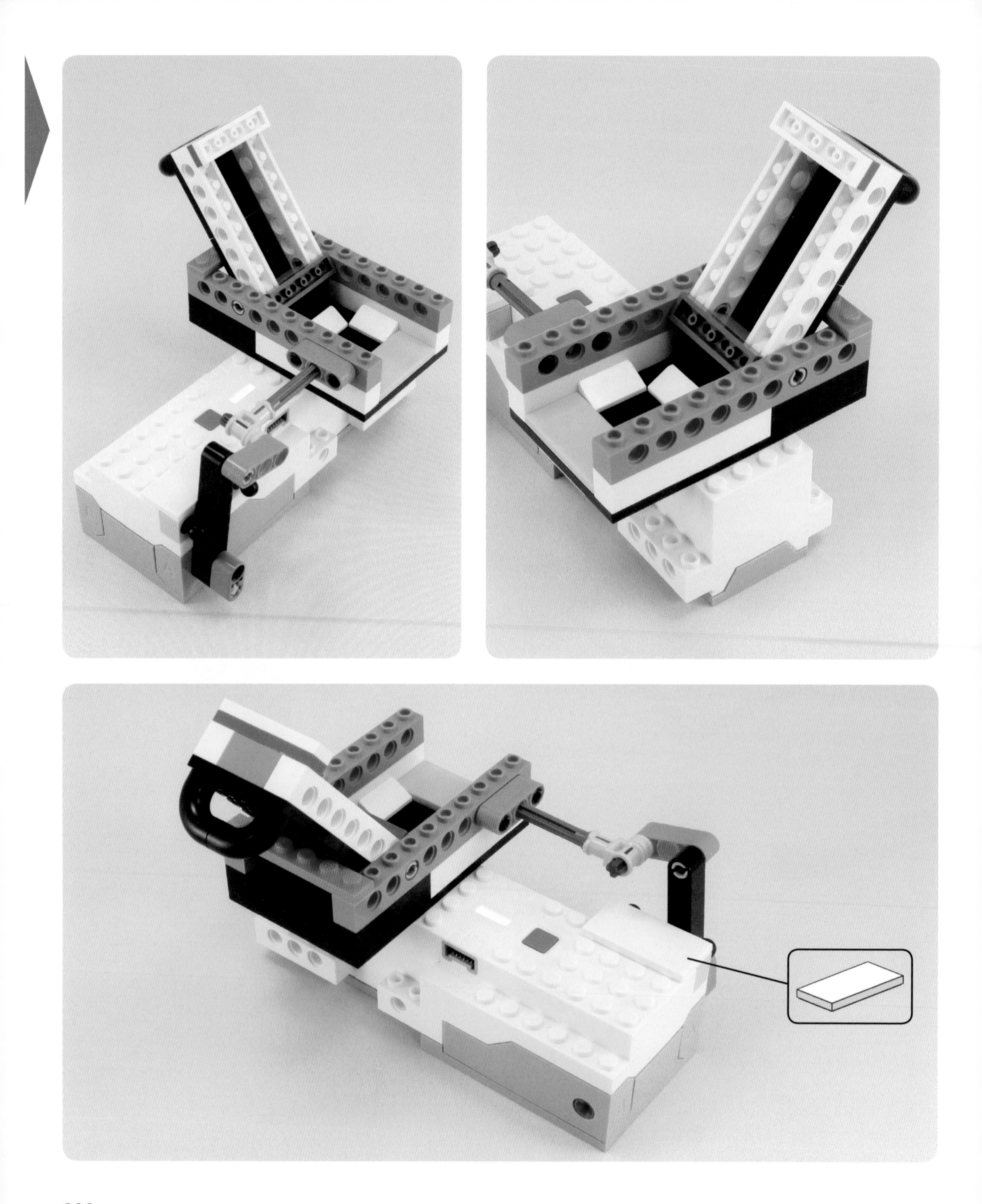

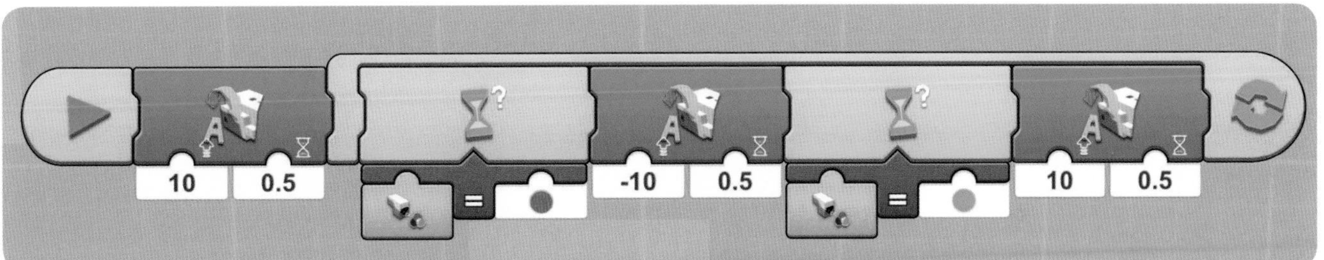
10
0.5
-10
0.5
10
0.5

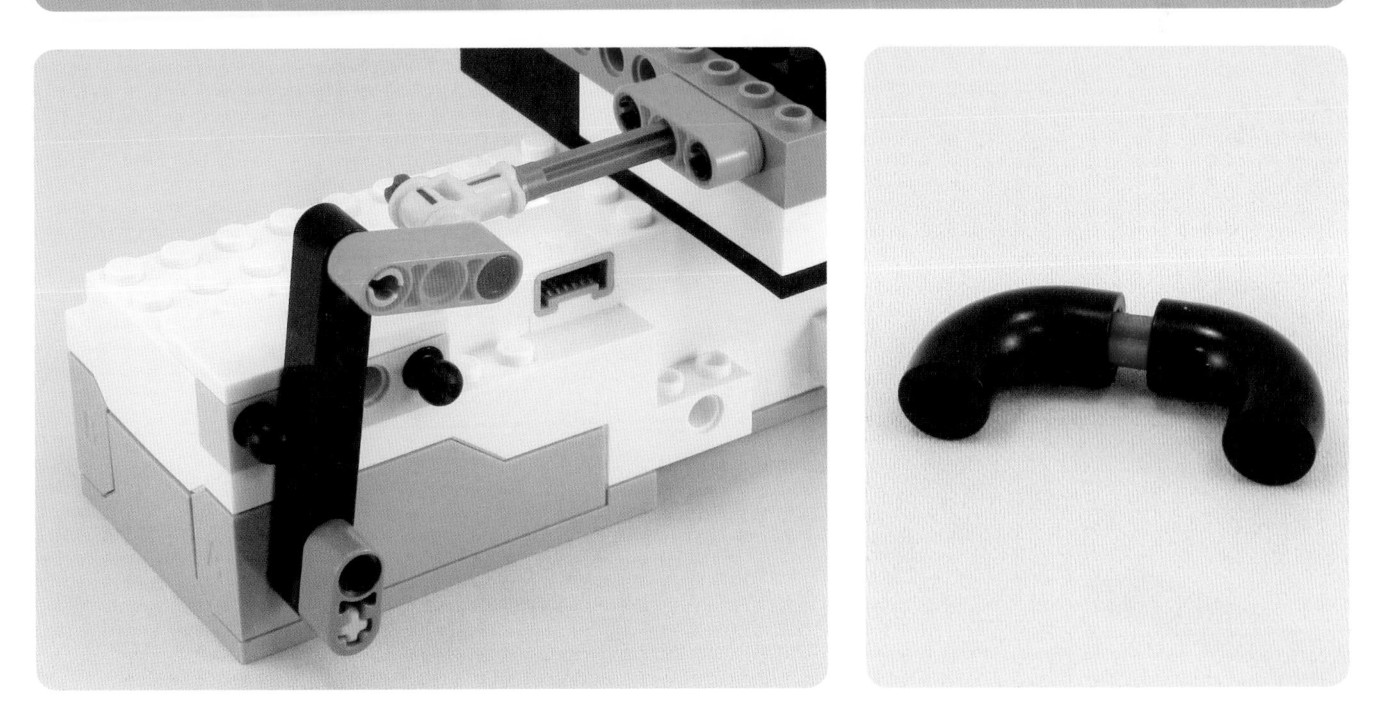

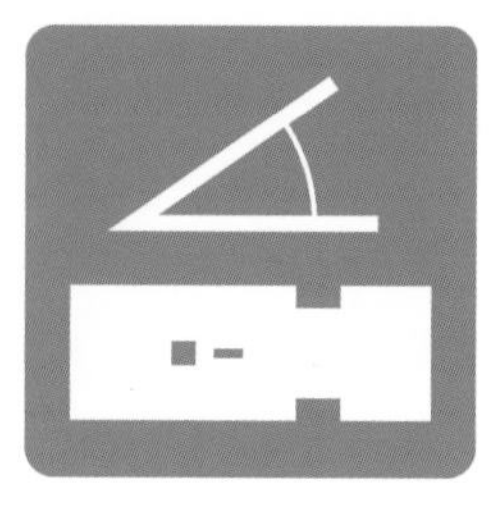

Using the Move Hub's Tilt Sensor

#90

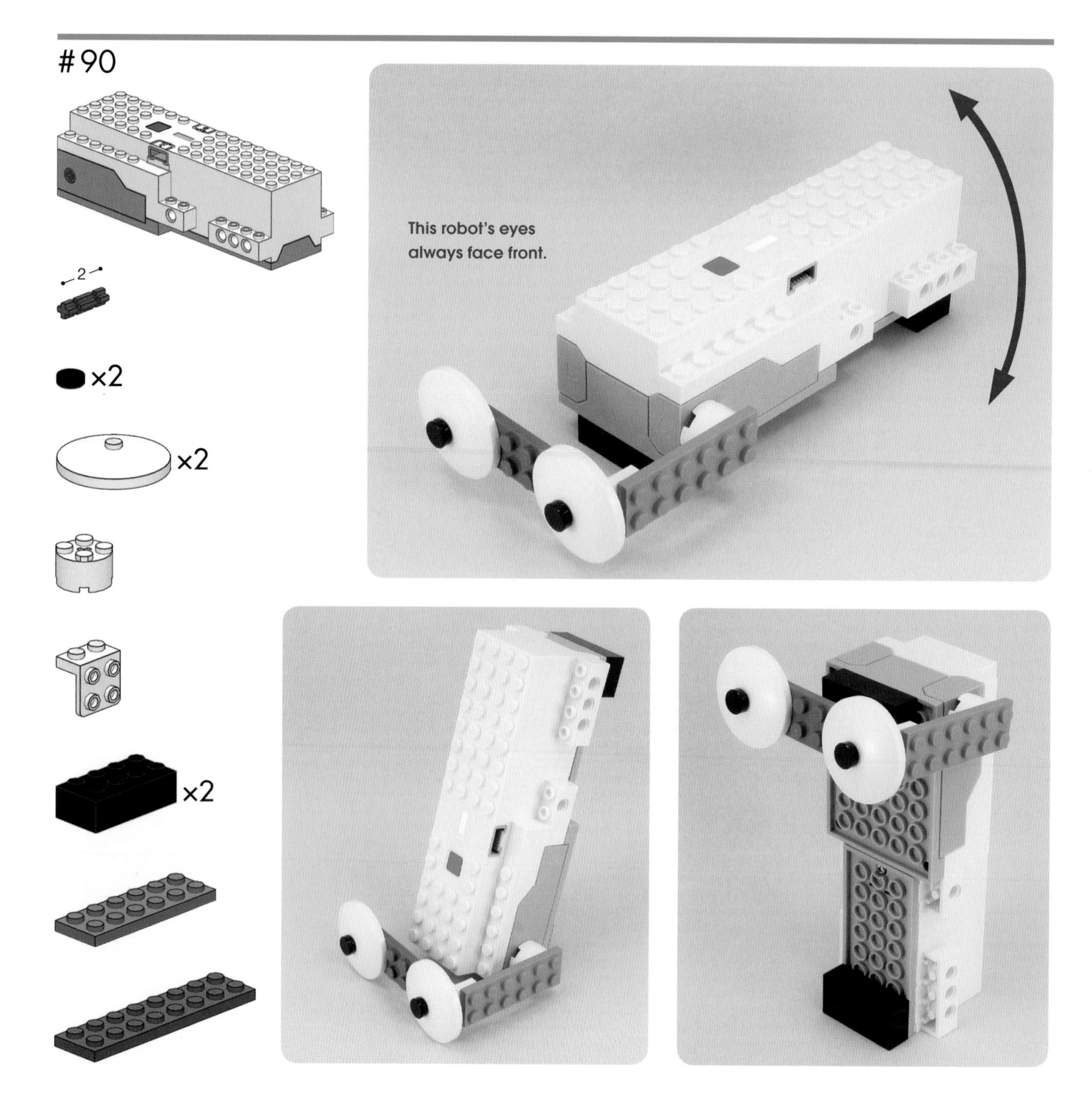

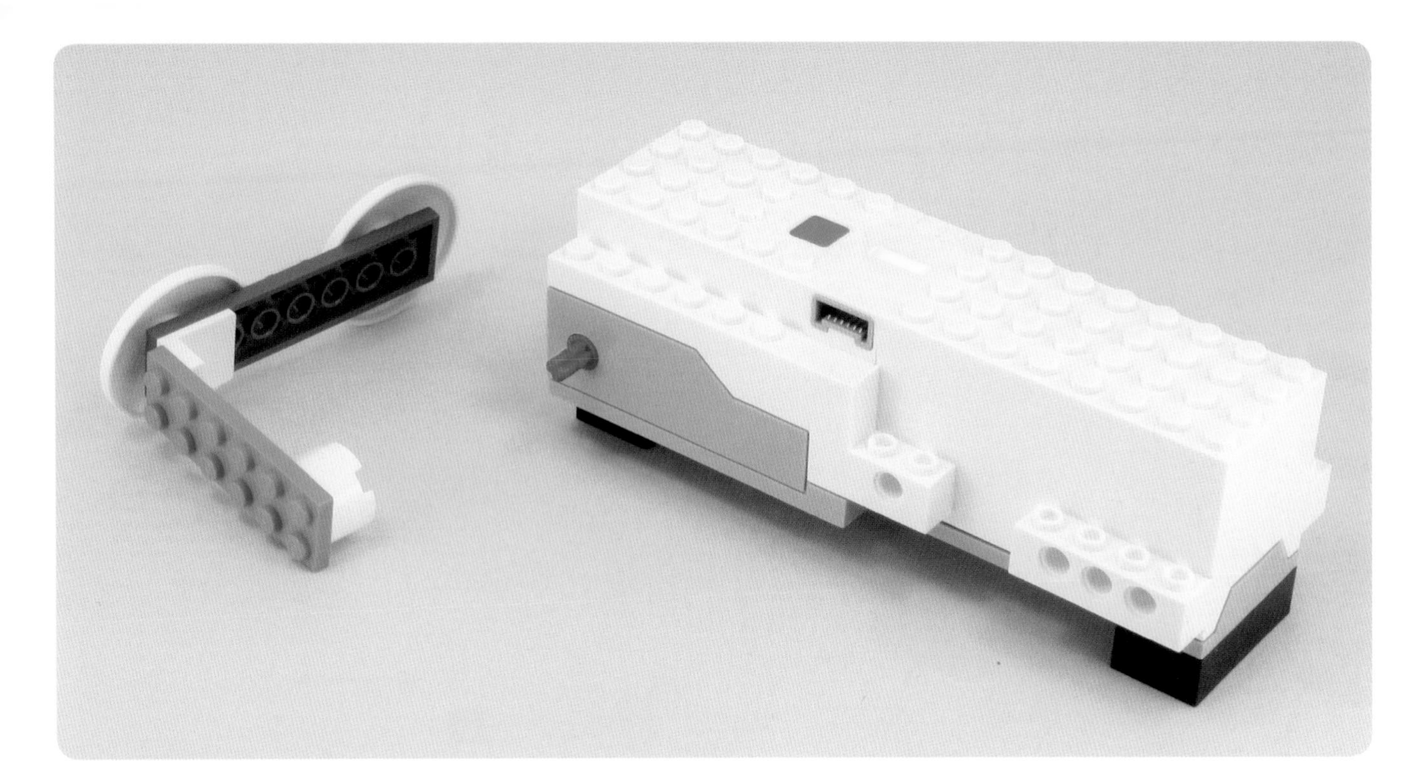

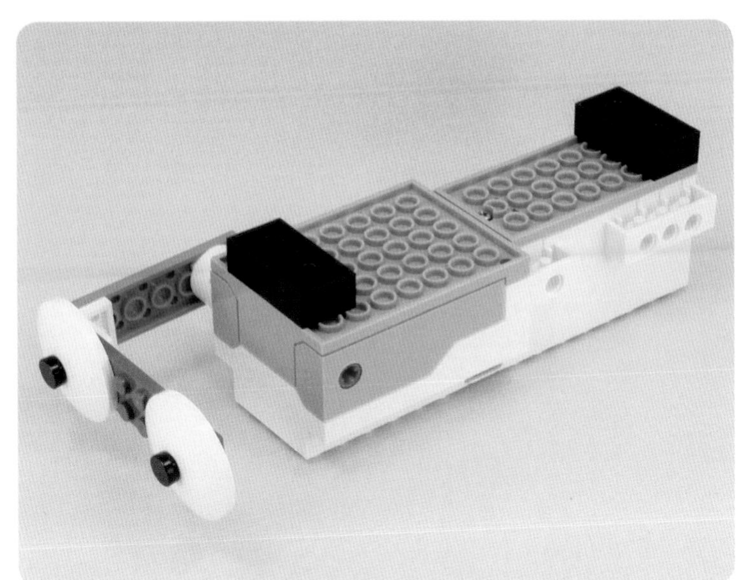

Start the program while the robot is in this position.

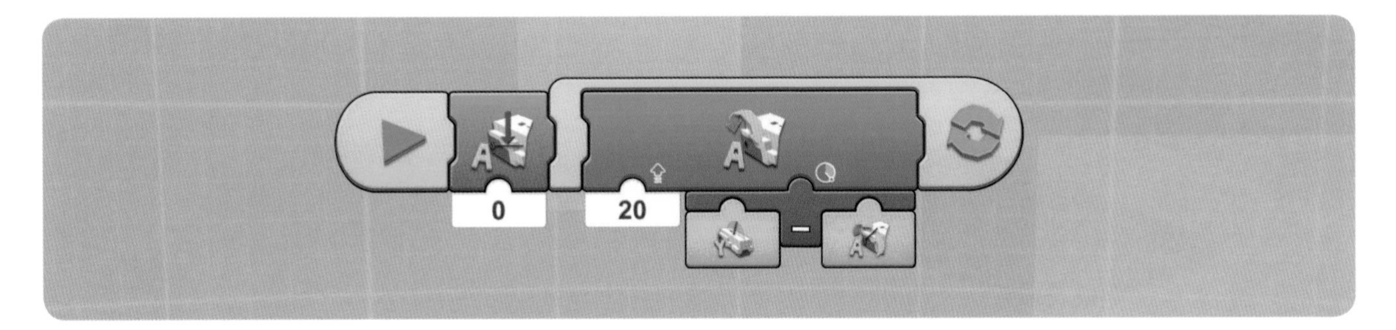

#91

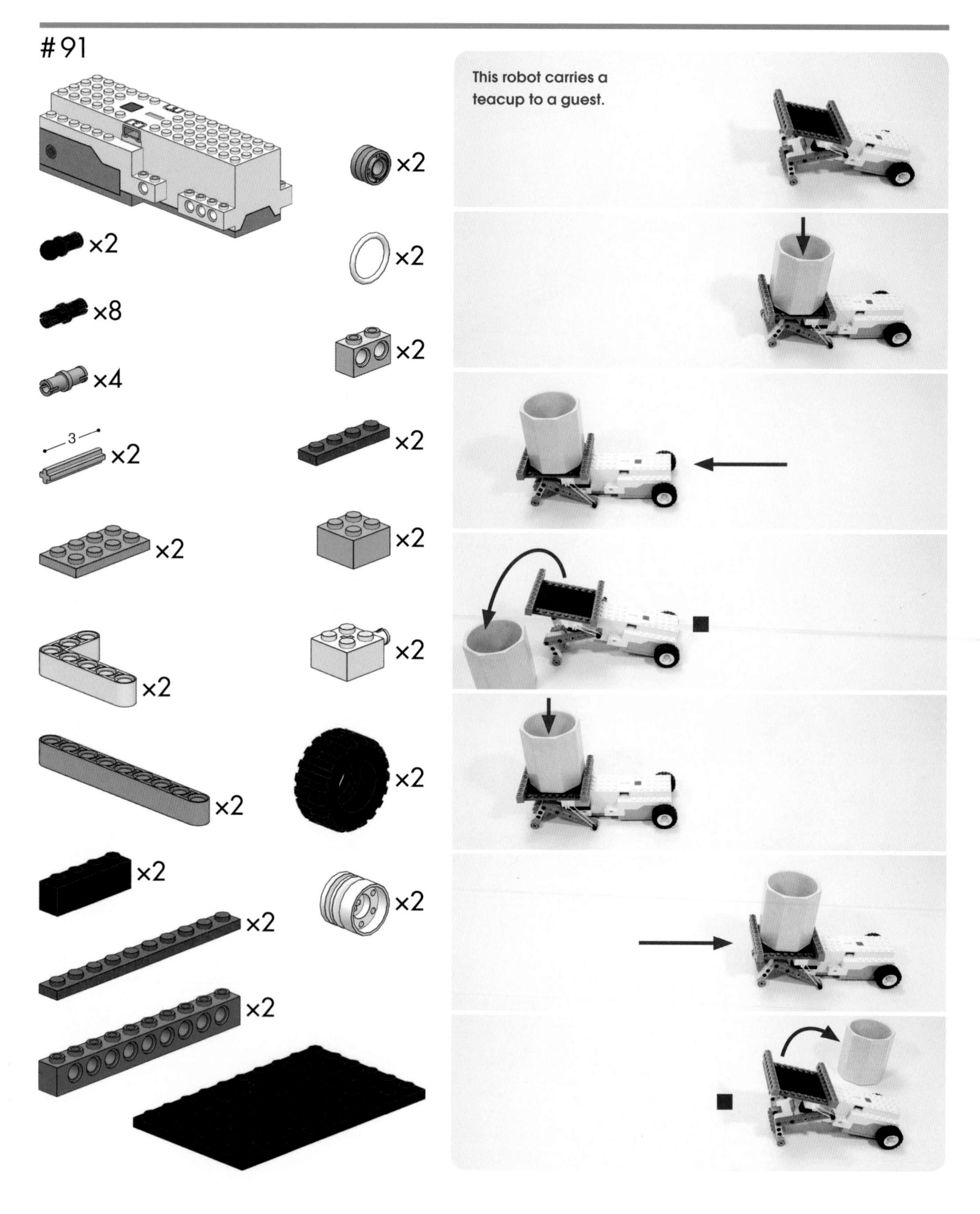

This robot carries a teacup to a guest.

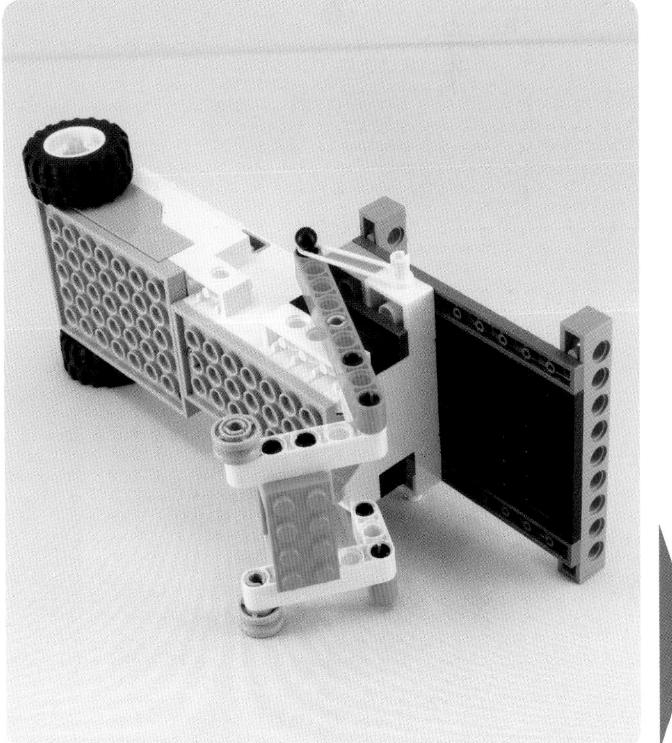

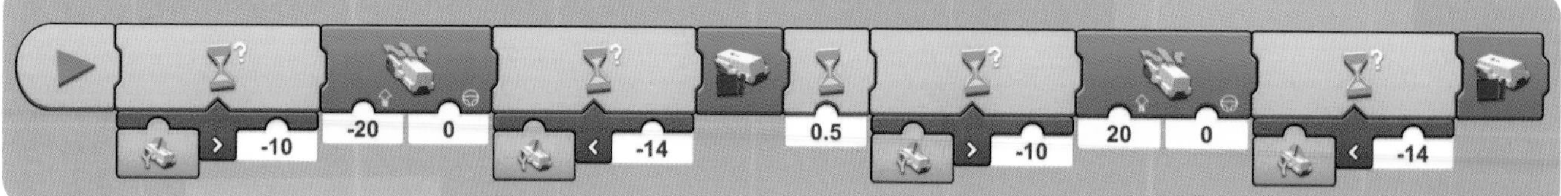

-10
-20
0
-14
0.5
-10
20
0
-14

Using Motor A and Motor B for different purposes

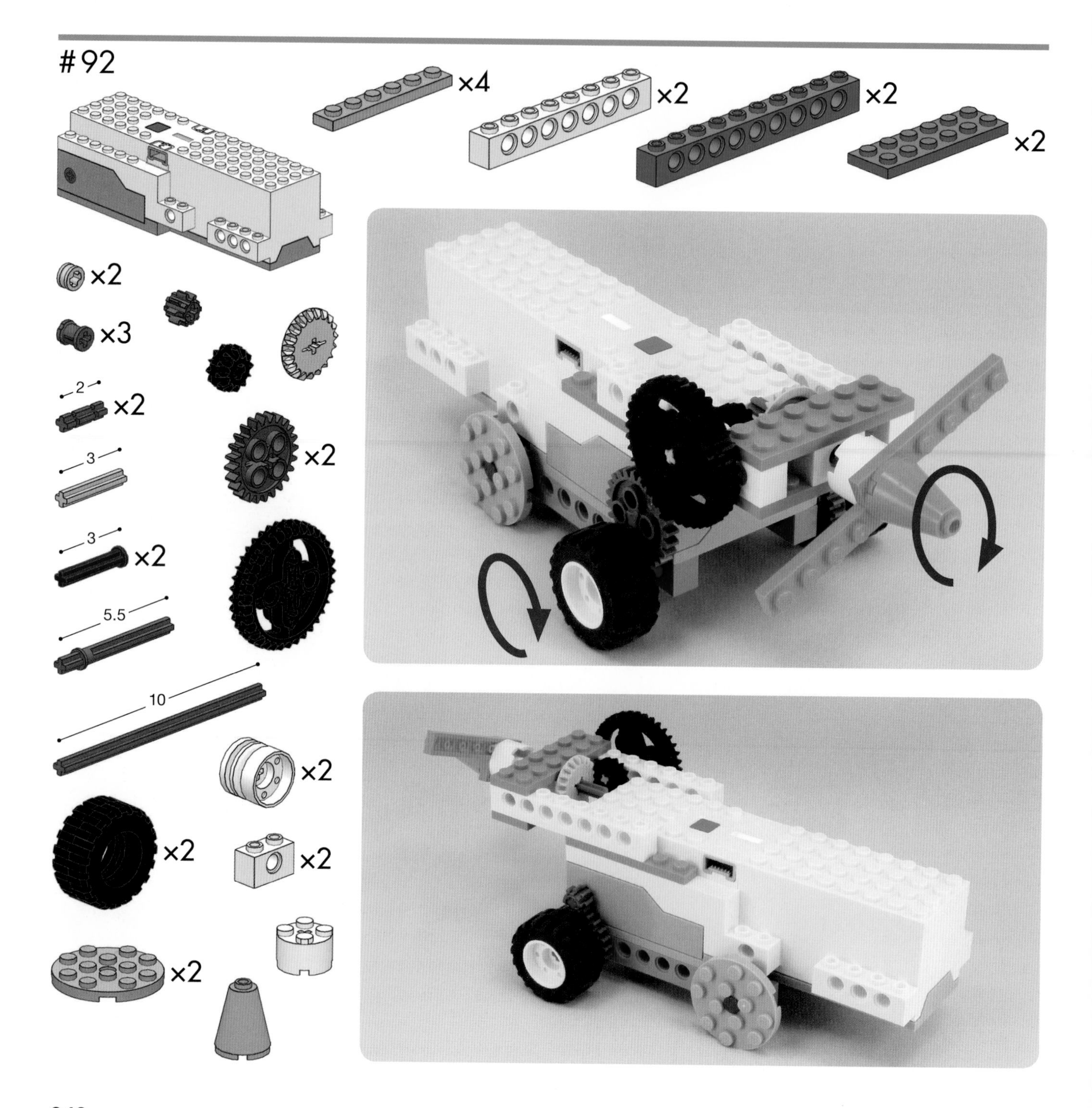

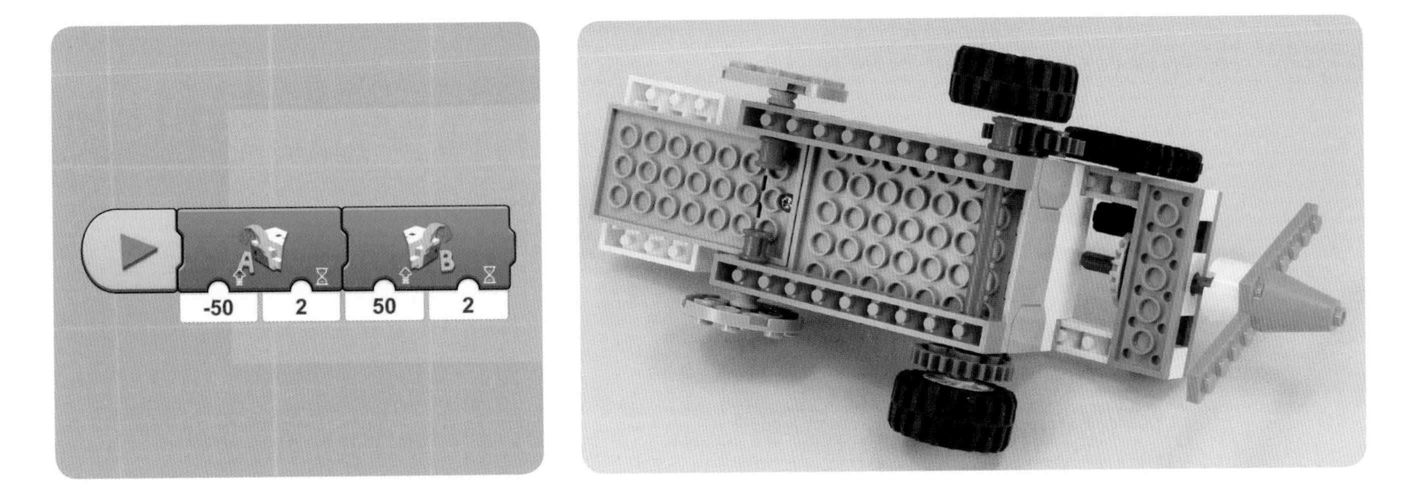
A
B
-50
2
50
2

More ideas

#93

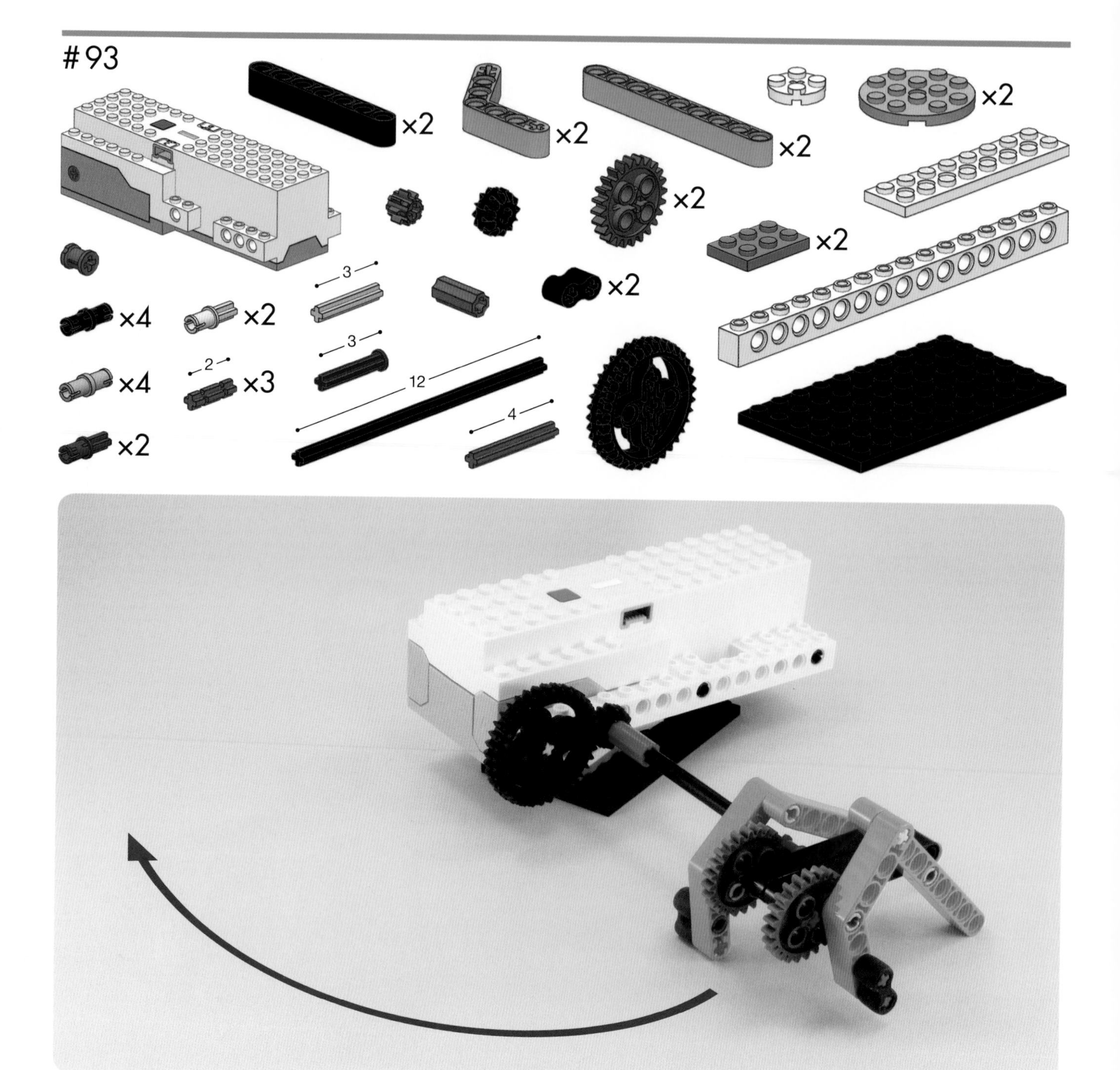

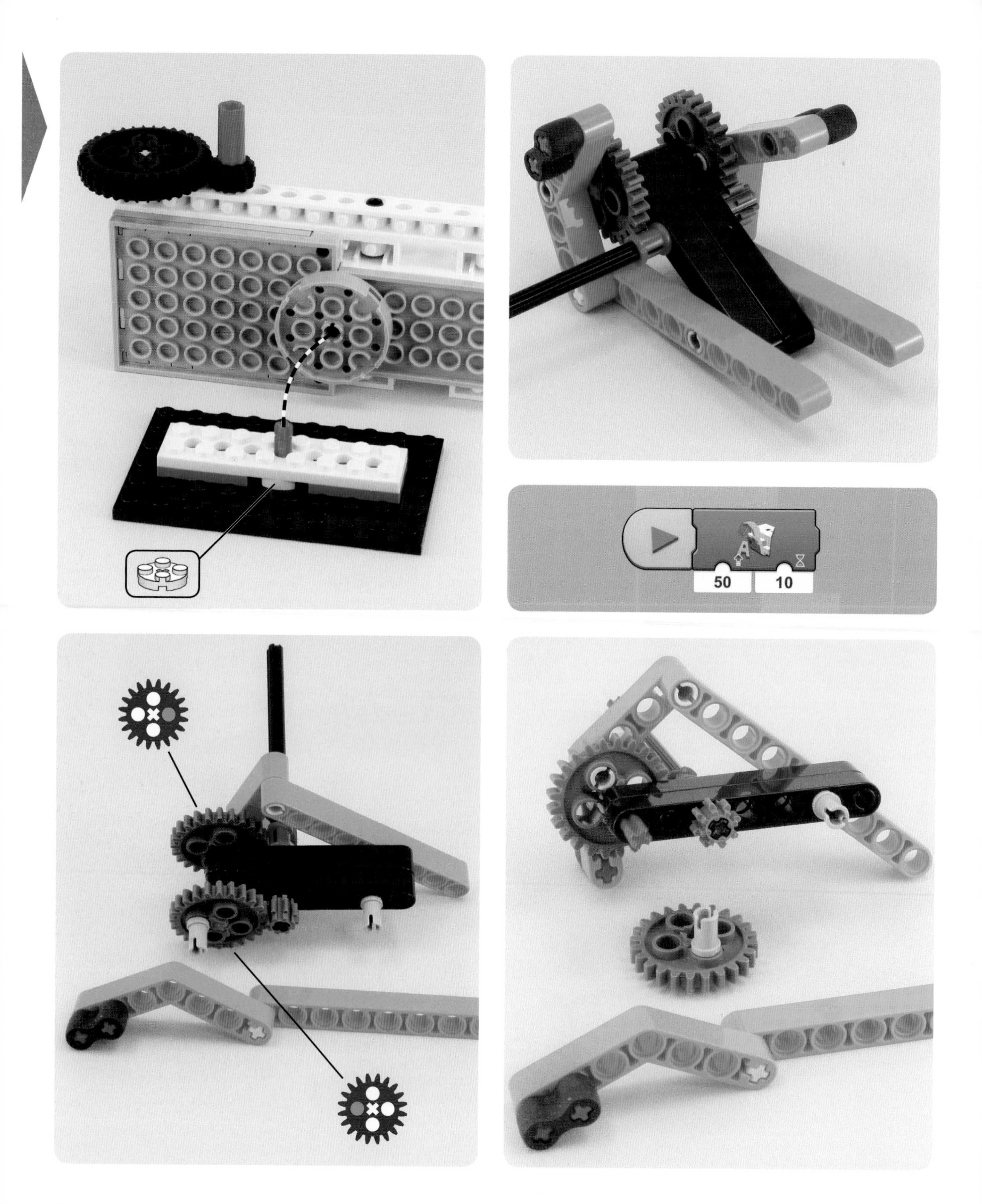
50
10

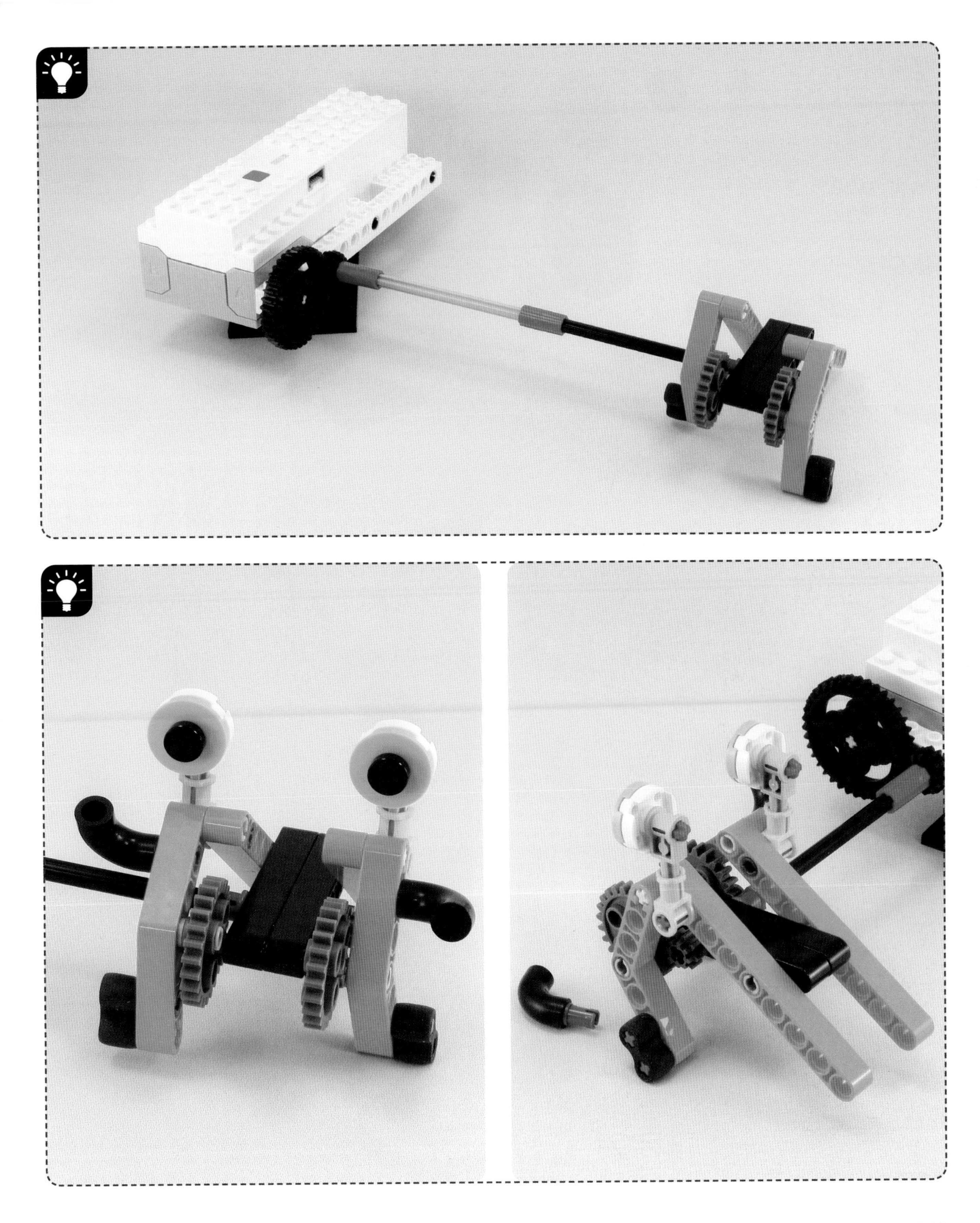

#94

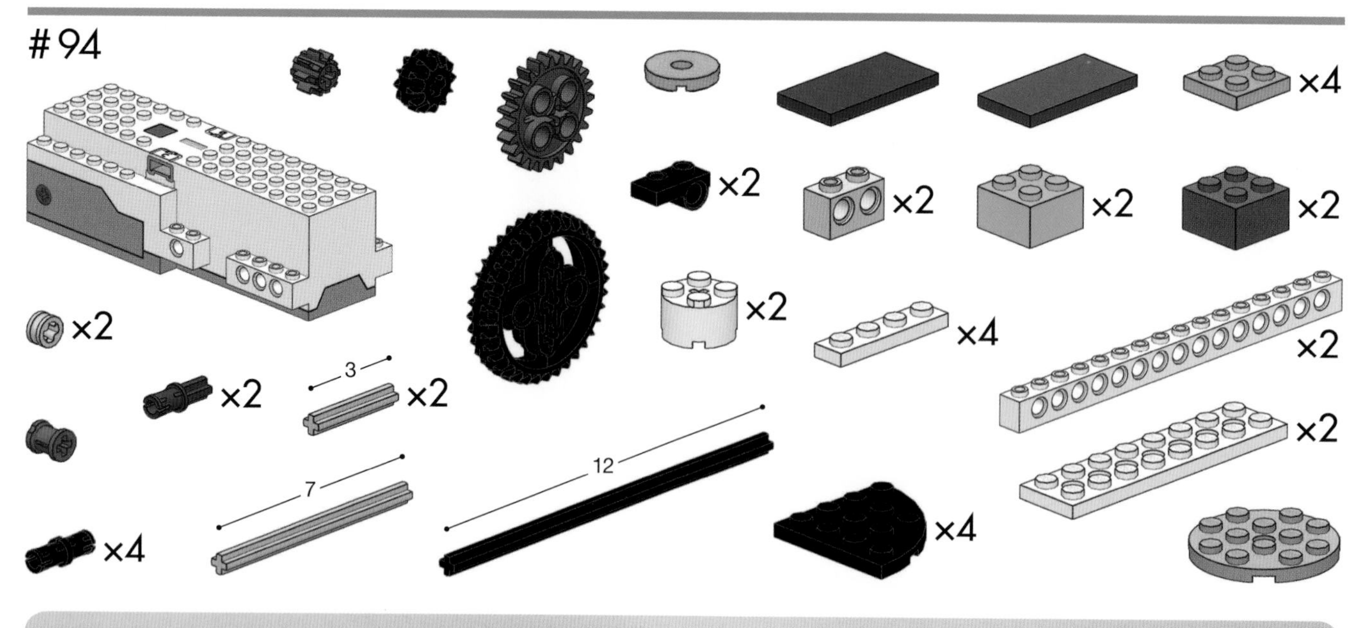

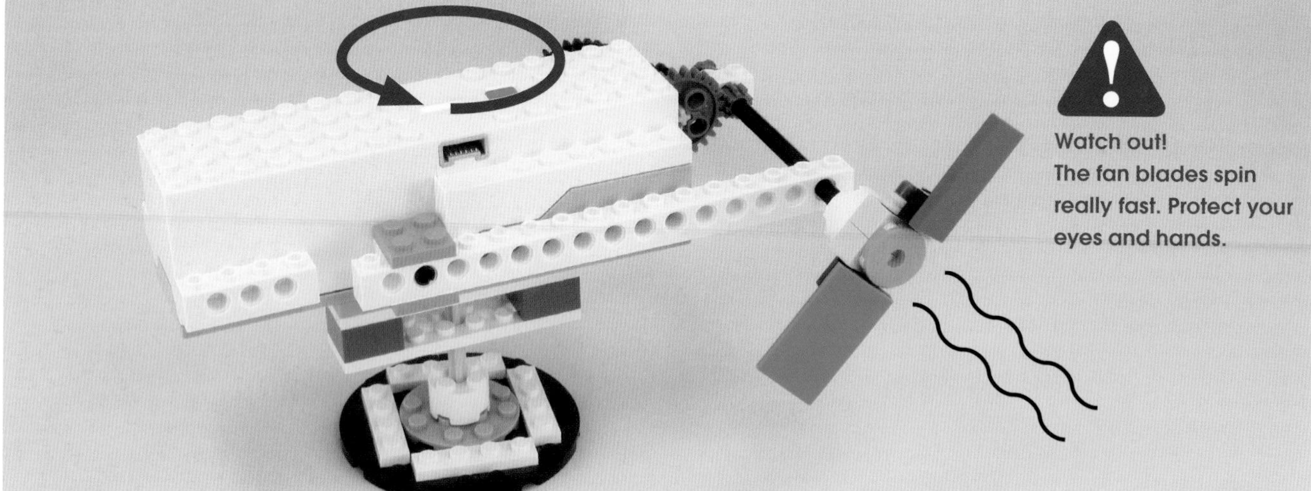

Watch out!
The fan blades spin really fast. Protect your eyes and hands.

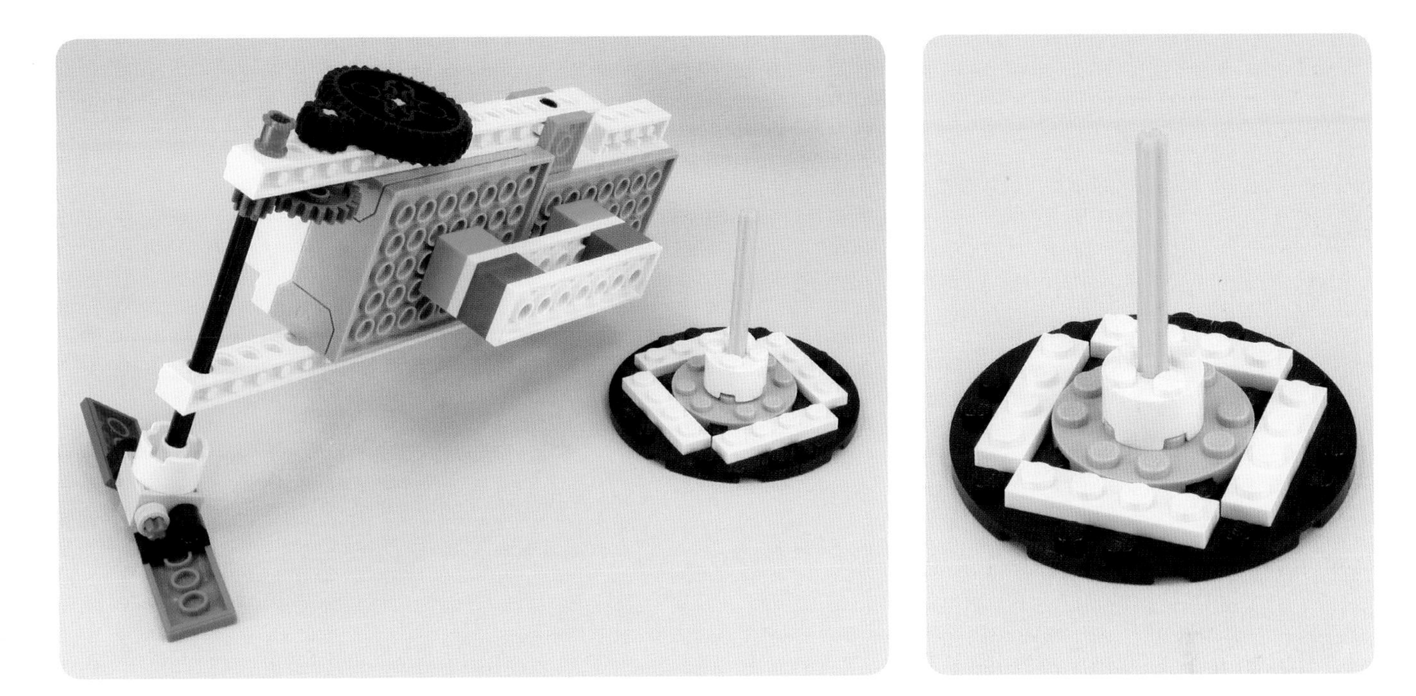

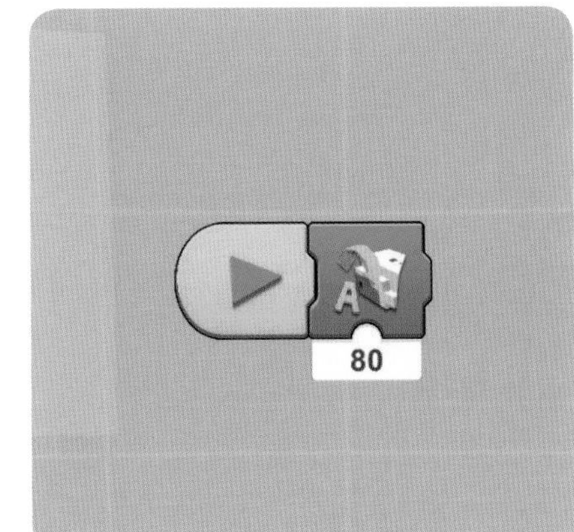
A
80

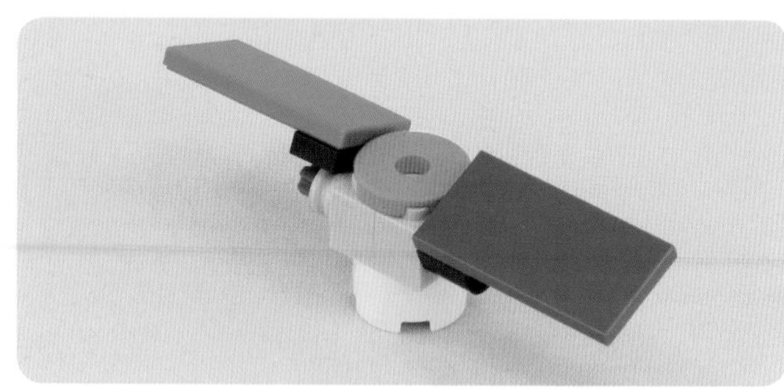

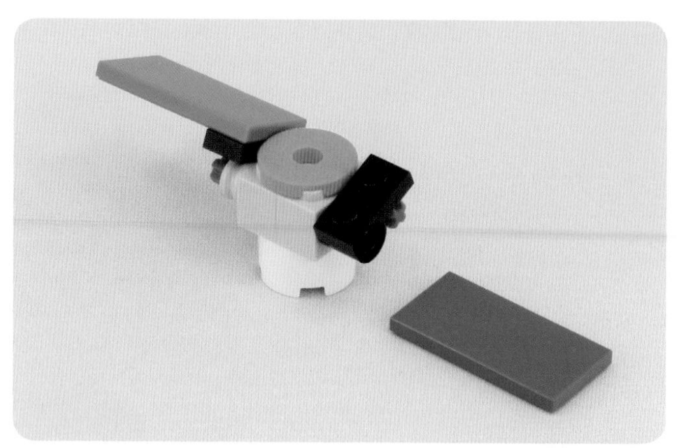

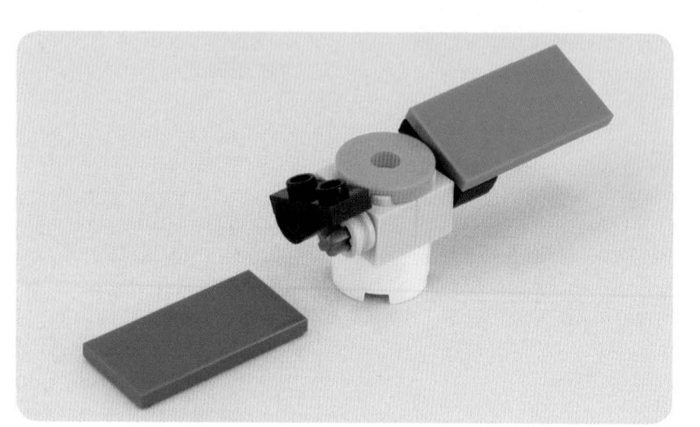

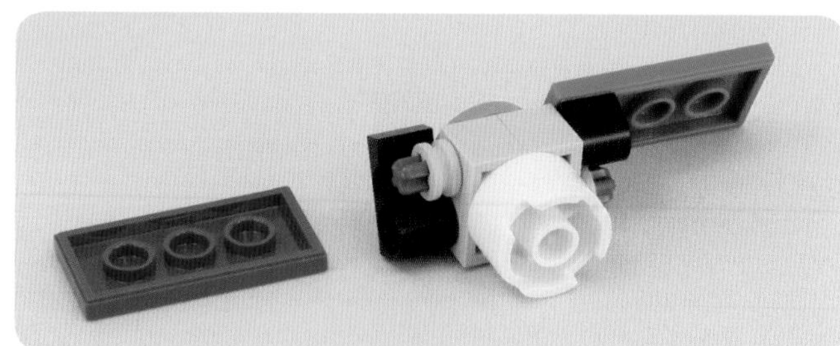

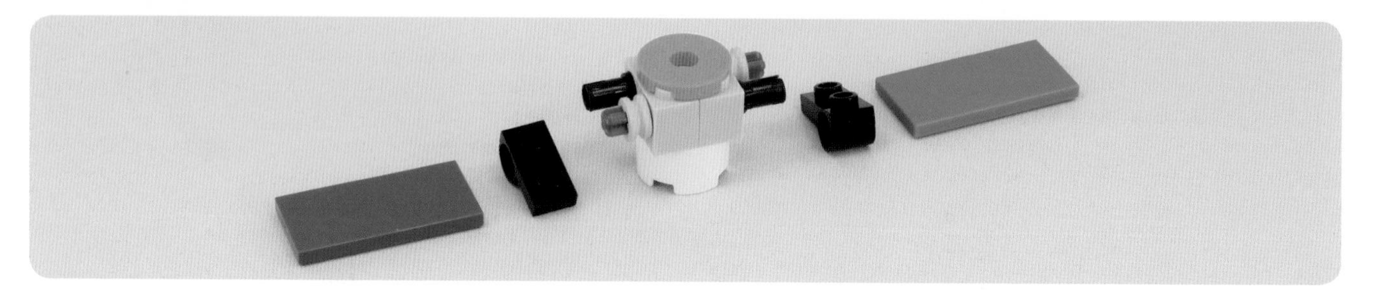

The weight of the batteries may affect balance. Find a good spot for the yellow axle so that the base holds the mechanism steady and the model spins without falling over.

#95

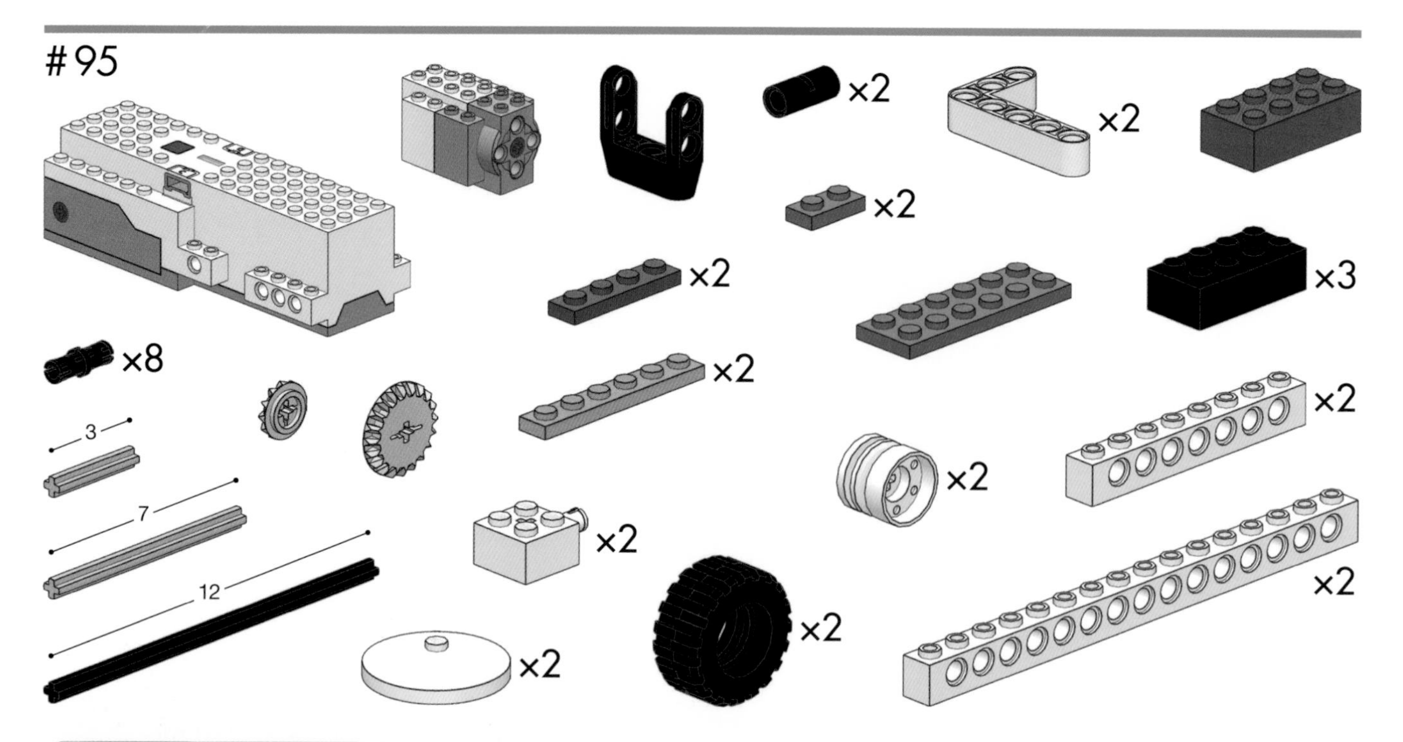

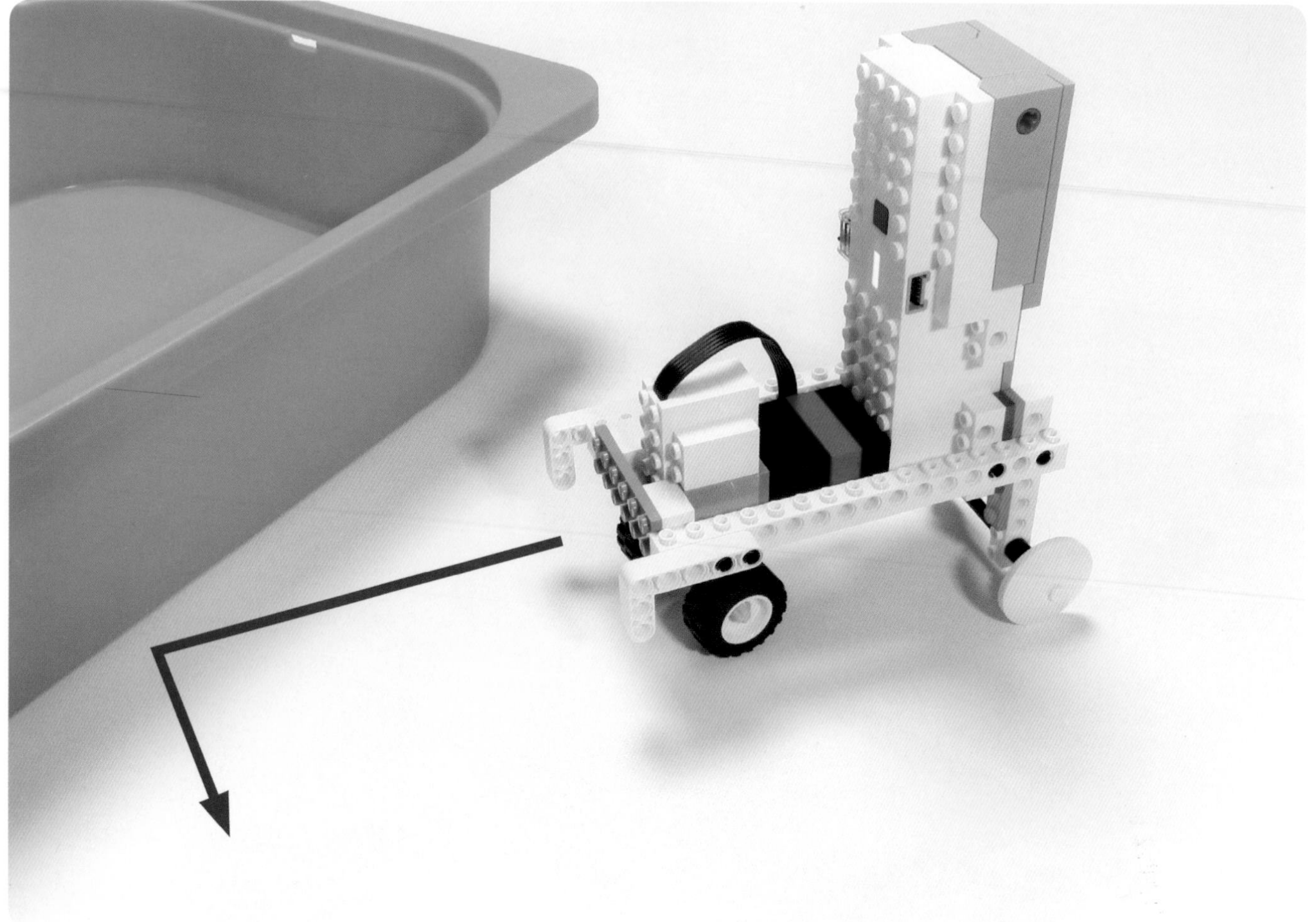

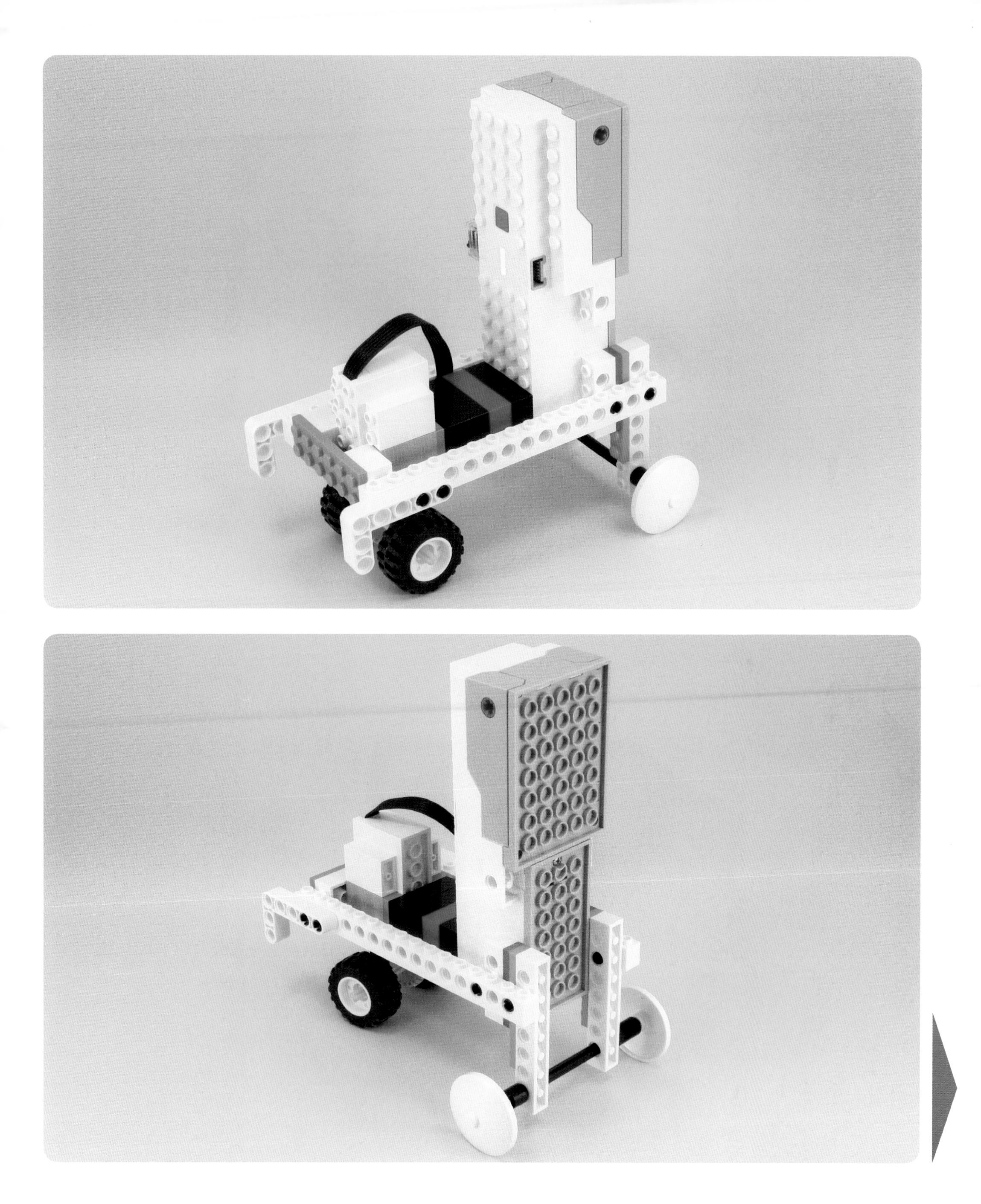

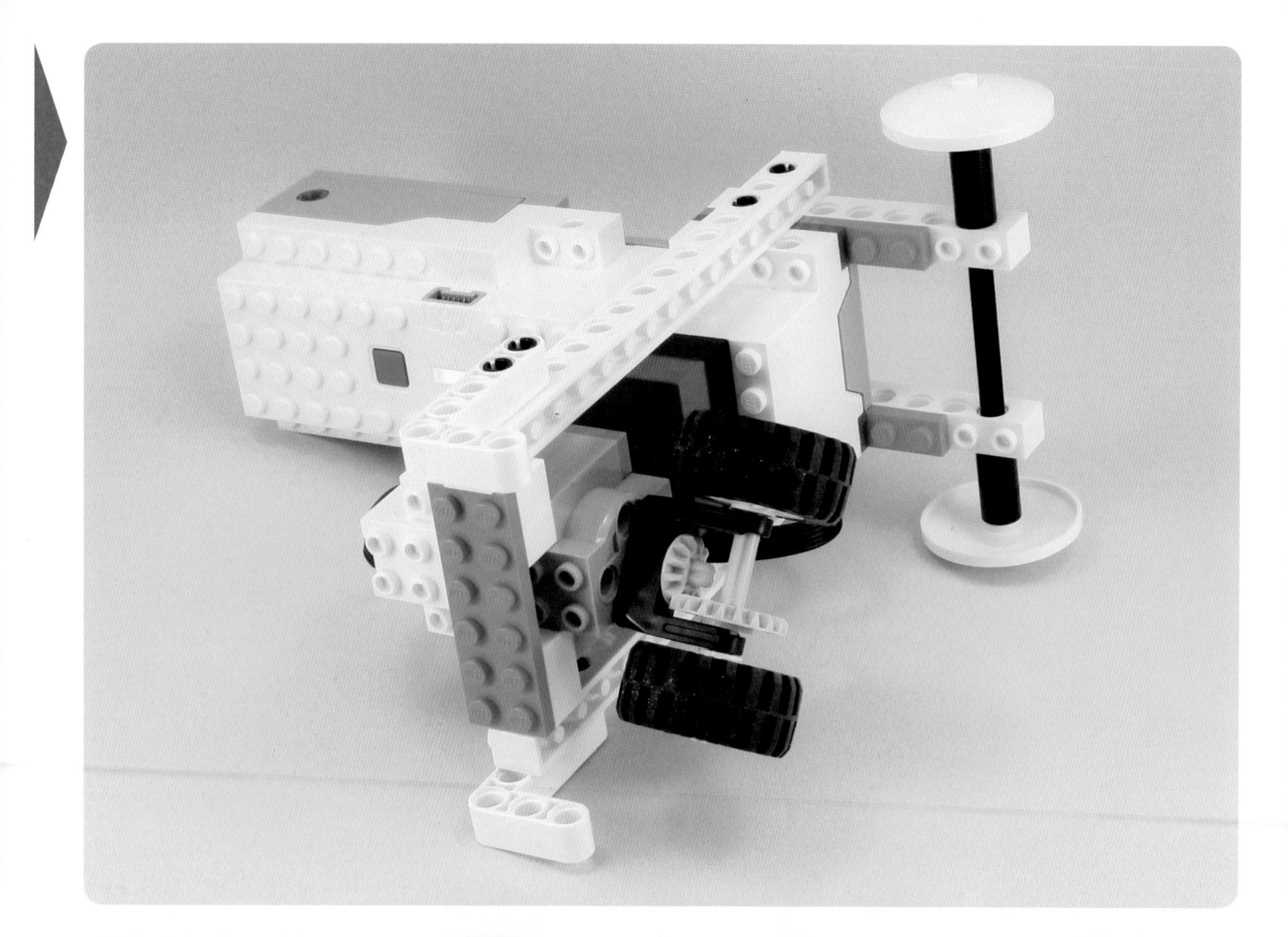

50

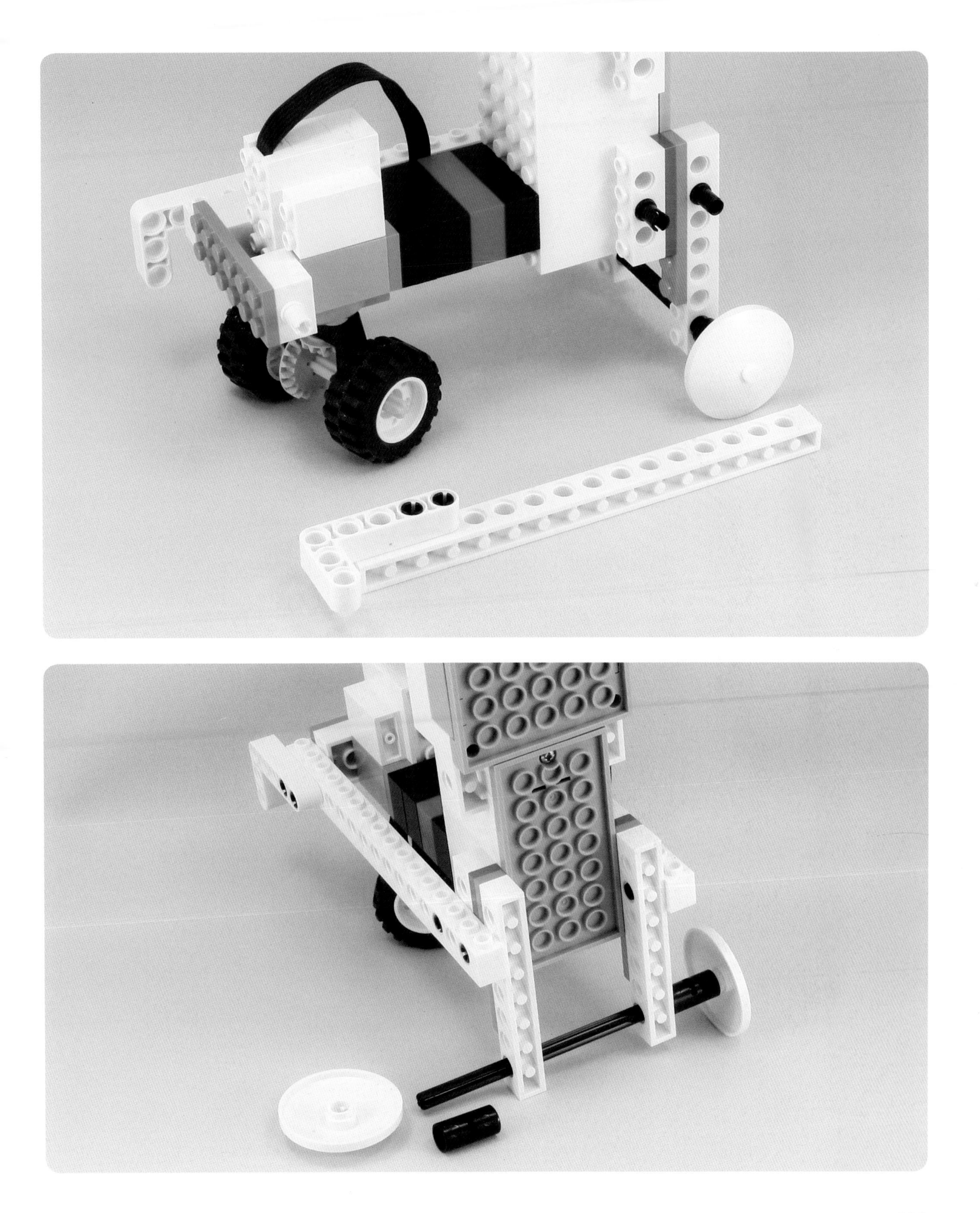

TILL NEXT TIME!